# An Environmental History of the Middle Ages

The Middle Ages was a critical and formative time for Western approaches to our natural surroundings. *An Environmental History of the Middle Ages* is a unique and unprecedented cultural survey of attitudes towards the environment during this period. Humankind's relationship with the environment shifted gradually over time from a predominantly adversarial approach to something more overtly collaborative, until a series of ecological crises in the late Middle Ages. With the advent of shattering events such as the Great Famine of 1315–22 and the Black Death of 1348–49, medieval people began to think of and relate to their natural environment in new and more nuanced ways. They now were made to be acutely aware of the consequences of human impacts upon the environment, anticipating the cyclical, "new ecology" approach of the modern world.

Exploring the entire medieval period from 500 to 1500, and ranging across the whole of Europe, from England and Spain to the Baltic and Eastern Europe, John Aberth focuses his study on three key areas: the natural elements of air, water, and earth; the forest; and wild and domestic animals. Through this multi-faceted lens, *An Environmental History of the Middle Ages* sheds fascinating new light on the medieval environmental mindset. It will be essential reading for students, scholars, and all those interested in the Middle Ages.

**John Aberth** is an independent scholar, wildlife rehabilitator, and medieval historian. He is the author of numerous books on the Middle Ages, including *From the Brink of the Apocalypse: Confronting Famine, War, Plague and Death in the Middle Ages* (2009) and *A Knight at the Movies: Medieval History on Film* (2003).

# An Environmental History of the Middle Ages
The Crucible of Nature

John Aberth

LONDON AND NEW YORK

First published 2013
by Routledge
2 Park Square, Milton Park, Abingdon, Oxon OX14 4RN

Simultaneously published in the USA and Canada
by Routledge
711 Third Avenue, New York, NY 10017

*Routledge is an imprint of the Taylor & Francis Group, an informa business*

© 2013 John Aberth

The right of John Aberth to be identified as author of this work has been asserted by him/her in accordance with sections 77 and 78 of the Copyright, Designs and Patents Act 1988.

All rights reserved. No part of this book may be reprinted or reproduced or utilized in any form or by any electronic, mechanical, or other means, now known or hereafter invented, including photocopying and recording, or in any information storage or retrieval system, without permission in writing from the publishers.

*Trademark notice*: Product or corporate names may be trademarks or registered trademarks, and are used only for identification and explanation without intent to infringe.

*British Library Cataloguing in Publication Data*
A catalogue record for this book is available from the British Library

*Library of Congress Cataloging in Publication Data*
Aberth, John, 1963-
An environmental history of the Middle Ages : the crucible of nature / by John Aberth.
 p. cm.
 "Simultaneously published in the USA and Canada"—T.p. verso.
 Includes bibliographical references.
1. Human ecology—Europe—History—To 1500. 2. Nature—Effect of human beings on—Europe—History—To 1500. 3. Europe—Environmental conditions. 4. Europe—Social conditions—To 1492. 5. Four elements (Philosophy)—Social aspects—Europe—History—To 1500. 6. Forests and forestry—Europe—History—To 1500. 7. Animals—Europe—History—To 1500. 8. Human-animal relationships—Europe—History—To 1500. 9. Europe—History—476-1492. 10. Middle Ages. I. Title.
 GF540.A24 2012
 304.2094'0902--dc23
            2012002332

ISBN: 978-0-415-77945-6 (hbk)
ISBN: 978-0-415-77946-3 (pbk)
ISBN: 978-0-203-10769-0 (ebk)

Typeset in Garamond
by Taylor & Francis Books

Printed and bound in Great Britain by the MPG Books Group

To our wonderful animals,
our horses Kit, Rob Roy, Esther, Jack, Chloe, and Star,
and our cats Midnight and George,
this book is dedicated,
even though they never seem to understand it,
no matter how many times I read it to them.

# Contents

*Illustrations* ix
*Acknowledgments* xii
*Preface* xiii

Introduction 1

PART I
Air, water, earth 11

   *In the beginning ...* 11
   *Worshipping the elements* 18
   *The Medieval Warm Period* 26
   *Harnessing the elements* 28
   *Collaboration, or exploitation?* 41
   *The Little Ice Age* 49
   *Earth, wind, and death* 51
   *Environmental causes of the plague* 56
   *Man-made pollution of the environment* 63
   *The poison thesis* 69
   *Weather magic* 73

PART II
Forest 77

   *Pre-Christian tree cults* 78
   *Surviving wildwood at the start of the Middle Ages* 84
   *The early medieval woodland* 87
   *An era of "great clearances"?* 92
   *A brief history of the royal forest of England* 97
   *The evidence of the eyre rolls* 105
   *Managing the king's woods* 111
   *Disafforestment and the rise of private woodland* 119

*The management of woods elsewhere    123*
*Shaping the idea of wilderness    127*
*A renaissance in regrowth of the forest?    137*


**PART III**
**Beast**                                                                                                                 141

*Animals on the farm: the Early Middle Ages    148*
*Animals on the farm: the High and Late Middle Ages    155*
*Animals as pets and companions    169*
*Animals of the hunt: origins of medieval hunting    176*
*Animals of the hunt: deer and other game    183*
*Animals of the hunt: romance vs. reality    195*
*Animals of the hunt: falconry and fishing    200*
*Animals and disease    206*
*Animals on trial    217*
*Animals in the bed    224*
*Animals and magic    228*


| | |
|---|---|
| *Afterword* | 233 |
| *Notes* | 235 |
| *Bibliography* | 280 |
| *Index* | 308 |

# Illustrations

1.1  MEDIEVAL WATERMILL, employing a vertical water wheel. Manuscript illumination from *Hortus Deliciarum*, Alsace, late twelfth century. The Granger Collection, NYC – All rights reserved  35
1.2  WATER-POWERED FLOATING MILLS. French ms. illumination, thirteenth century. The Granger Collection, NYC – All rights reserved  37
1.3  CARRYING WHEAT TO A WINDMILL. "Obituaire de Notre Dame de Pres", after 1270. French School, Bibliotheque Municipale, Valenciennes, France / The Bridgeman Art Library  38
1.4  WINDMILL BENCH END. English School, All Saints' Church, Thornham, Norfolk, UK. Photo © Neil Holmes / The Bridgeman Art Library  39
1.5  MILLER AND HIS WINDMILL. Misericord carvings, French School, fifteenth century. Musée National du Moyen Âge et des Thermes de Cluny, Paris / Giraudon / The Bridgeman Art Library  40
1.6  WINDMILL & WATER WHEEL. French ms. illumination, c.1270. The Granger Collection, NYC – All rights reserved  40
1.7  WITCHES BREWING UP A HAILSTORM. German woodcut, 1489. The Granger Collection, NYC – All rights reserved.  74
1.8.  AN INCANTATION TO OBTAIN A FLYING THRONE, conjured by invoking 33 demons of the air, from a fifteenth-century necromancer's manual – BSB Clm 849. © Bayerische Staatsbibliothek  75
2.1  GREEN MAN, an example of anthropomorphic representation of perhaps a woodland deity. Roof boss, English School, Norwich Cathedral, Norfolk, England. Photo © Neil Holmes / The Bridgeman Art Library  81
2.2  HARVESTING ACORNS FOR THE PIGS in the month of November, from the "Breviarium Grimani", c.1515,

x  Illustrations

|     | | |
|---|---|---|
| | Flemish School, Biblioteca Marciana, Venice, Italy. Giraudon / The Bridgeman Art Library | 89 |
| 2.3 | FELLING TREES in the month of March, from a Book of Hours by Simon Bening, c.1540. British Library, London, UK / © British Library Board. All Rights Reserved / The Bridgeman Art Library | 95 |
| 2.4 | COPPICED TREES in the foreground, with oak standards left standing in the cleared area behind, from a Book of Hours by Simon Bening, c.1540. British Library, London, UK / © British Library Board. All Rights Reserved / The Bridgeman Art Library | 112 |
| 2.5 | THE "BIG BELLY" OAK in the Savernake Forest of Wiltshire, England, a nearly 1000 year old tree that is thought to have taken root during the time of William the Conqueror. © Jim Champion http://www.geograph.org.uk/photo/153419 | 113 |
| 2.6 | MODERN COPPICING in Bradfield Wood © Robin Chittenden/Frank Lane Picture Agency/Corbis | 114 |
| 2.7 | TRISTAN AND ISOLDE, misericord carving from Chester Cathedral, Chester, England, c. 1390. An illustration of the Tristan and Isolde romance legend, with King Mark spying out from a tree as Tristan and Isolde tryst in the forest. © Dominic Strange www.misericords.co.uk http://www.misericords.co.uk | 129 |
| 2.8 | TREES, GROWING NATURALLY AND IN POTS, woodcut from Konrad of Megenberg, *Buch der Natur* (Augsburg, 1481). U.S. Library of Congress. | 136 |
| 3.1 | ADAM NAMING THE ANIMALS, from the Aberdeen Bestiary, composed in the north Midlands of England during the twelfth-century. Aberdeen University Library, Scotland / The Bridgeman Art Library | 154 |
| 3.2 | FOUR-FOOTED BEASTS, woodcut from Konrad of Megenberg, *Buch der Natur* (Augsburg, 1481). U.S. Library of Congress | 159 |
| 3.3 | BUTCHER SLAUGHTERING OX, misericord carving from Worcester Cathedral, fifteenth-century. © Dominic Strange www.misericords.co.uk http://www.misericords.co.uk | 164 |
| 3.4 | STAG HUNT, misericord carving from Gloucester Cathedral, fourteenth-century. © Dominic Strange www.misericords.co.uk http://www.misericords.co.uk | 184 |
| 3.5 | A FALCON WITH A PREY BIRD IN ITS CLAWS, misericord carving from Chester Cathedral, c. 1390 © Dominic Strange www.misericords.co.uk http://www.misericords.co.uk | 200 |
| 3.6 | BIRDS, woodcut from Konrad of Megenberg, *Buch der Natur* (Augsburg, 1481). U.S. Library of Congress | 202 |

3.7 FISH AND OTHER AQUATIC CREATURES, woodcut from Konrad von Megenberg, *Buch der Natur*. Johannes Hartlieb, Kräuterbuch (Cod. Pal. germ. 311), fol.160v, Universitätsbibliothek Heidelberg  203
3.8 MISERICORD CARVING OF FISH, probably pike, at Exeter Cathedral, thirteenth century © Dominic Strange www.misericords.co.uk http://www.misericords.co.uk  204
3.9 THE TALE OF REYNARD THE FOX, misericord carving from Bristol Cathedral, England, early sixteenth century. Here Reynard is about to be hanged for his crimes, with the execution presided over by the lions, King Noble and his queen, while the bear, wolf, goose, cat, and squirrel all prepare the gallows. © Dominic Strange www.misericords.co.uk http://www.misericords.co.uk  222

# Acknowledgments

The publisher would like to thank The Granger Collection, Bridgeman Art Library, Bayerische Staatsbibliothek, Corbis, the Library of Congress, Universitätsbibliothek Heidelberg, and with particular thanks to Jim Champion and Dominic Strange for their permission to reproduce the images within this book.

# Preface

My interest in environmental history dates back to 1967, when I was just four years old. It was in that year that my parents bought our farm in Roxbury, Vermont. The natural history of that farm can be considered, in microcosm, a natural history of the whole state, which is one of the most heartening environmental success stories in the United States. Simply put, Vermont, over the course of about a century, reversed the all-too-familiar tale of deforestation and environmental degradation. Whereas as much as 80 percent of Vermont was cleared in 1880, now the complete opposite is true, with nearly 80 percent of the state forested. This happened as farming went into decline in Vermont, particularly the growing of wheat and corn and the raising of sheep on upland fields and pastures scattered along the state's hillsides. In the case of our family's farm, it can be seen in the abandoned stone walls and foundations that now enclose a forest wilderness. And it is a history that has not stopped being written, ever since, as a boy, I remember frolicking in an "upper meadow" that no longer exists, while even on our lower field the forest has been steadily encroaching on its inexorable march at the edges. It is also a history that sometimes has been aided by man, as I remember my father planting dozens of white pine trees as part of the state's reforestation program during the 1970s.

So environmental history, like my interest in the Middle Ages, exerts a very personal hold over me, and this book represents an opportunity to happily combine both my fascinations. It is probably true that Vermont's example of environmental reversal that so inspires me was driven, at least initially, by largely economic motivations. Vermont's ravaged land was allowed to heal mainly because, after the Civil War, many farmers left to make their fortunes elsewhere, as more agriculturally viable land became cheaply available out West, great cities in southern New England beckoned, and as Australia began to corner the worldwide market for wool by the mid-nineteenth century. Nonetheless, a distinct change in environmental outlook and culture began to overtake Vermont, particularly during the 1960s and '70s, when many so-called "hippies" and back-to-the-landers began a great migration into the state, but whose path had been blazed even earlier, during the 1930s by the "good life" advocates, Helen and Scott Nearing. All this has transformed

Vermont politically from a stolidly Yankee bastion of conservatism to a liberal-leaning outpost of cutting-edge conservationism. As embodied in Act 250, Vermont's famous land use and development law (enacted in 1970 under a Republican governor, Deane Davis), the state is now uniquely associated with an environmental ethic and ethos, where natural aesthetics and integrity are supposed to trump economic considerations.[1] And yet, such ecological achievements are fragile, as a new report finds that forests in New England are beginning to decline once again, although Vermont's at a much slower pace than in the rest of the region. Moreover, the Great Flood of 2011 that occurred in our state from Tropical Storm *Irene* at the end of August, which rivaled or even surpassed damage from the last, "Great Flood" of 1927, brought home to us how the power of nature, perhaps in response to long-term climate change, can alter our landscape rather quickly.

Nonetheless, Vermont stands as an example of how exploitative attitudes towards nature can be changed, and changed mostly for the better. But if you ask someone what was medieval people's outlook on the environment, you will probably get a quizzical look. Some might have an inkling that medieval man had a naturally close relationship with his physical surroundings, since the more primitive technology of the age and the more rural lifestyle of the average person made encounters with nature far more likely. But many others might well wonder if medieval people ever stopped to think about their environment, and hence if there even is such a thing as an environmental history of the Middle Ages?

I hope in the pages that follow to show that the writing of a book like this one is more than justified. For I believe that the thousand years of medieval history between roughly 500 and 1500 were among the most crucial for determining man's future relationship with the natural world. This was a period, I believe, that especially in its latter centuries saw a significant shift in attitudes towards the environment brought on by a series of ecological catastrophes—to include the "Little Ice Age" and the Black Death, to name the most obvious—that have a distinct parallel in our own present crises, such as global warming. And yet there were also silver linings, as there are today, such as partial reclamation of Europe's forest in the wake of depopulation from the Black Death, just as the forest came back in Vermont and New England since the mid-nineteenth century. There is thus much we can learn from a study of how the environmental attitudes of our medieval forebears were forged in their own "crucible of nature."

Although the meaning of "environment" can be spectacularly all-encompassing, embracing man-made, urban environments, for example, this book will confine itself to exploring only the natural world and medieval man's relationship with it. We will also be approaching the natural world primarily from a cultural perspective, albeit in the context of interrelated developments in technology, climate change, and population pressures. Some have argued that a unified view of the environment or nature simply did not exist in the Middle Ages, but that contemporary attitudes towards the natural world must instead be culled from a variety of peripheral commentary on seemingly unrelated topics.[2] Others locate the beginnings of a modern philosophical approach to the environment firmly after the medieval period, during the early modern era starting in the sixteenth century.[3] I hope to show in this book that the beginnings of what we today call the "environmental movement," or a more nuanced and integrated approach to nature, can indeed be traced back to at least the late Middle Ages. The overall evolution throughout the Middle Ages of these attitudinal changes towards the environment will be sketched out in the introduction, while the remaining sections of the book are organized along the lines of the basic, fundamental aspects of man's environment with which he has had to deal in any age. Indeed, these are the same themes—the elements of air, water, and earth we use to live in and build our world, the forest that shelters and provides for all things wild, and the beasts that share our natural world—which continue to shape my own, personal relationship with the environment right here in Vermont. High up on our mountain, for example, is an old talc mine whose gash can still be seen in the earth and whose tailings still yield talc. Alongside the nearby Flint Brook are the stone foundations of a water mill that sawed the lumber used to build our nineteenth-century farmhouse. Our mountain is part of a range known as the Northfield Ridge, which is one of a number of ridgelines in Vermont that are proposed as sites for industrial-scale wind turbines which, ironically, would devastate the natural landscape in order to purportedly save it from global warming. Our tree farm was one of the first to enroll in Vermont's Current Use program, which grants tax breaks to landowners in exchange for maintaining a working, productive forest, which we do in consultation with our local forester. And at the Vermont college I taught at most recently, there is an annual Animal Blessing Service held on October 4, the feast of St. Francis of Assisi, considered the "patron saint of animals," where members of the college community are invited to bring their pets to the front steps of the chapel (but not actually into it) in order to receive the officiating priest's blessing.

For so many of us moderns, the medieval world seems to be an alien place, particularly with regard to its environment. To the degree that our ancestors' world was more "natural" or more wild than our own, our alienation from them is perhaps a function of our modern alienation from nature, where a visit to more rustic surroundings, such as a state or national park or even to a farm, seems like a visit to a foreign country or a strange land. I'll never forget the time when some inner city kids from Boston were brought to our Vermont farm for a weekend outing. The first few hours were spent trying to coax the kids to leave the comfort of the car and venture outdoors. Their reluctance to do so, they explained, was due to the fact that they had never seen such a profusion of blood-sucking bugs!

But even such visits to the unfamiliar can be instructive, and one may be surprised to find out how much there is that *is* familiar. For example, while much of what men thought about the environment during the Middle Ages was obviously informed by Christianity, there also was much that was seen as magical in the natural world by our medieval forebears. If some of the properties ascribed to plants and animals in medieval magical recipes may strike us as ludicrous or bizarre, nonetheless their sense of the sacredness, of wonder and enchantment, to be found in nature is something with which almost any modern nature lover can identify. By the same token, there also was considerable cruelty shown by medieval men towards their fellow creatures, which finds its own parallels in our present world. Bear baiting, for example, in which dogs were pitted against bears for the public's amusement, may seem like something that can safely be consigned to the distant past. Yet even in our own day and age, right here in Vermont, the sport of hunting bears with dogs is still sanctioned and defended, though it be but a small remove from what most would consider to be a barbaric medieval practice.

To understand medieval attitudes towards the environment, therefore, is to gain an invaluable insight into our own.

# Introduction

> May this stand as a perpetual reminder to everyone, now living and yet to be born, how almighty God, king of heaven, lord of the living and of the dead, who holds all things in his hand, looked down from on high and saw the entire human race slipping and sliding towards all kinds of wickedness, enmeshed in crimes, pursuing numberless transgressions, immersed up to their bowels in every kind of vice out of an unfathomable malice, bereft of all goodness, not fearing the judgments of God, and chasing after everything evil. No longer able to bear so many abominations, so many horrors, God called out to the earth: "What are you doing, Earth, held captive by gangs of worthless men, soiled with the filth of sinners? Are you totally helpless? What are you doing? Why do you not demand human blood in vengeance for this wrongdoing? Why do you tolerate my enemies and adversaries? When confronted by such wantonness, you ought to have already swallowed my opponents. Make yourself ready to exercise the vengeance which lies within your power.
>
> "And I, the Earth, having been established at your command, will open my veins and swallow up an infinite number of criminals once you give the order. I will deny the usual fruits of the earth, I will not bring forth grain, wine, and oil."
>
> And when the very irate Judge gave the word, with thunder sent down from the heavens, He marshaled the elements, the planets, the stars, and the orders of the angels against the human race with an unspeakable judgment and He armed every single animate being in order to exterminate the sinners and He called them forth to execute His justice with quite a cruel stroke.[1]

Thus opens Gabriele de Mussis' *Historia de Morbo* (History of the Disease), a chronicle of the Black Death of 1348–49 written from an eyewitness' point of view in Piacenza, Italy. Mussis provides a familiar explanation for the occurrence of a disease, now identified as plague, that wiped out as much as half of the population of Europe and the Middle East within a few years during the mid-fourteenth century. That humankind is being punished for its sins through deadly epidemics is an idea that goes back in the Western tradition to at least the Old Testament (the ten Egyptian plagues in Exodus, for example) and the ancient Greeks (Apollo's plague that opens the history of the Trojan War in *The Iliad*). What seems new, however, is that Mussis is now providing a concurrent, *natural* explanation alongside the divine one: A dialogue ensues between God and the earth in which the latter is the one that actually carries out the sweeping, punitive sentence through its environmental influence and power over man. Mussis therefore enunciates a rather sophisticated, cyclical understanding of humans' relationship with the natural world: Nature is not merely a backdrop to human history but instead plays an integral part in it, since even in the Middle Ages man's actions were understood to be able to shape and impact the environment, which then redounds back onto him. These days this view has been christened the "new ecology" to express the mutual, two-way dialogue between humans and nature that drives historical change, and it is one that will be adopted in this book.[2]

It is my belief that the new ecology expressed by Mussis represents a seismic shift in attitudes towards nature that came about in the late Middle Ages, beginning in the fourteenth century. This was no mere coincidence: Not only was this the time of the Black Death, often described as the greatest demographic disaster and disease event in the history of humankind, but it was also the start of another ecological catastrophe, the "Little Ice Age," a time when, in contrast to our present-day crisis of global warming, average temperatures became significantly cooler, thus curtailing growing seasons and areas of Europe where agriculture could be pursued. More than that, the weather also became wetter and wilder, swinging from one extreme to the other, so that crop failures and famines became more frequent, including the Great Famine of 1315–22 that encompassed almost all of the northern half of Europe. Seas became stormier and fisheries disappeared, glaciers advanced, and rivers and seas flooded or froze over, destroying coastal communities and trapping ships in the ice. It is also possible that earthquakes became more frequent at this time in Europe, as they are attested by some contemporary observers to be one of the causes of the Black Death.

All this ecological upheaval and uncertainty was bound to change how medieval men and women approached their environment. But although historians have certainly recognized the general significance of ecological crises such as the Black Death and the Little Ice Age and the parallels that they provide for our own, game-changing scenarios such as global warming, I believe that they have yet to grasp the unique power and influence that such a collective series of crises had upon contemporary society, especially with respect to its attitudes towards the environment. For example, the Black Death is often interpreted as the product of a Malthusian dilemma whereby Europe's ever expanding population and consequent over-exploitation of natural resources in order to feed itself resulted in a necessary correction that happened to be expressed in the famines and plagues that were so prevalent in the fourteenth and fifteenth centuries.[3] This is an attractive interpretation, if only for the convenient and salutary warning it provides for modern man's ecological sins, yet when examined closely, it does not really stand up within the context of the Middle Ages itself. Why, then, if medieval population and exploitation was in need of a correction, did a demographic decline or stagnation continue until at least the middle of the fifteenth century, long after the first, great "check" occurred in the middle of the fourteenth? Why were sentiments very close to those of Malthus expressed by Latin Christian authors such as Tertullian around the turn of the third century C.E., when the Roman Empire was in fact entering a period of decline that was the exact opposite of the Malthusian dilemma of the high Middle Ages? Certainly, examples can be found of medieval technology's precocious impact upon the environment, which can now be recovered by isotope analysis of ancient pollution trapped in deep-drilled ice cores, but one has to wonder if all this adds up to a man-made catastrophe on the scale of our current phenomenon of global warming? Isn't it possible, after all, that ecological events like the Black Death and the Little Ice Age happened for arbitrary reasons, largely independent of man's actions, the product of a happenstance microbial invasion mediated by rats and fleas or an unforeseen and inexplicable turn of the weather? How can we know for certain anyway that medieval communities on the eve of the Black Death or the Little Ice Age had in fact outgrown their self-sustaining capacities and were therefore ripe for a fall? Couldn't instead a state of demographic "deadlock" have persisted indefinitely absent an "exogenous" variable like plague or famine? And doesn't the Malthusian attempt to project the causes of the Black Death and the Little Ice Age back into preceding decades diminish the sudden impact that these disasters surely made upon medieval Europeans and instead reflect the wish-fulfillment of modern historians to understand them on their own terms?

All this is by way of challenging the overmighty role that historians too often accord to human actions in their own affairs, albeit some find it a comforting notion in that it gives us greater control, as well as greater responsibility, in our own destinies. But too often, especially when applied to the distant past of the Middle Ages, this view has been overemphasized and is by now rather dated, since we are far less sure that it accords with actual reality. The latest scholarship on the Black Death and the Little Ice Age, for example, has been at pains to swing the pendulum the other way and stress the technological and other societal changes that were forced upon Europeans by cataclysmic forces such as plague and famine, rather than the other way around. Indeed, if the agricultural and technological capacity of the high Middle Ages to exploit its environment leading up to the late medieval crises has previously been underestimated, this then at the same time calls into question the inevitability of a Malthusian dilemma in the first place. For example, the idea that "marginal" agricultural land reclaimed from forest or waste was quickly exhausted is really dependent on a cultural definition of what constitutes marginal: Land can be classified as "waste" just as easily if it is too distant from the market to be able to be exploited effectively, as much as due to the fact that it is no longer agriculturally productive. If intensive cultivation and wasteland reclamation are assumed to have eventually failed to keep pace with population expansion by the fourteenth century, then this is at the expense of neglecting other contemporary achievements, such as crop rotation, fertilization, and legume enrichment, that may have helped maintain or even increase production. Above all, the Malthusian assumption downplays the significance of autonomous ecological crises such as the Black Death and the Little Ice Age, which instead some historians have been trying recently to reintegrate back into our mainstream history of the Middle Ages.[4]

In sympathy with the latter approach, I believe that major shifts in attitudes towards the environment by and large only come about in human societies as the result of major ecological catastrophes, such as the Black Death, the Little Ice Age, or global warming, which prove themselves uniquely capable of impinging upon human consciousnesses in myriad and wide-ranging ways. Such earth-moving crises obviously only come around once in a great while, and therefore are all the more important for that. On the other hand, one must beware of succumbing to the temptation of the too-easy, mono-causal explanation for environmental catastrophes such as the Black Death, which would eliminate any role for human beings in their history. For example, a new theory posits that undersea outgassing of ancient carbon stored in the oceans and recorded by tree-ring growth records (dendrochronology) killed large numbers of people at the time of the mid-fourteenth century, corresponding to the believed outbreak of plague; this theory is attractive because it meshes well with chronicle accounts of deadly miasmas or vapors leaving a trail of corruption in their wake, which are usually dismissed by historians as apocryphal. (Other explanations for the higher carbon levels in the atmosphere include volcanic eruptions and cosmic ray fluxes from outer space.) Yet the consensus seems to be that if such environmentally apocalyptic events as outgassing did occur, that they were local in character and impact which fail to account for the Black Death persisting over wide geographical areas and for long periods of time.[5] In other words, this can only be one factor among many, and perhaps an eccentric anomaly at that. The same can be said for the dust-veil event of 536 C.E., the "year without a summer," which it is argued contributed to the decline of the Roman Empire.[6]

For quite some time now historians have advanced their respective interpretive models to account for the overall views of nature in medieval Europe and any changes thereof that may have occurred. An older theory took the position that the Judeo-Christian inheritance of the Middle Ages inaugurated an exploitative attitude towards the environment that persists down to the present day: It was medieval Christianity, so this argument goes, that alienated human beings from their natural world and in turn deprived all other living things of their humanity, reserving this exclusively for man as the closest likeness on earth to an anthropocentric God. Ultimately, however, medieval exploiters of nature traced their lineage back to the Hebrew tradition of the Old Testament, which had placed man at the center of a God-created universe and where progress was measured in terms of humans' divinely ordained dominance and control over his surroundings. Moreover, medieval Christians broke definitively with older pagan religions that had breathed life and spirit into all things, even what we regard as inanimate objects (a belief system known as animism), and eventually embraced a burgeoning interest in science and technology that was to have a nugatory impact on the environment.

The only discordant note was sounded in the thirteenth century by the founder of the mendicant tradition, St. Francis of Assisi, who should be regarded as the "patron saint" for modern-day ecologists because of his espousal of an alternative to the "orthodox Christian arrogance toward nature." St. Francis allegedly preached an "equality of all creatures" that veered on the heretical, since it assumed that even animals had souls.[7] Though it goes unmentioned in this argument, an even greater rebuke than St. Francis to the "arrogance" medieval Christians supposedly assumed towards their environment would undoubtedly have been the famines and plagues of the late Middle Ages, when God seemed to be punishing man in a particularly severe manner for all his manifold sins, chief among which in the medieval view of things was pride.

In more recent decades, this overly broad and rather stereotypical interpretation of medieval attitudes towards nature has been replaced by a more nuanced one, which proposes four sets of attitudes towards the environment, each one roughly succeeding the other as the Middle Ages progressed.[8] The first of these was the "eschatological" view, which supposedly prevailed during the transition period from classical times to the Middle Ages, here defined as between the third and fifth centuries. The Church fathers who wrote at this time had indeed inherited an "ecological triumphalism" from the Judeo-Christian tradition that urged man to "increase and multiply" and "take command of ... all living things that move on the earth". But they had also, under the leadership of perhaps their greatest theologian, St. Augustine, enunciated a doctrine of original sin in which Adam and Eve were condemned to eke out a hardscrabble existence from their former Eden, as well as a linear view of time in which the earth was entering its "senility" of maximum bounty in preparation for the imminently awaited apocalypse or end times. Consequently, this view was, in the main, a pessimistic one that urged fellow Christians to actually reign in their procreative urges, perhaps with the aid of a contraceptive or abortifacient plant lore that was inherited from the ancient world.

The next stage was an "adversarial" one, extending through the early Middle Ages from the sixth to the tenth centuries, in which Europeans, confined to small population "islands" isolated from each other in a sea of wilderness where trade and commerce and other forms of communication had severely declined from the heady days of the Roman Empire, were now characterized as having an excessive fear or awe of nature, as if besieged by it. Emblematic of this attitude were Christian tales of heroic contests with monstrous or demonic forces believed to be embodied in hostile and "uncivilized" creatures of nature, such as the "fierce spirit" called Grendel who roams the moors and fens and attacks the Danes in their mighty mead hall in the Anglo-Saxon epic poem, *Beowulf*. Otherwise, such epic struggles are played out in the saints' lives featuring early Christian missionaries who strive to convert by demonstrating their power and dominance over nature, as represented by pagan holy sites in wood, rock, or spring that are worshipped by Celtic druids and other sects as the locus of the divine. Even so, an alternative view sees the early medieval Church assimilating the mystical, integrated view of nature in Anglo-Saxon and Germanic culture, whereby the physical and natural world is honored and respected for its ability to channel the spiritual and supernatural realms. Such a synthesis supposedly came under assault by the high Middle Ages from a growing scientific and rational strain in scholastic thought, yet still remained influential.[9]

The high Middle Ages, from about 1000 to 1300, constitutes the "collaborative" stage of this evolution in medieval attitudes towards the environment. This is when medieval man demonstrated his ability to impact his surroundings, instead of simply being subject to nature's whim, as both the number and size of population centers expanded, along with the arable support network needed to maintain them, at the expense of the wild and untamed landscape. A good example is the twelfth-century expansion of the Cistercian order of monks, who carved out their monasteries from swamps, forests, and other desolate "wastes" as "God's partner" in managing his creation.[10] If ever there was an "ecological triumphalism" in medieval society, it was surely at this time, when humans proved the power of their contemporary abilities at environmental exploitation. In place of the pessimism and trepidation that characterized the early Middle Ages, there was now unbounded optimism and hopeful expectancy of what lay beyond the next horizon. This was also apparently a time of discovery of the intrinsic value of nature for what it could reveal of the divine plan for creation.[11] A sign of the times was Alain of Lille's twelfth-century allegorical poem, *De Planctu Naturae* (The Complaint of Nature), which was far less a celebration of the natural world than a hectoring lecture to mankind to fulfill his God-given duty to procreate by leaving off his depraved and distracting "sins against nature," such as sodomy (encompassing any non-procreative act such as homosexuality, bestiality, oral sex), incest, adultery (i.e., prostitution), etc.[12]

Meanwhile, in the emerging universities and cathedral schools, there arose the idea of collaboration or balance between humans and nature, in which each played their part to bring harmony to the natural world order. In this scheme, God was the first principle who had created the essential building blocks of the universe, nature was what maintained or kept them going in a never-ending cycle of birth and decay, and man was the artisan who, imitating both God and nature, put his own stamp on things by fashioning them to his needs, such as when he makes shoes or cheese. In the words of Bernard Silvester, one of the twelfth-century exemplars of this school who was heavily influenced by Platonic philosophy, man is the "ruler and high priest of creation, that he may subordinate all to himself, rule on earth and govern the universe" and who actually improves on what is provided him by God and nature in order to bring it to perfection. Whereas brute beasts "bear their heads towards the earth from downturned faces," man, in accordance with the greater "dignity of his mind," directs his attention heavenward towards the stars. This speech was put into the mouth of Noys, a goddess personifying Divine Mind, who along with her fellow female deities, Physis, Natura, and Endelechia, or the World Soul, are portrayed in Silvester's *Cosmographia* as the ones who, acting on behalf of God, actually preside over man's birth. The work ends with a celebration of the male penis as a "joyful" (*jocundus*) and "fitting" (*commodus*) member by whose procreating semen men "may fight unconquered with death armed with [most] genial weapons / they may repair Nature and perpetuate his species."[13] A greater contrast with the rather prudish views of early Church fathers like St. Augustine cannot be imagined. Sexual and phallic overtones also abound in the twelfth- and thirteenth-century Occitan love lyrics dedicated to Lady Natura, who is likewise seen as a figure of healing, harmony, renewal, joy, hope, fertility, and, of course, love.[14]

Naturally, all this optimism and harmony comes to a crashing end with the advent of the Black Death, and perhaps even earlier with the Great Famine. Humans were forcefully reminded that nature could still make war on man and that an adversarial relationship was quite possibly the norm between the two. A good example of the way in which the environmental tectonic plates had shifted is the *Iudicium Iovis* (The Judgment of Jupiter), a prose work written in both the vernacular and Latin by the German humanist, Paul Schneevogel, going by the pen name of "Paulus Niavis," which he produced at the very end of the Middle Ages in the 1490s. This was part of a genre of such astrologically-minded allegorical works: A poem entitled *De Judicio Solis in Conviviis Saturni* (On the Judgment of the Sun at the Feasts of Saturn) was composed by Simon of Corvino of Liège in 1350 in order to explain the planetary conjunctions that produced the great plague of 1348. In Schneevogel's version, a hermit living in the Erzebirge or Ore Mountains of Bohemia has a dream vision in which *homo* or man accompanied by his advisers, the Penates or household gods of Rome, disputes with Mother Earth, her advocate, Mercury, and all the classical deities of nature including Bacchus, Ceres, the Naiads, Minerva, Pluto, Charon, and the Fauns, in Jupiter's court of justice on a charge of "parricide," or despoiling the environment, chiefly through mining for silver and gold. Mercury accuses man of grievously wounding Earth, despite her generous bounty, and spurning the gods by digging without rest, day and night, toiling by torchlight, probing ever deeper into the Earth, boring into her stomach and striking "against her very veins" in his insatiable greed for precious metals. In reply, man denounces Earth as behaving more like a stepmother than a mother to him, covetously hiding and concealing the metals man needs for his currency within "her inward parts" or bowels, even though "everything was created for the sake of man and man alone." In the end, Jupiter decides to refer the case to Lady Fortune, who decrees that "men ought to mine and dig in the mountains, to tend the fields, to engage in trade, to injure the Earth, to throw away knowledge, to disturb Pluto [god of the underworld], and finally to search for veins of metal in the sources of rivers." But at the same time, men's "bodies ought to be swallowed up by the Earth, suffocated by its vapors and intoxicated by wine, and afflicted with hunger and remain ignorant of what is best," for "these and many other dangers are proper for men." Thus, no final judgment and resolution is really rendered except that man and the earth are destined to engage in perpetual conflict, each inflicting their respective harms upon the other, in an ongoing war of mutual attrition. This supposedly returns us full circle back to the "adversarial," pessimistic view of nature with which we had opened the early Middle Ages.[15]

One other environmental attitude that is identified in the late medieval period and projected back into the high Middle Ages, when man was beginning to become alienated from his natural surroundings by living in ever greater numbers in towns, is a "recreational" view of nature that celebrates its restorative powers for the human spirit. This is evidenced in the Robin Hood ballads in England, the goliardic poems and troubadour culture associated especially with France and Germany, and in the works of Italian humanists such as Francesco Petrarch and Leon Battista Alberti. It is also identified with the writings of St. Francis of Assisi, such as his famous *Canticle of the Sun*, in which he celebrated the harmonious possibilities between man and the forces of nature as our "brothers" and "sisters," without the "ecological imperialism" that was typical of the high Middle Ages.

All of the above observations with regard to the general trends in environmental attitudes during the Middle Ages have, I believe, considerable merit and will be duly incorporated into our more detailed examination of the medieval history of the environment in the pages that follow. But each of them also fails, I believe, to explain the unique and important contribution that the Middle Ages made to environmental history. For example, the classical world of Ancient Greece and Rome provides a precedent for both the exploitative and conservationist attitudes towards nature that have been located in the Middle Ages. Sacred precincts or groves dedicated to the gods were set aside for preservation, while the Orphic school of philosophy championed by Pythagoras and Empedocles preached a universal harmony and kinship among all living things in nature on the grounds that human souls were reincarnated in plants and animals. At the same time, Hippocrates and Aristotle pioneered the study of ecology by demonstrating the complex interrelationships that existed between humans and their physical environments. But even though Aristotle introduced a more scientific study of nature based on classification of organisms, he also adopted a hierarchical view of the natural order that placed man at the apex, thereby justifying a "triumphal" or imperialistic approach despite divorcing his system from the gods. While a host of classical authors deplored the myriad impacts mankind made upon his environment, to include air and water pollution, deforestation, soil erosion, and the wanton hunting and slaughter of animals, they could also extol man as "the orderer of nature" and "the finisher of the creation."[16]

However, I believe that the unprecedented ecological crises of the late Middle Ages forced a radical rethinking of environmental attitudes, one that anticipates the "new ecology" of today. Rather than simply coming full circle to the pessimistic, "adversarial" or "eschataological" approaches, the later medieval period saw a more nuanced, sophisticated view of the environment than ever before. While nature was certainly proven capable once again of waging war on man, humans understood more completely their own role in provoking this war. In this way, then, man was not simply at the whim of nature but played a significant part in his own subjection to it, a sort of combination or summation, if you will, of the preceding views from both the early and high Middle Ages. It should be emphasized, however, that human interference in the environment was comprehended here on completely different terms than our modern understanding. For medieval people, environmental culpability was measured not just directly, in terms of human pollution for example (which in the Middle Ages was of rather limited impact when compared to modern capabilities), but also indirectly through human behavior in other, seemingly unrelated spheres, such as morals and ethics, which could then bring down ecological punishment, as we have seen in Mussis' account of the Black Death. At the same time, man's interconnected relationship with the environment was also being understood on a much more scientific basis than ever before, relying on both ancient authorities and empirical observation, which was likewise a direct result of the late medieval ecological crises and which was being paralleled in other fields, such as medicine and eschatology.[17] Paradoxically, the late Middle Ages was also a time when beliefs in the magical ability to manipulate the natural elements for human purposes and in the occult properties of nature were on the rise. Necromancers devised rituals to command the winds to bear them aloft through the skies, local witches were accused of raising storms or fashioning charms and potions from plant and animal parts in the first stages of the witch-hunt, and Renaissance "magi" such as Marsilio Ficino and Pico della Mirandola mined the works of Neoplatonic philosophers, especially Plotinus, for their notion that all of nature is pervaded and interconnected by magical influences, a belief ably summed up by Ficino's maxim, "Nature is everywhere a magician."

Moreover, there are two other ways in which this book will amend and improve upon previous scholarly interpretations of medieval attitudes towards the environment. One is that, although scholars have given sufficient notice of the breadth or scope of different views about the natural world during the Middle Ages, they have focused less often on the *depth* of those views, or how widely-held such notions were among the population at large. In other words, how deeply did then-current environmental attitudes penetrate the popular consciousness? I believe that the late Middle Ages was a turning point in this regard as well, again largely owing to the ecological crises of plague and famine, which naturally impinged upon the awareness of large numbers of people, if not the entire population, at a time. Second, I am convinced that even those interpretations that adopt a multi-faceted approach to medieval views of the environment—such as the four-part eschatological, adversarial, collaborative, and recreational paradigm narrated earlier—are still overly broad surveys, in that they fail to differentiate among human attitudes towards different *facets* of nature. For example, the chief concern during the later medieval period with regard to natural elements such as the air, water, and the earth seems to have been over their possible corruption or contamination giving rise to disease, such as the Black Death, an idea that medical authorities naturally inherited from Hippocrates' *On Airs, Waters, and Places*. However, the woodlands or forests present a rather different scenario: Since these had largely been cut down or devastated during the expansion of the high Middle Ages, there was perhaps a renewed appreciation for the greenwood as trees started to grow back in the wake of depopulation from the plague. Human attitudes towards animals, both wild and domestic, also have their own, unique history. While medieval bestiaries record a view of animals that was mired in fantasy and religious allegory, the animal trials of the late Middle Ages seem to accord a rationality and near human equivalence to animals that is reminiscent of the Orphic school of Pythagoras during classical times. Consequently, each of these aspects of the natural world—the elements, the forests, and the beasts—will receive its own section in this book, tracing the history of human attitudes towards them throughout the Middle Ages.

As mentioned in the preface, this book will primarily be a cultural survey of medieval attitudes towards the environment. Our main focus will be on what medieval people *thought* about their natural surroundings, rather than what they *did* to them. While the "hard numbers" of mining outputs, fish catches, wood harvests, or ice core readings can indeed help set the contemporary context, most of our recourse will be to evidence of a more literary kind, particularly the genre of natural history such as the *Buch der Natur* (Book of Nature) by Konrad of Megenberg. However, a wide variety of works ranging in seemingly unrelated fields, such as plague treatises, will be mined for the nuggets they reveal of attitudes to nature, and archival records will certainly not be excluded from this effort. Although this may yield a more anecdotal and impressionistic history that disappoints the more scientifically minded, it is my hope that a more entertaining, as well as informative, read will be the result, one that can both blaze a path for future research and uncover its own truths, inconvenient or otherwise.

# Part I

# Air, water, earth

We don't usually think about the air we breathe, the water we drink, or the earth on which we trod and till to grow the food we eat. It is in fact quite natural to take all these most basic elements for granted—that is, until they are taken away from us through naturally occurring or man-made pollution and contamination. It is then, and perhaps only then, that we begin to truly appreciate the elemental building blocks of nature.

### In the beginning …

The first rational, systematic thought in the West on the natural elements and human interactions with them is invariably to be traced back to the Ancient Greeks. The pre-Socratic philosophers, such as Thales of Miletus (c.624–548 B.C.E.) and Anaximander (c.611–547), speculated on the make-up of all living things and how they could be bound together by a single element. (Thales said it was water; Anaximander called it the *apeiron* or the "infinite.") Such thought culminated with the classification system of Aristotle (384–322), who in the *Physics* divided all the elements into five categories, four of which—water, earth, air, and fire—were to be found here below in the earthly realm, while a fifth element, ether, was a different substance entirely that made up the heavens. This mistaken dichotomy was to hold true in the West right up until the Scientific Revolution of the seventeenth century. But the great contribution of the Greeks to man's thinking about nature is to have taken the gods out of the equation. Other, older civilizations had, of course, thought about the elements, yet this was only in the context of a divinely ordained universe. In the Book of Genesis, Noah was able to survive a flood that otherwise wiped out humankind (a story that goes further back to the *Gilgamesh* epic of ancient Mesopotamia), while in Exodus Moses bested pharaoh's army by calling down a pillar of fire and parting the Red Sea. The natural elements could certainly affect man on a cataclysmic scale, and man in turn could sometimes harness and control them, but only at the behest of a supernatural being who was behind it all. The Greeks changed all that.

Perhaps the most important contribution the Greeks made to environmental history was in explaining the relationship between the natural elements and human disease, therefore showing how man was intimately interconnected with his environment. The pioneer of this approach was the Greek physician, Hippocrates of Cos (460–370 B.C.E.), who is credited with the classic work, *On Airs, Waters, and Places*. Hippocrates was primarily interested in how the elemental forces of nature, such as the winds, the stars, changes in the seasons, and water and soil quality, could affect human health. For example, Hippocrates opens *On Airs, Waters, and Places* by admonishing that any physician who wishes to make a study of medicine must pay attention to the following: the special characteristics of the seasons of the year and their changes; a locality's exposure to the sun and to the winds, whether these be hot or cold, that blow through it; and the kinds of waters, whether "hard" or "soft," that people drink from and the topography of the ground from whence the waters find their source, such as "elevated" and "rocky" places or "marshy," low-lying, and sheltered locations. Each of these environmental conditions is related to specific diseases: Those exposed to hot winds will experience discharges from the anus, such as diarrhea and hemorrhoids, as well as fevers, while those living in the face of cold winds will suffer from lung and eye diseases. Those who drink from stagnant waters such as in lakes and marshes will be prone to spleen problems and to dropsy. Almost equally as bad are hard waters that bubble up from rocky places where there are a lot of minerals, such as iron, copper, silver, gold, sulphur, and the like; these tend to cause difficulties in urination. The best waters are those that flow from "elevated grounds and hills of the earth," since these are sweet, light, and clear. Rainwater is also recommended, as Hippocrates explains how the sun evaporates water lying in seas and lakes and thereby separates it from its impurities, such as salt; nonetheless, it should be boiled before drinking because it spoils easily. Water melted from snow and ice, however, is bad for humans, for once frozen and congealed, such waters "never again recover their former nature." River waters are also to be avoided, since in them flow together impurities from various sources, so that they give rise to kidney stones and bladder problems. People should also watch for patterns of the stars and changes in the seasons for what these might portend about their health: If the autumn is rainy, the winter mild, and the spring and summer seasonable, then the year will be a healthy one; on the other hand, a cold, dry winter followed by a hot, rainy spring forecasts a summer full of fevers, eye diseases, and dysenteries.

Since the natural environment also determined humans' humoral make-up or constitution, Hippocrates could even make the case that a region's climate determined national character. Because the climate in Asia was generally mild and temperate and the continent was abundantly supplied with good waters and soils, its inhabitants, whether these be humans, animals, or plants, were "beautiful and large" as well as vigorous and prolific, and their dispositions were "gentle and affectionate"; on the other hand, a placid environment also made for a character lacking in courage, spirit, and enterprise. Europe, by contrast, was marked by great diversity in its people that reflected the vagaries of its climate and seasons: Those who lived on stony soils and in mountainous and exposed places tended to be blond, rugged, enterprising, and war-like; those who lived on low-lying, ill-ventilated meadowlands fed by stagnant waters were more likely to be dark haired and complexioned as well as cowardly and sluggish. In general, though, Europe's inhabitants were characterized as being more courageous and industrious than those in Asia as a result of their more "changeable climate"; this was also reflected in their governmental institutions, which in Asia tended to be more despotic and enslaving. Here, perhaps, Hippocrates is governed more by the ethnic bias that the Greeks felt towards their "barbarian" neighbors in Persia, which also shows up in the histories of Herodotus, than by environmental science. Nonetheless, Hippocrates explains the difference in character and physical features between the two continents as owing to the different "coagulation of semen" that was effected by the differing climates.[1]

Hippocrates thus provided a model for how air, water, and the earth could impact humans, on both an individual and regional scale. The transmission of his ideas to medieval Europe came about gradually, largely through Persian Islamic intermediaries who wrote on medicine and made a study of Hippocrates and Galen (129–c.217 C.E.). The connection between disease and the air, known as the "miasmatic" theory from miasma, referring to foul vapors or exhalations in the atmosphere and which has its origins in the Greek word for "pollution," was further developed by Ali ibn Abbas al-Majusi, known as Haly Abbas in the West. A Persian physician, Haly Abbas, who lived during the tenth century and wrote a celebrated textbook, the *Kitab al-Maliki* or the *Complete Book of the Medical Art* (also known in its Latin translation as the *Liber Pantegni*), explained that air can undergo a change in substance and qualities such that it is turned into decay and putrefaction, thereby causing epidemic illnesses. This happens when the miasma, or bad vapors and exhalations, arise from such sources as rotting fruits and plants, stagnant waters in ditches and marshes, garbage and filth in town sewers, and decomposing animal carcasses and dead human bodies such as those who fell on the field of battle; the miasma then mixes in with the air and alters its uniform, simple composition. When humans breathe in this bad, "pestilential" air, they then fall ill to epidemic diseases such as the plague. Haly Abbas also related disease to unnatural changes in the seasons of the year, an idea obviously indebted to Hippocrates.[2] Here he was joined by his contemporary colleague from Persia, al-Razi, known as Rhazes in the West (865–925): According to medieval redactions of his most famous works, such as the *Kitab-al-Mansuri* (Book of Medicine for Mansur) and the *Secret of Secrets*, he forecast that a foggy and rainy autumn would result in pestilences and skin diseases characterized by swellings or *apostemes*, or that future plagues were indicated by unnatural changes in the weather within a given season, such as the air alternating between hot and cold or rainy and dry for several days in succession.[3]

Plenty of other Arabic authors in this period wrote on how air and water pollution can arise in a given region and how it can impact human health and how this can be prevented. Qusta ibn Luqa (820–912), a Melkite Greek

Christian writing under Abbasid patronage in Baghdad, explained in his treatise, *On Contagion*, how disease can be spread by the "surrounding air," which is spoiled by pollution sources arising from the earth, such as swampy vapors, smoke from furnaces and cremated bodies, and decomposing things left to rot out in the sun. Far more extensive is al-Tamini's tenth-century treatise on treating "air spoilage," which discusses different types of polluted air in the lands of Islam depending on weather and geographical conditions and the various diseases that can arise from this pollution. He also gives methods of treating polluted air, such as by burning incense (whose recipes are derived from an earlier author, al-Kindi), and by treating stagnant water sources. Later authors focused on specific, municipal sources of air and water pollution in some of Islam's largest cities: Ali ibn Ridwan (c.988–1061) and Abd-el-Latif (1162–1231) wrote on Cairo; Ibn Jumay' (twelfth century) wrote on Alexandria; and Yaʻqūb al-Isrāʼīlī (twelfth century) on Damascus. Abd-el-Latif in particular noted how various aspects of urban planning in Cairo, such as ventilation shafts in houses, sewage construction, and street layout and orientation, could affect human exposure to sources of air and water pollution.[4] Here, too, the Muslims could be said to be continuing a tradition going back to ancient authors, particularly among the Romans, who complained of pollution coming from their very own marvels of engineering that occurred especially in urban environments, such as the dangers of drawing water from lead pipes.[5]

Perhaps the most famous commentator in Arabic on the relationship between environment and disease was the great Persian medical authority, Ibn Sina, known as Avicenna in the West (c.980–1037), whose encyclopedic *Al-Qanun fi al-Tibb*, or *Canon of Medicine*, rivaled Haly Abbas' *Kitab*. Eventually translated into Latin during the thirteenth century by Gerard de Sabloneta, the *Canon* became the standard medical textbook in schools throughout Europe and was a chief authority for plague doctors during the later Middle Ages, such as Gentile da Foligno, who wrote a commentary on nearly all of its five books. In the fourth book of the *Canon*, Avicenna follows Haly Abbas in explaining that air, like water, is a simple element that does not putrefy into a pestilence-causing substance unless bad vapors are mixed in with it, changing the entire quality of the air to bad, just as water does not corrupt unless mixed in with bad "earthy bodies." In the case of air, this can happen whenever winds bring with them "bad fumes from fetid places," such as deep valleys or battlefields strewn with unburied corpses. In addition, Avicenna connects all three elements—air, water, and earth—to disease by explaining that sometimes pestilences arise from "putridities" that occur "in the bowels of the earth" that then "cause harm to water and air" but that the reason for why this happens is unknown. However, pestilential fevers are specifically caused by "unsettled and humid air"; moreover, the ultimate cause of this is a synergistic interaction between "celestial figures" such as heavenly bodies or planets, and "terrestrial dispositions" such as occur here on earth, which then cause a "vehement moisture" that releases "vapors and fumes" into the air.

When breathed in, the putrefied air "comes to the heart where it corrupts the complexion of the vital spirits that are in it and rots them" by surrounding the heart with its moisture. There then occurs an "unnatural heat" that spreads throughout the body, resulting in pestilential fever.[6] It was crucial that Avicenna establish this connection between the celestial and terrestrial causes, as Islamic tradition dictated that the ultimate cause of any disease be traced back to God, even if there be secondary causes that can explain the disease's immediate appearance here on earth.

Avicenna's other main contribution in this regard was to instruct his readers in how to look for signs that would herald plague: In terms of the air, Avicenna advised scanning the skies for celestial portents such as shooting stars, or *asuhub* in Arabic, a notion derived from Aristotle's *Meteorics*. These would typically occur at the beginning of autumn; in winter, one should look for the arrival of southerly and easterly winds, or else an unsettled and nebulous quality to the air, such as when it seems that it is about to rain and yet it does not rain. A paucity of rain in the spring, accompanied by cold nights and stiflingly warm days, portends a summer plague. Like Rhazes, Avicenna also forecast pestilential fevers when the weather changed many times in one day, or was clear one day and stormy the next. On the other hand, Avicenna also believed that plague portents could come from the earth below as well as from the air above: A perhaps unique contribution of his was to find signs of a coming pestilence in frogs and snakes suddenly multiplying over the ground, which for him signified a putridity in the earth generating such creatures. At the same time, "you may see mice and [other] animals that live underground flee to the surface of the earth and manifestly exit [from there]." Obviously, modern commentators have speculated that this may refer to a quite real phenomenon whereby rodents, when infected by the plague, come out of their holes to stumble around as if in a drunken stupor and eventually perish in a mass die-off, or what is termed an epizootic, that generally heralds a corresponding epidemic among humans. However, one should beware of an element of fantasy that creeps in at the end of this chapter: Avicenna warns his readers to watch out for the *alalzalilzi*, an "animal of an evil nature" that seems to be a cross between a serpent and a bird, perhaps akin to the mythological basilisk of the Western bestiary tradition, since this creature and other animals like it "flee their nests and go [far] away from them and perchance abandon their eggs" when a pestilential corruption emanates from the earth. While some of these signs may well have originated in a close observation of nature, others seem to have had their ultimate inspiration in classical augury, such as the divination of future events by noting the flight of birds.[7] As is typical of much of medieval literature, the prosaic and the marvelous mix here almost imperceptibly and are granted equal credence, presenting a challenge to the modern reader in terms of teasing out which is which. By the time of the late Middle Ages during the Black Death of 1348–49, plague treatise authors were to greatly elaborate on these signs, explaining how birds migrated from the mountains to the plains or vice-versa in order to escape the corrupt air emanating from the heavens or the earth, and likewise frogs, mice, serpents, and other ground-dwelling creatures would either flee from their holes or burrow deeper into them (see "Animals and disease" in Part III). Yet there is no evidence that this was anything but a fanciful elaboration on Avicenna's authority, rather than direct testimony of phenomena actually observed in nature.

In the first book of the *Canon*, Avicenna discusses the natural elements and the relationship of changes of the seasons and climate with disease.[8] These sections are obviously much indebted to Aristotle and Hippocrates, although Avicenna also had access to other authorities not available to the Greeks, such as the Qur'an and Indian and Chinese concepts. In part two of the first book, Avicenna makes the observation that a physician must pay close attention to the natural efflorescence of the seasons, such as when trees leaf out and flowers and fruits begin to form, more so than other observers such as astronomers, since this is what has immediate impact on human health. He then discusses the specific qualities of each season and their impact upon the human body, particularly in terms of heat and moisture that are the source of putrefaction. Avicenna notes how each season produces its own diseases and affects people of different humoral temperaments differently. The autumn time is the worst for humans in terms of disease for a number of reasons, some of which are reprised over three centuries later by Gentile da Foligno in order to explain why the Black Death will wax during the autumn (even though Gentile himself died in June): These include that fruits (which are moist, leading to putrefaction) are now in abundance and that the bodily humors, which have been fermented by the heat of the preceding summer, are now suddenly coagulated by the onset of cooler weather in the fall.[9] A region's particular climate, which also greatly affects human dispositions and health, can vary by a number of factors, including the types of winds it is exposed to, its terrain (whether mountainous or low-lying), and the characteristics of its soil and water. In discussing changes to the air caused by winds, Avicenna follows Haly Abbas in distinguishing between substantial and qualitative changes to the air, only the former of which can result in air that is described as "pestilential," or liable to cause putrefaction and disease (which tends to occur at the end of summer and during the autumn). Since the element of the air is a simple one that cannot putrefy, Avicenna emphasizes that he is referring here to the "atmosphere," a composite substance composed of air, watery vapors, and terrestrial and fiery particles. Avicenna discusses the various kinds of countries and their implications for human health, whether these be hot, cold, wet and damp, mountainous, low-lying, maritime, southerly, northerly, etc. He even gives advice on environmental factors to take into consideration when choosing one's dwelling, such as soil quality, altitude, wind exposure, water supply, and so on. With regard to the last category, water, Avicenna devotes a whole separate section that is heavily dependent on Hippocrates' *On Air, Waters, and Places*. For instance, Avicenna notes, like Hippocrates, that rain water and spring water are the best kinds of waters, while stagnant, marshy waters and melted snow can be harmful. However, Avicenna also comments on how humans can artificially modify and thereby improve the quality of their waters, such as by boiling and distillation, which separates out impurities and sediments.[10] In another work, *Repelling General Harm from the Human Body*, Avicenna classifies various types of air and water that have the potential to harm humans: "Coal smoke" seems an obvious example of man-made air pollution, while "arsenic water" or "ammonia water" are sure to be poisonous if drunk. Avicenna's contemporary, Abū Sahl al-Masīhī, and a later author from the thirteenth century, Ibn al-Quff, likewise wrote preventative treatises on preserving human health from environmental pollution: The former focused on abnormal changes to the air, while the latter addressed water quality and treatment.[11]

The contribution of the Arabs to medicine and science is often characterized as merely passing on to posterity the innovation of the Greeks, without making their own observations and experiments. (The same, however, *cannot* be said of philosophy, where the Arabs had to adapt Greek ideas to the religious precepts of the Qur'an.) It is therefore said of the Middle Ages that "man's attempts to observe nature herself were very weak."[12] We will see, however, that with the late Middle Ages and the advent of the Black Death, direct, personal observations of nature will once again take their place alongside appeals to authority in discussions of environmental causes of disease. The environmental catastrophe of the Black Death thus seems to have convinced many of the importance of paying renewed attention to contemporary natural phenomena, which perhaps comprises a radically different attitude towards nature. But before we take our leave entirely of the Arab contribution to environmental literature, we should note that the tenth century also saw works of comparative geography, such as by al-Khwarazmi and al-Mas'udi, that likewise have an environmental component. Based largely on Ptolemy, Muslim geographical treatises divided the world into seven latitudinal zones or "climes," each of which predisposed their inhabitants to possess attributes that reflected their particular climate. For example, since the European "Franks" lived in the northern climes, where the weather was cold and damp, their characteristic attributes tended to include being dull-witted, coarse and crude in their mannerisms, and in physical terms to have a lumbering stature and an excessively pale complexion. These prejudicial stereotypes based on geography lingered on into the next century, even in Spain, where presumably Muslims had much more direct, intimate contact with Christian Europeans than elsewhere.[13] This is really no different than Hippocrates' climactic profiling of Asiatic "barbarians" in *On Airs, Waters, and Places*.

A standard part of encyclopedic compendiums of knowledge, even those with a medical focus such as the *Liber Pantegni* of Haly Abbas and Avicenna's *Canon*, was a description of the elements of nature, including air, water, and the earth, a tradition that again goes back to the classical authors of ancient Greece and Rome. In the Christian West, the first work to address natural history on the model of ancient classics such as Lucretius' *De Rerum Natura*

(On the Nature of Things) or Pliny's *Historia Naturalis* (Natural History) was the *Etymologies* of Isidore of Seville (c.560–636). In Book XIII, Isidore addresses the natural elements of the world, which, following the Greeks, he lists as fire, earth, air, and water, declaring them to be interdependent upon each other as well as interchangeable as they seek out their true origins. His main focus here is on air and water, explaining airy phenomena such as thunder, lightning, rainbows, and winds, and dividing waters into oceans, seas, lakes, and rivers which are liable to such things as tides and floods. The earth is discussed in Book XIV mainly as a geographical entity in terms of its three continents, Asia, Europe, and Africa, while Book XVI discusses the precious products that issue from the earth, namely precious stones and metals. Once again, there is a curious mixture of the fabulous and the factual in Isidore's account. For example, in the section on different kinds of water (XIII.xiii), Isidore accurately describes the origins of hot springs, noting how the water "boils" underground due to the sulphuric heat of the earth, but his claim that the rivers of Thessaly turn the sheep drinking them either black, white, or mixed and other such tales must be put down to pure legend. In an earlier section on winds (XIII.xi), Isidore presciently describes the environmental impacts of the airy elements, such as that the south wind (*auster*) sometimes corrupts the air and so brings with it pestilences, while the northern wind (*aquilo*) cleanses the air of its corruption; while not accurate from a modern epidemiological point of view, this was to become a standard feature of many a plague treatise when discussing the first "non-natural" in a health regimen, namely the air, in the aftermath of the Black Death during the late Middle Ages. Isidore likewise describes how floods can be destructive to man, not just in Biblical times but also in his own, due to the heavy rains that gave rise to swollen rivers, an assertion that would come to pass during the Great Famine of 1315–22 caused by excessively wet weather (section XIII.xxii).[14] Over a century later, much of this information in the *Etymologies* was to be re-issued almost word for word by the Carolingian scholar, Rabanus Maurus (780–856), in Books IX, XI, XII, and XVII of the *De Universo* (On the Universe, also known as *De Rerum Naturis*, or On the Nature of Things), while during the high and late Middle Ages natural history was to be taken up by encylopedists such as Vincent of Beauvais (c.1190–1264), Bartholomaeus Anglicus or Bartholomew the Englishman (1203–72), Thomas of Cantimpré (1201–72), Albertus the Great (c.1200–80), and Konrad of Megenberg (1309–74).[15]

## Worshipping the elements

The elements of nature were also the battleground in the early Middle Ages for epic contests between Christian missionaries and pagan cults that worshipped their gods as part of the natural landscape. This was a fertile field of endeavor: Hundreds of healing water shrines at lakes, rivers, and springs, some of them hot thermal ones, are attested throughout the former Roman Empire, 80 of them in Gaul alone, one of which was still in use as a pagan center as late as 700 C.E. The water was either drunk, selectively applied to the ailing part of the body, or subjected to a full immersion, and in return coin offerings were usually thrown into the water, which if recovered allow us to date the shrine.[16] Some of the most famous of these water shrines were the hot springs at Bath in England dedicated to the Celtic water goddess Sulis, whom the Romans conflated with their healing goddess, Minerva, and the *Fontes Sequanae*, or "springs of Sequana," at Dijon in Burgundy, the source of the River Seine. Both sites were older Celtic shrines that were taken over by the Romans and "monumentalized" by building an elaborate temple complex. At *Fontes Sequanae*, votive offerings took the form of wooden and later stone carvings of the pilgrim suppliants in the dress appropriate to their station bearing gifts to the goddess, or else depictions of their afflicted limbs or bodily organs, which they believed would be replaced by the Sequana goddess with a whole or healthy substitute. At Bath, in addition to coin offerings to Sulis-Minerva, lead *defixiones* or curse tablets have been recovered which asked the goddess to take revenge for any kind of ill done to the devotee. Other offerings besides the usual items of coins, food, or clothing could include more durable, "prestige" goods worked in metal that have been recovered from archaeological sites, such as chariots, weapons, or tools, and even animal and human sacrifices could be deposited in sacred rivers and lakes; these indicate a more propitiatory intent towards the body of water that associates it not just with healing or beneficial properties but also with potential destruction that hopefully could be averted, such as from flooding or drowning. Sites at Llyn Cerrig Bach and Llyn Fawr in Wales have yielded numerous bronze and iron votive objects of high quality workmanship that have been preserved in the peat bog sediment that was formerly at the bottom of their lakes, while the Lindow Man who had been buried face-down in a swamp after consuming a ritual meal that included mistletoe and after having been axed, throttled, and had his throat cut may have been an offering to the Celtic god Teutates, who was appeased with drowned victims.[17] Another use for water was in magical rain-making rituals, such as Burchard of Worms witnessed among women of the Rhineland in the early eleventh century.[18] Since water shrines were virtually indestructible, the Church, if it wished to end pagan worship there completely, usually had to appropriate them and assign them to one of its local saints, such as the "holy wells" dedicated to St. Brigit one comes across quite often in Ireland. As late as 1410 a "certain well and stone" in the parish of Turnastone in the diocese of Hereford in England was still being worshipped with "genuflections and offerings of diverse things," while the worshippers, described as a "copious multitude," "carried away with them parts of the said stone and well and, if water was lacking, the mud of the same, and kept and guarded them like relics."[19]

Less popular but also hard to eradicate completely, unlike the sacred groves where trees could simply be cut down, were mountain and hilltop shrines and idol-bearing pits, rocks, and stones. Celtic deities often made their homes on high peaks: These included Vosegus of the Vosges Mountains and Albiorix on

Mount Ventoux in France, and Latobius on Mount Koralpe in Austria; likewise in the East, parallel cults existed such as to Dolichenus on Mount Doliche in Syria. In the Central Pyrenees of southern Gaul, the names of the gods Andeis, Arpeninus, Aereda, and Arixus have been recovered from inscriptions found on hillocks or heights of land on the mountains themselves, while the goddess Baeserta was appropriated by the Christian cult of Our Lady of Basert, to whom travelers prayed before traversing this lonely and sinister mountain. In the last example, Baeserta was held not to reside on the very summit of the mountain but along a difficult and dangerous passage through it. The Pyrenean deities were often associated with the Roman god Jupiter as lord of the skies, and in Celtic mythology a sky-god was also given command of the weather, particularly sunlight, storms, and thunder and lightning. More down to earth, Celtic gods were also worshiped in specially dug pits or shafts in the ground, where modern excavations in Britain, France, and Germany have uncovered the remains of animal and human sacrificial victims as well as carved figures of the gods themselves. Above ground, single stones ranging in size from huge boulders and pillar-like menhirs to more moveable monuments, some in their natural or "undressed" state and others dressed or modified with carvings, inscriptions, and other markings, were also worshipped, especially in the British Isles and Gaul. As at water shrines, sacrifices and offerings were made at these sites, sometimes in the form of food, a lighted candle, or even a small stone pitched at a larger one or added to a cairn. Again, the main purpose of the idol seems to have been a healing one, a power that was especially associated with the earth as the source and sustainer of all life. This even extended to ingesting dirt for its medicinal value. The Celts also worshipped the earth in the form of a Mother Goddess—equivalent to the chthonic deities of ancient Greek and Roman mythology—that were often depicted in triplicate images, whose cult ensured fertility either in plant, animal, or human form and which had to be appeased in order to maintain the essential harmony or balance of nature.[20]

The air was likewise the pagan locus for the divine: As far back as the first century C.E., Roman authors such as Seneca and Pliny the Elder were commenting upon how "hail guards" and "cloud-drivers" would be on the lookout for hail or destructive winds and storms and who would then signal the people to offer up sacrifices of lambs, chickens, or even their own blood to keep the weather at bay from their fields.[21] Seven centuries later, in 815–16, Agobard, archbishop of Lyons, encountered similar beliefs among a broad cross-section of his flock, only this time people believed that certain "storm-makers" could raise thunder, lightning, and hail at will, against whom they had to utter counter-curses. Moreover, it was believed that the storm-makers conspired with "cloud-sailors" who came down in their ships from a sky kingdom called *Magonia* (literally, "Magic Land") in order to carry off grain and other produce that had been beaten down by the hail. All this was in addition to "defenders" who claimed to have the power to magically ward off such storms in exchange for regular tribute in the form of a percentage of crops, which is more along the lines of what was described by Seneca.[22]

Nonetheless, it is impossible to know exactly what the natives believed in and why, since Agobard is concerned above all not with describing such "absurd" beliefs, but only in refuting them. His concern with storm-makers was apparently widely shared among authorities at the time: In 789 Charlemagne in his capitulary, *Admonitio generalis* (General admonition), ordered that storm-makers, among other enchanters and diviners, be "corrected" or else condemned, while his Church council gave more explicit instructions on how such transgressors were to be brought to heel. Wherever they were apprehended, those suspected of raising storms or performing other *maleficia* were to be subjected to a "most diligent examination" by the chief priest of the local diocese, all the while to be kept in prison, yet not under such harsh conditions that "they lose their life," but still until they confess and promise "to amend their sins."[23] Although it is not mentioned, it is likely that torture was used, since this seems one of the few instances in Charlemagne's reign when inquisitorial procedure, which did not operate by the normal laws of accusation, applied; moreover, Germanic law codes did allow for torture in cases of capital crimes and other serious offenses, based on the precedent provided by Roman law.[24] At the Council of Paris convened in 829 under Charlemagne's son and successor, Louis the Pious, storm-makers found guilty by the Church were to be handed over to the lay ruler for punishment, a policy later adopted for heretics under the Inquisition. Storm-making was also the target of penitentials designed as manuals for confessors when probing for the sins of their flock, which from the eighth century authorized priests to impose penances of up to seven years for the practice.[25]

At this stage, it does not seem that the powers that be in Frankish society took seriously the common people's belief that human beings had the power to control the weather and the elements: Agobard, for example, characterized it as a "great madness" or "foolishness" that could easily be disproved through rational arguments, such as that storm-makers seemed incapable of making it rain or hail when their crops were dry or enemies were at hand. Likewise, when the Anglo-Saxon courtier Cathwulf advised Charlemagne in 775 to act against storm-makers, witches, diviners, and other "evil-doers," he did so on the grounds that the king needed to defend his Christian "honor" by upholding the laws and eliminating injustice and other vices from his kingdom, not out of a sense that such practices posed a real and present danger to his subjects. For Cathwulf, Agobard, and others, the real offense behind the belief in storm-making was that it was a form of pagan backsliding and therefore an implicit denial of the Christian faith; to accord any reality to such superstition would be to succumb to the "foolishness" itself. And yet, the joke may have been on the authorities themselves, because in some circles, the belief in storm-makers may have been a charade to hide newly harvested grain from inquisitive tax collectors.[26]

On the other hand, there may have been something to all the superstitious fears of bad weather, because some modern climatologists believe that the weather of the ninth century was stormier than before, and even today the region of France where Agobard tracked his storm-makers is known for its violent thunder and hailstorms. But instead of countering storm-making with appeals to naturalistic explanations for hailstorms such as we would employ today, Agobard peremptorily declared that all such phenomena must come from God alone, an assertion he backed up with ample quotations from the Old Testament. Nor was a naturalistic explanation the intent of treatises on thunder and bad weather during the medieval period, but rather the focus was on what thunder portends for the human condition, based on what direction the thunder came from and in what month it occurred.[27] The early Church father, St. Augustine of Hippo (354–430), had laid down the law here by pronouncing that all natural magic, or the belief in occult powers in nature, was the work of superstition and worship of demons, which had no real validity when compared to Christian miracles, even though Augustine did acknowledge certain wondrous properties in nature, such as the power of a magnet to attract iron or of goat's blood to shatter adamant.[28] But we will see that, since monks and saints were accorded power to affect the elements, eventually by the end of the Middle Ages the Church came round to the position that witches, too, could perform real and genuine weather magic, a reversal that was made possible by the view that witches were agents of the devil and that the devil's power to rival God's was so evidently at work in the world.

The rich but biased hagiographical literature of the Church offers us glimpses into the struggle between Christian and pagan forces for control over the natural elements. Invariably, such contests were portrayed as unequal ones in which the "demons" residing in the earth, water, or air were conquered by their Christian adversaries and driven out. A classic example comes in the *Life of St. Gall*, in which the Irish monk, Gallus, traveled in the company of St. Columbanus to the Lake Constance region of present-day south-west Germany and Switzerland in c.612 to convert a pagan Germanic tribe who worshiped "idols of gilded metal." One night, while laying fishing nets in the lake, Gallus overheard a conversation between two demons, one residing in the mountain, the other in the water. The demon of the mountain called out to his colleague in the lake to help him drive the Christian strangers out of the land, but the demon of the lake responded that he had no power to harm Gallus, since his sign of the cross protected him and he never slept. It is not clear whether these demons normally resided among these natural elements, since previously Gallus had smashed some idols and "cast them into the depths of the lake," while the demon of the mountain complained that he had been driven out of his temple. Whatever the case, it is clear that nature was seen, at least by the Christian monks, as a place of refuge for pagan gods and therefore something to be watchful of, until Columbanus was able to drive out the demons from even this sanctuary, accompanied by their "wailing and lamentation" that "echoed from the mountain top." Nonetheless, the real threat to Gallus of drowning in the lake may not have come from demonic water but from man, since some pagans who witnessed Gallus' earlier destruction of their idols were reported to be "angry and enraged" at this sacrilege and "departed in wrath." Although the Germanic pagans clearly could find religious significance in the natural elements, it is not clear whether in this particular instance they perceived the Christians as enemies of nature, or simply the enemies of their gods, wherever they might reside. From the Christian point of view, however, it is clear that a saint's power over the elements constituted proof of the supreme power of the Christian god over rival gods or "demons."

In other saints' lives, Christian missionaries conducted exorcisms of demons causing floods, crop failure, or bad weather not only in order to demonstrate their benevolent power over nature but also to thereby win over the hearts of their converts. A large degree of assimilation with prior pagan practices is assumed here. A number of saints in both the Eastern and Western parts of the late Roman Empire were reputed to have the power to turn back hailstorms: One of them, the seventh-century St. Theodore of Sykeon in Anatolia in present-day Turkey, was appealed to by the villagers of Reake just like a pagan "cloud-driver" whenever "a fierce cloud from time to time visited the land and brought the hail upon the vines when the grapes were full." Theodore would plant *horoi* or markers such as wooden crosses at the corners of the field and then retire to his monastery for prayer, in return for which the grateful men of the village would bring a yearly tribute to the monks of wine and grapes, akin to a sacrificial offering left at the shrine of the pagan gods. According to the text of one exorcism that survives from this region, the demon of the skies was adjured "by the power of the God of Hosts and the throne of the Lord" and "by the letters of the planets" to stay outside the *horoi* of corner-stones, while the archangels Raphael, Ragouel, Israel, and Agotheol were asked to seal the boundary.[29] Another example of this assimilation between Christian and pagan adjurations of the elements, this time of the earth, is the Anglo-Saxon "*æcerbot* charm" or field ceremony that was performed before plowing and planting of the fields in order to restore their fertility or counteract witchcraft. Here the charm's enchanter, usually the village priest acting in cooperation with the local farmer, called upon both Father God and Mother Earth to ensure bountiful crops, beginning and ending the charm with the "Be fruitful and multiply" passage from Genesis 1:28 and a Trinitarian blessing or Our Father prayer. In between, the ceremony has rituals that seem to combine biblical (Old Testament) and pagan elements, such as anointing four pieces of sod taken from the four corners of the field with oil, honey, milk, and wine, or placing incense, fennel, soap, and salt in a hollowed-out hole in the plough-tail, or burying bread made from every kind of flour in the first furrow. What made these rituals "Christian" was simply the addition of holy water and wooden crosses placed over and under the sods, with accompanying prayers.[30] How all this is different from pagan magic so fulminated against by Augustine is hard to discern.

Instead of contesting the elements, saints could also collaborate with them for the benefit of the Church and the entire monastic community, as well as constituting proof of their sanctity. This is foreshadowed in stories of the early Christian martyrs, where the elements saved Christians from persecution or at least eased the pain of their passing, as in the fourth-century life of Bishop Marculus, who was gently cradled by the rocks as he was thrown down to his death from a high cliff, or where rain put out the funeral pyres of those sentenced to be burnt alive.[31] In Bede the Venerable's early eighth-century *Life of St. Cuthbert*, the holy man was able to call forth a spring out of dry land in his hermitage on the Farne Islands, and even the sea obeyed him by throwing up on the shore a piece of wood exactly corresponding to the desired length—twelve feet—needed for construction of a shed attached to his dwelling. Bede's explanation of Cuthbert's power over the elements is quite illuminating: He said that it was "hardly strange that the rest of creation should obey the wishes and commands of a man who has dedicated himself with complete sincerity to the Lord's service," since such human "dominion over creation" was "ours by right" but which was lost through "neglecting to serve its Creator."[32] Evidently, Bede was harking back to a biblical age, perhaps when Adam and Eve were still in the garden of Eden before the fall as told in Genesis, which implies that man's intended state in nature was to direct and control it at his will. This certainly anticipates the "collaborative" stage of the high Middle Ages, when medieval man became more confident in his ability to tame and even exploit his environment, even though from Bede's perspective this ability was as yet limited to a privileged, select few.

On the other hand, an alternative tradition emerged at the dawn of the high Middle Ages in which nature expressly refused to obey the human will and there were clear limits to what man could command of the natural elements. We see this already in some of the Latin verse produced during the "Carolingian Renaissance," such as Sedulius Scotus' letter to Einhard in 848, in which he personified the north wind as "white faced Borea" or "Aquilon" who greeted travelers to France with a howling cold blast that tore into its victims with "cruel talons."[33] The autonomy of the elements, however, is best represented by an incident in the life of Cnut the Great, king of England, Denmark, and Norway during the early decades of the eleventh century. As told in the thirteenth-century chronicle of Henry of Huntingdon, the story goes like this:

> When at the summit of his power, [Cnut] ordered a seat to be placed for him on the sea-shore when the tide was coming in; thus seated, he shouted to the flowing sea, "Thou, too, art subject to my command, as the land on which I am seated is mine, and no one has ever resisted my commands with impunity. I command you, then, not to flow over my land, nor presume to wet the feet and robe of your lord." The tide, however, continuing to rise as usual, dashed over his feet and legs without respect to his royal person. "Let all men know how empty and worthless is the power of kings, for there is none worthy of the name, but He whom heaven, earth, and sea obey by eternal laws."[34]

It is true that Cnut was no saint; he was suspected of bigamy for keeping a handfast wife, Aelfgifu of Northampton, at the same time as he wedded

Emma of Normandy, and his native country of Denmark was but recently converted to Christianity under his father, Svein Forkbeard. But the legend has all the trappings of a Christian morality tale, even though it has obvious parallels with the pagan history of Herodotus, who told of how the Persian King Xerxes futilely punished the waves of the Hellespont with 300 lashes for destroying a bridge over which he hoped to cross his army into Greece in 380 B.C.E. In both cases, these are cautionary tales against pride, which the Greeks called *hubris* and Herodotus blamed for starting the Persian Wars and which the Christians denounced as one of the seven deadly sins responsible for Lucifer's rebellion against God. To presume to command the elements, then, is to be guilty of assuming god-like powers, one of the worst possible sacrileges in almost any of the world's religions or mythologies. The only reason Christian saints can do it is by humbly acting as a conduit for God's power. The author of Cnut's tale would presumably interpret a saint's ability to do what a king could not as a miraculous exception that proves the rule of ordinary human impotence before the forces of nature.

Yet Cnut's legend may also owe something to pagan Viking sensibilities with regard to the natural elements on the eve of their conversion to Christianity. The Viking pantheon, particularly the chief god, Odin, were known for being fickle and unreliable, even towards their most devoted worshippers, so that calls upon the gods to control the elements might not always produce the desired result. This is well demonstrated in *Egil's Saga*, by the tenth-century Icelandic skald or poet, Egil Skallagrimsson. In one of his poems, Egil mourned his favorite son, who drowned at sea. He cursed Odin for "stealing" his son from him by allowing a "rough storm" to capsize his son's ship, and even though he continued to make offerings to Odin, he did so "not in eagerness" and with but small compensation in the form of his poetic gifts. Either the gods could not always master the world they had created, or else they indifferently heeded the appeals of their devotees. Normally, of course, it was the Norse god Thor, the son of Odin and Jord, or the earth, who was most associated with the elements: According to the Christian author Adam of Bremen, writing in 1080, Thor was at the center of the pagan shrine at Old Uppsala in Sweden and was the god of the air, controlling thunder, lightning, wind, rain, fair weather, and crops. He therefore seems somewhat akin to the Celtic thunder god, Taranis,[35] but by contrast Thor's cult, judging from the extent of hammer amulets uncovered from excavated Viking graves, sometimes right alongside Christian crosses, was apparently quite popular throughout the Viking world, especially among the farmers, fishermen, and other common folk who made up the bulk of society, and it endured right down into the era of Christian conversion. For it was Thor who in Viking mythology was the god most devoted to combating the deadly chaos of giants, serpents, and other monsters that threatened to engulf Midgard, or Middle Earth, where men dwelled as well as the entire Viking cosmos. Appropriately enough, Thor was worshipped in open-air sites such as hill shrines and sacred groves, where perhaps Vikings felt closest to the elements. In an age before sophisticated navigational instruments, Viking sailors also had to rely on close observation of memorable features of the natural landscape in order to complete their astounding oceanic voyages, such as to "Vinland" on the eastern coast of North America around the year 1000.

## The Medieval Warm Period

The North Atlantic voyages of the Vikings are a good introduction to the next, "collaborative" stage of Europeans' relationship with the elements, which must have seemed more cooperative with the advent of the "Great Warming" in the world's climate, which generally extended from 800 to 1300 C.E. The Vikings' explorations and settlements in the western North Atlantic around the year 1000, such as in Iceland, Greenland, and Vinland, were made possible at this time by the retreat of sea ice and the warmer summers and winters that made human habitation and cultivation possible. Similarly, when colder weather began to set in with the advent of the "Little Ice Age" in the fourteenth century, the settlements in Greenland were doomed. A fascinating source for the study of climate change and human impacts at these sites are the "insect fauna" that have been discovered buried among the archaeological artifacts. By tracing the appearance and disappearance of various insect species scholars can track the modifications imposed upon the environment by climate and pollution.[36]

The Medieval Warm Period, sometimes also referred to as the "Little Optimum," can be deduced from a number of records. While instrument readings, such as from thermometers and rainfall gauges, only date back as far as the late seventeenth century, there are plenty of other sources to hand. These include documentary data such as chronicles, annals, grain price and grape harvest records, and sometimes even weather journals; but the problem with documentary sources are that, although descriptively fresh and immediate to their time period, their records tend to be impressionistic and of limited historical range and reliability.[37] Consequently, they need to be supplemented with proxy data such as oxygen isotope and radiocarbon readings from ice cores and lake sediments, pollen and peat bog analyses, tree rings (dendrochronology), and, as already mentioned, insect fauna remains; while proxy records are more indirect evidence, they can also yield more precise readings and for a longer duration.[38] All these confirm that Europe was indeed experiencing a significant warm period especially between 1100 and 1300, when summer temperatures in England and on the Continent were an average of one degree Celsius, or almost two degrees Fahrenheit, hotter than averages for the twentieth century.[39] (The current phenomenon of global warming, it should be pointed out, is not really comparable to the Medieval Warm Period.)

Such large-scale climactic changes are now thought to be due to the interplay between arctic air and ocean seawater at two "downwelling" sites off the coasts of Iceland and Greenland that power thermohaline circulation, otherwise known as the great ocean conveyer belt, which acts as a kind of heat pump for Europe, largely through the circulation (or lack thereof) of warm ocean currents (caused ultimately by cyclical shifts in the earth's orbit and axis).[40] In this way, then, the elements of air and water could be said to have determined the extent of humankind's exploitation of the earth. During the "Great Warming" period of the high Middle Ages, farmers were able to push cultivation 60 meters, or nearly 200 feet, higher than it is today and grow hardier grains such as oats, barley, and rye at the coolest margins of the Continent, such as in the northern districts of Norway, the highlands of Scotland, and in Iceland; grape cultivation extended as much as 500 kilometers, or roughly 300 miles, north of today's commercial vineyards, such that England was able to challenge the French trade in wine; and a cod fishing industry sprung up off the coast of western Greenland in waters that are estimated to have been 4 degrees Celsius, or a little over 7 degrees Fahrenheit, warmer than at present.[41] But just because a warmer climate made greater exploitation of the earth's natural resources possible, is not the same as saying that it *caused* such changes in man's behavior. Inevitably, a whole host of other factors was involved.

This hasn't stopped climatologists from making extravagant claims for the weather's role in history, however. A colder, less agriculturally productive climate regime (the "Dark Age Cold Period") that descended upon Europe beginning in the fifth century C.E., for example, supposedly explains the Germanic migrations that led to the fall of the Roman Empire by 476. Likewise, the rapid expansion of Islam beyond the Arabian peninsula beginning in the seventh century is allegedly due to drier weather and droughts that made Arabian deserts even more inhospitable. Viking raids out of Scandinavia from the end of the eighth century onwards were apparently driven by a warm, "climate amelioration" in the preceding centuries that resulted in overpopulation and a quest for more arable land. The Crusades launched at the end of the eleventh century at the height of the Medieval Warm Period were seemingly a product of the expansion-inducing climate of Europe and an opposite, contracting one in Egypt and Syria. The problem with all of these scenarios is not only in establishing a convincing cause-and-effect correlation, but even in firming up each side of the equation, i.e., in demonstrating that a crucial change in climate did occur or that the weather was consistently influential for a given region or period of time.[42] On the other side of the coin, oftentimes the political, social, and economic explanations for an historical event are not so easily explained nor mesh so well with a climactic causation. Many historians these days, for example, reject a dramatic decline of the Roman Empire in favor of an anti-climactic transformation, or question whether Islam was its own, distinct religion during Muhammad's lifetime and during the reigns of his successor caliphs in the decades thereafter. It is now thought that the true explanation for the Viking raids is to be found not in an expanding population within Scandinavia but rather in their integration into commercial and patronage networks outside it. Nonetheless, this still leaves a role to play by the climate, for example in making rivers and ocean passages more or less navigable through summer storms or winter ice floes, even if

Scandinavia's weather pattern did not always match that in the rest of Europe. If the Crusades are seen these days as more of a defensive operation on Europe's part, perceiving itself to be under assault from an expanding Islam on both Eastern and Western fronts, which is the exact opposite of what the climate tea leaves say, then it still may be true that a "coolish and rainy clime" in Palestine at the close of the eleventh century may have made it more comfortable for European knights to campaign in their chain-mail armor and to forage for crops.[43] But those historians who discount the climate altogether from history, such as on the grounds that a 1 degree Celsius change in the weather makes no difference in the grand scheme of events, are missing the obvious point that humans are inevitably part of nature and, therefore, will be ineluctably influenced by their climate.[44] Moreover, it ignores both the local variability and impact of dramatic weather events and the long-range scope of major climate shifts, which all too often go unnoticed by contemporaries living through them. The compromising consensus nowadays seems to be that climate necessarily *did* play a role in history, but only one among many influences in a closely interrelated web, and often less as the motivating factor behind events and more as having a catalyzing or else retarding effect upon other, more conventional causes.[45]

## Harnessing the elements

What is undeniable, though, is that an impressive expansion of human interactions with and impacts upon the environment was taking place at this time. In terms of the elements of air, water, and earth that are the focus of this chapter, this supposedly "collaborative" exploitation of the environment mainly occurred in the fields of agriculture, mining, and water and wind milling technology. Since population in England, and probably the rest of Europe, is estimated to have increased two-, three-, or even fourfold over the course of the high Middle Ages from 1086, the year of the Domesday Survey, up to c.1300, it almost goes without saying that agricultural production must have somehow kept pace and increased also. Logically, the easiest way of increasing crop production was by expanding the amount of arable land, such as at the expense of the forest by cutting down trees (see Part II) or at the expense of pasture land for livestock, even though such choices of allocation of environmental resources posed some conundrums for medieval farmers, such as whether expansion of arable would compensate for reducing the amount of manure available for fertilizer, reducing the pannage available for pigs, reducing good grazing land which in turn could lead to animal diseases, etc.[46] Aside from the direct impacts of arable farming—clearing of forests and draining of wetlands—there were indirect ones as well, such as increased spring run-off and topsoil erosion into rivers and streams, which did not go unnoticed at the time and which demonstrate the age-old truism that the ecosystems of air, water, and earth are closely interconnected.[47] And yet this land "reclamation" process, known in the Middle Ages as "assarting," is likely to have increased the area of cultivation only by about a third—obviously not enough by itself to do the job of feeding the many more mouths that were becoming available. Therefore, a concurrent process of intensification of agriculture—whereby farm yields were dramatically improved—is also likely to have taken place at some point during the Middle Ages, even though an older generation of scholars assumed that medieval society was inevitably headed for a Malthusian dilemma of being unable to have its food supply match its population growth.[48]

What was behind this "agricultural revolution" of the high Middle Ages? An older interpretation has been that it was achieved through some technological innovations and new agricultural techniques that, after percolating from the sixth through to the ninth centuries, finally were able to synergistically transform society starting in the tenth century, perhaps around the turn of the millennium—the rigid horse collar, the heavy mouldboard plough, nailed horseshoes, etc., as well as a switch from a two-field to a three-field rotational system, in which only a third, rather than a half, of the total arable land was left fallow and unproductive in any given year, formation of nucleated village settlements (leaving more room for arable and pasture), and greater adoption of the speedier horse over the ox as the preferred plough beast.[49] However, recent research has questioned many of the assumptions behind the medieval "agricultural revolution," challenging in particular the supposed interdependence of its various innovations, which are now thought to have been more widely scattered in space and time—extending from roughly 900 to 1300—and in general shifting its most important phase, in which grain yields in some parts of Europe increased exponentially compared to averages elsewhere and from an earlier time, to the thirteenth century and later. Moreover, the mix of innovations and adopted techniques across Europe were highly variable and dependent upon local conditions, and many social and economic factors, including the willingness of seigneurial regimes and peasant communities to innovate, the rise of towns and urban markets, the stability and peacefulness of state systems, etc., helped determine the pace of technological change.[50] For example, one strand of revisionism has been to show through archaeological excavations that there was a great deal of continuity in agricultural knowledge from Roman to medieval times, and in fact some of the more intensive techniques such as collection and distribution of manure and heavy tilling of soils involved simple manpower (of which there was now much more available) and updated familiar implements such as the iron-shod spade and hoe.[51] Three-field crop rotations were not all the rage even in Flanders, perhaps one of the most advanced agricultural areas in Europe, where high yields—sometimes as much as 20–30 grains for each one sown, akin to those during the modern Agricultural Revolution—were instead achieved by a more fragmented and flexible cropping system known as *Flurzwang*.[52] (Average yields during the high Middle Ages are assumed to be 3–4:1, and 2–3:1 during the early medieval period.) Nor were horses universally adopted but rather, their diffusion was more locally sporadic and piecemeal, being embraced more often than not by small peasant farmers rather than large manors or monastic estates, since by necessity peasants with their limited resources had to opt for a more versatile beast of burden.[53] Where horses were used, they were often paired with the lighter ard plough, to which they were better suited, and therefore this older instrument continued to be used alongside the heavy, mouldboard variety down through the late Middle Ages and into the early modern period, as in parts of Denmark.[54]

In short, the underpinnings of Europe's agricultural revolution were much more complicated and hard to pin down than was previously thought: It seems to have been accomplished through a multi-faceted network of "small improvements" that included selective cropping and grazing; mixed farming that balanced arable and pastoral husbandry; greater use of available labor; soil improvements through better crop rotations, manuring, seeding and weeding; investment and exploitation of new or improved tools and equipment; and finally, the incentive of surplus-devouring cities nearby that could absorb and repay higher yields.[55] Even the technological breakthroughs in agriculture so trumpeted in the past instead came about through a "technological complex," whereby each individual innovation only became meaningful in the interdependent context of others, such as the rigid horse collar being used as part of a whole harnessing package that included the whippletree, traces, bridles, reins, and so on.[56] In terms of humans' relationship with the environment, the "technical determinism" of scholars such as Lynn White, Jr., who saw the technological innovations of the Middle Ages, particularly in agriculture, as heralding a new "estrangement between man and nature, in which man became the master," or, in White's own words, the "exploiter of nature," can now be challenged. Rather, historians these days see an exact opposite trend—that the intensification of agriculture that was a concomitant or even afterthought of population growth during the high Middle Ages was in fact made possible by a greater involvement on an individual level with the land, rather than an alienation from it. Instead of an "agricultural revolution," therefore, what we instead should speak of is a long and sporadic process of "diffusion and adaptation," in which change was more gradual than radical. This makes it far more likely that the real shift in environmental attitudes only came during the later medieval period, when ecological catastrophes such as the Black Death of 1348–49 did indeed necessitate drastic reconfigurations in human lifestyles, including the development and use of more labor-saving devices, but which also created a "silver lining" of new opportunities, such as more land and greater surpluses available to peasant landholders that encouraged initiative.[57]

The other great exploitation of the earth aside from agriculture at this time was in mining: Once again, there was some continuity from the classical era, especially in the mining of iron ore, which was much in demand during the Middle Ages for agricultural implements and tools, armor, weaponry, even as building and sculpture material. Another crucial mining industry for medieval society was in the quarrying of stone, which took off especially after the year 1000 with the construction of massive new castles and soaring Gothic cathedrals.

Gold, silver, copper, lead, tin, and zinc were mined extensively in Central Europe by the eleventh and twelfth centuries, with the silver mines at Rammelsberg in the Harz Mountains near Goslar and at Freiberg in Saxony and Kutna Hora in Moravia becoming famous. England, especially in the southwest of the country in Devon and Cornwall, was an important source of silver, lead, and above all, tin, which was used in the manufacture of pewter tableware and the mining of which reached a height of production of more than one and a half million pounds of metal per year during the early decades of the thirteenth century, when it was shipped from English ports to all over Europe and beyond. The thirteenth century was likewise the beginning of a "great expansion" in the production and burning of coal, which was mined chiefly in the north of England and in the Low Countries. The English also dug extensively for clay—used for pottery, brick-making, and other building material—as well as for marl, a chalk-like clay used for fertilizer. France was known for its supply of building stone, which was quarried most intensely from the eleventh to the thirteenth centuries and was also shipped abroad, such as to England, at great expense. Towards the end of the Middle Ages, iron mines and forges appear extensively in the Othe Forest in eastern France, and the mining of saltpeter, a crucial component of gunpowder, became another important industry.[58]

The Romans had been experts at water engineering and manipulation, building aqueducts that brought water to cities from miles away across deep gorges, and public baths where water was piped to different rooms in which it was heated to specific temperatures. Some, but not all, of this water know-how survived the transition to the Middle Ages: Aqueducts continued to supply water to early medieval cities, particularly in Italy, where it was probably a point of pride as well as practicality to keep the reminders of ancient civilization in good repair or to build modern exemplars of them. However, many cities and households also had to supplement their water supply with cisterns, wells, and springs. Public baths, on the other hand, eventually fell into decline as a result of Christian moralizing against recreational bathing, even though the merits of cleanliness and hygiene were maintained, now largely through private (or in the case of monasteries semi-private) baths or public fountains and basins where only the extremities could be washed. Another feat of water technology in which the Romans excelled, the sewer, was also occasionally maintained by some Italian cities, but most seem to have made do with cesspits or drains and gutters that eventually ran directly into rivers, where it was hoped it would be flushed away. This was obviously not an ideal system, as contamination of underground and downstream water supplies could all too easily occur.[59]

It was in the monasteries where much of the water engineering and technology was kept alive, concerned as they were with maintaining a self-sufficient community and many of them planned and sited beforehand with a view to access to supplies of fresh water, which the monks were careful to preserve by keeping these sources well separate from their waste streams. Water was essential not only for drinking, cooking, and bathing but also for religious ritual, such as baptisms. Monasteries often went to great lengths to secure their water supplies, building aqueducts or conduits and lead and wooden pipelines that spanned long distances, sometimes in tunnels under rivers or across bridges; other monasteries built dams or dug trenches and ponds to hold and store water, which could also be used to house fish or service a mill. Once water was used it continued to serve the community by being flushed out through sluices that drained the monastic *reredorter* or latrine.[60] Cities likewise were obligated to provide reliable supplies of water to their citizens for a variety of uses besides drinking and cooking, to include brewing, tanning of leather, manufacture of woolen and linen cloth, butchering of animals, and so on. For many cities, especially in Italy and Germany, fountains that gave public access to a bountiful display of water became a source of civic pride, and as in the monasteries elaborate pipe networks, complete with settlement tanks or washout valves to help filter the water of sediment, were constructed to maintain the supply of what was regarded as a rightfully free commodity. English cities apparently did not invest so highly in public water systems, perhaps reflecting a lower cultural priority accorded to securing safe, fresh supplies of water; the exceptions were the cities of Gloucester and Southampton, which benefitted from the water pipes already laid down by religious houses. However, it is true that throughout medieval Europe, water as a drink was much less preferred to other liquids that had been processed or treated in some fashion, such as by fermentation, but this seems to have been for reasons of status just as much as hygiene. At the same time, medieval people did have the expectation that their water at least be clear and free from odor. As urban populations rapidly expanded starting in the high Middle Ages, cities also began to pass laws that attempted to regulate pollution of rivers and other water supplies from human and animal waste as well as from the discharges of various industries that used water in the processing of their products, such as the tanning of leather or the washing, fulling, and dying of woolen cloth. And yet there are modern claims that the alum, tartar, and tannin used in the manufacture of cloth and leather garments may actually have helped purify rivers when discharged into them, or at least have had a neutral effect.[61]

Two areas of water technology and manipulation where the Middle Ages probably exceeded what had been achieved during the classical period, even if it inherited some of the Roman techniques and know-how, were in drainage/ irrigation and milling. Undoubtedly the most advanced drainage systems were developed in Holland, some two-thirds of which lie at or below sea level. Once the Dutch—perhaps beginning in the ninth or tenth centuries— started down the path of land reclamation from natural wetlands, which nowadays are appreciated for their own unique ecology, they were then caught in a vicious cycle, or "technological lock-in," in which they were required to devise ever more sophisticated techniques of water engineering— comprising not only drainage ditches and canals to draw water off but also a system of dikes and dams to keep water out—just to maintain the status quo of their newly cultivated fields. This is because the bog peats of Holland, some 80–90 percent of which can be made up of water, greatly reduce in volume once they are drained, dried out, trampled, and oxidized: Modern experience has shown that peat can subside on average by nearly an inch a year; nearly thirteen feet was lost in the 130 years between 1848, when the Holm Fen Post, an iron rod 22 feet long, was first sunk into the ground in Huntingdonshire, England, and 1978, when the latest measurement was made. Usually drainage was accomplished piecemeal, starting with a main reclamation ditch, often dug along the bed of a naturally occurring stream issuing from the raised center of the peat-bog wilderness, and then with other ditches radiating out at right angles from the main one and parallel to each other to form individual settlements; the dirt raised up from the digging also served as embankments to keep in the water. Eventually, by the mid-twelfth century, such reclaimed lands began to be prone to serious flooding from the sea, helped along by both the subsidence created by man and the increased incidence of storms from nature. An area known as the Zuiderzee emerged in the lowland coastal zone where the sea made substantial inroads into the drained basin and which now had to be held back by a network of dams and dikes. This necessitated a coordinated system of drainage and flood control among the various *polders* or communities enclosed by their own water control embankments, since each *polder* was at a separate level of subsidence. Collective agencies such as the Regional Drainage Authority of Rijnland, which emerged during the thirteenth century, demonstrate the amount of cooperation and coordination that could be achieved, even in the absence of a central state institution, when everyone's livelihood and indeed very existence was at stake. These drainage works also changed the topography of the land, literally relocating Rijnlanders further inland from the coast and orientating them towards the heartland of European identity to the east and south, making them "Hollanders" instead of "Frisians". By the end of the Middle Ages, water hydraulics had advanced to the point that windmill pumps were now able to defy gravity and keep even the lowest-lying lands of the Netherlands dry. It is no exaggeration to say that one of the most densely populated regions of Europe was made possible by its ability to master, or "dominate," the element of water.[62]

Elsewhere, an expanding population's need for more arable soil was satisfied by other drainage and irrigation projects that either drew water off from land that had too much or brought water in to where there was too little. The lowland regions of England, such as East Anglia and Kent, mirrored Holland in that drainage ditches and canals were built primarily in the twelfth and thirteenth centuries in order to reclaim thousands of acres of fens and marshland, while at the same time defenses were constructed against incursions of the sea. In the case of the Fens of East Anglia, medieval drainage works took advantage of earlier, Roman engineering, such as the "Roman Bank," a 60-mile earthwork that ran along the Wash or estuary of the North Sea along the coasts of Lincolnshire and Norfolk, while the Car Dyke was a 90-mile long canal that ran inland between Lincoln and Cambridge. Maintaining or rebuilding these works also required collaboration among separate communities and private landowners, as in Holland. Other drainage and dam projects occurred during the high Middle Ages in the inland swamps of northern and southern France and along the Atlantic coast, as well as on the plains of northern and central Italy and in the Carpathian Basin of Hungary. In both France and Italy the lead was often taken by ecclesiastical institutions, such as the Cistercian monasteries, while Italy had a system of *magolato* or fields demarcated by ridges and drainage furrows that was very much akin to the Dutch *polders*. But in addition many cities in France as well as in Italy invested heavily in canals and other hydraulic works, such that they became known as "mini-Venices". Both France and Italy also had irrigation systems, especially in the southern, drier regions of each country. Hungary's interconnected complex of canals allowed for both flood control, or drainage of excess water, as well as for irrigation, or intake of water and sediment needed for agriculture and fishponds. The most advanced irrigation techniques, however, were in Islamic Spain, or al-Andalus, which benefitted from Persian and even Indian traditions. Already before the Muslim takeover in 711, the law code of the Christian Visigothic rulers of Spain evinced concern for the conservation of water resources rendered even more precious in an arid climate, for they penalized anyone who did damage to mills and ponds or who stole water from the streams of others.[63] In Muslim Spain, both valley-floor and slope irrigations systems were used, the latter taking advantage of gravity water feeds either perpendicular or parallel to the inclines of river valleys. Hydraulic techniques included dams, cisterns, *qanāts* or tunnels that filtered and channeled water as they ran through a naturally occurring water table, and *norias* or water wheels that scooped up water from a river or well and deposited it into an irrigation canal or reservoir. The last innovation, widely distributed in al-Andalus beginning in the tenth century, was not to come to England until the seventeenth century and was little known in Italy, despite the classical legacy of such water-lifting technology as the Archimedean screw.[64]

Mills harnessed the kinetic energy and density of flowing water in order to turn millstones that ground grain into flour, and as well harnessed water for a host of industrial uses, including the sharpening of blades and instruments; crushing of tanner's bark, dyes, and minerals for cloth and metal ore manufacture; and the fulling of wool, shredding of paper, sawing of wood, and forging of iron and steel using trip hammers and bellows worked by camshafts. Watermills were known to the Romans at the height of their empire, but it could be argued that medieval millwrights added their own refinements to the technology and that mills became far more diffused throughout Europe during the Middle Ages. Moreover, milling technology developed concurrently, if not necessarily independently, in areas that lay outside the Roman Empire: The existence of mills in early medieval Ireland, some dating to as early as the seventh century, have been uncovered through the archaeological and written record, with various terms for mills and their components devised in Gaelic. The Irish mills, which are among the earliest known to have arisen during the Middle Ages, employed both horizontal and vertical water wheels and developed some important refinements in milling technology, such as a special bore fitted to the penstock orifice of the flume in order to produce a more powerful jet of water onto the paddles of the wheel, and scoop-sectioned or dish-shaped paddles on the wheel that better absorbed the impulse energy of the waterjet from the flume. Vertical water-wheel mills, with either an undershot or overshot millrace, made more efficient use of both the weight and motion of the water to produce a more powerful and faster turning millstone, but because of their gearing they were more expensive and time-consuming to construct and maintain and thus were typically associated with seigneurial or ecclesiastical institutions and urban centers that could afford such an investment; eventually more and more watermills were of the vertical variety, but the horizontal-wheeled mills never disappeared completely since they were better suited to more humble peasant communities and to smaller, weaker-flowing streams and tributaries.

Tide mills, usually referred to in the records as "sea mills," which were sited in estuaries along the coast and harnessed the energy of the ebb and flow of the tides, also made their appearance at an early date, but they were

*Figure 1.1* MEDIEVAL WATERMILL, employing a vertical water wheel. Manuscript illumination from *Hortus Deliciarum*, Alsace, late twelfth century. The Granger Collection, NYC – All rights reserved

especially prone to damage from storms, which only grew worse as the Middle Ages progressed.[65]

Why medieval people should have constructed such mills has been variously explained: To save time and labor which could be appropriated to other tasks; to obtain a finer, whiter flour for the baking of more refined breads; to enable lords to squeeze more profits from their peasants; and as a symbol of pride (perhaps in the mastery over nature?), technological achievement, and continuity with the glorious ancient past. The true answer may well be a combination of all of the above. Whatever the reason, the high Middle Ages saw the high water mark of the distribution and variety of the medieval watermill. In England, the Domesday Survey of 1086 recorded a little over 6000 watermills, which probably increased to about 9000 by 1300, when they were also undoubtedly larger and more efficient than two centuries ago. England also made the mill a monopoly of the manor by the twelfth century, even though free peasants could still mill where they liked, either at home using a handmill or horse mill or at a neighboring "independent mill" that might compete for their custom. Mill rents, which could range from one-thirteenth to one thirty-second of the grain brought to the mill to be ground, made up a significant proportion of manorial incomes and helped perpetuate the manorial mill monopoly by providing the wherewithal to update mills or build new ones.

In France and Italy, which also had an ancient tradition of water-milling technology, the number of mills likewise picks up greatly starting in the eleventh century: Not only do we find both vertical and horizontal watermills but in addition a hybrid variety popularly known as the "French mill," which combined an overshot millrace with a horizontal wheel. In the countryside our picture of watermills is dominated by monastic institutions, but both countries also had a strong tradition of urban mills, which typically were sited under the arches of bridges where the water grew more accelerated and was harnessed by either floating mills, where the millwheel was mounted on boats, or by hanging or suspension mills, where the wheel was suspended across the entire width of the arch.

In the contado of Florence, over 700 grain mills were recorded by tax records in the early fifteenth century, but a significant percentage, 22 percent, were listed as destroyed or in disrepair. These mills were generally of three types, the suspension and terrestrial mills having a horizontal wheel, while the floating mills having a vertical one, and were widely distributed throughout the Florentine countryside, most of them in the mountainous district of the Apennines; they also had a variety of owners, including individuals and families, communes and consortiums, and monasteries and churches. By the late Middle Ages, the number of mills began to decline with population as fewer mouths needed to be fed with mill-ground flour: Florence's mills seem to have declined by twelve percent from the end of the thirteenth to the first quarter of the fifteenth centuries. But a shortage of labor also created opportunities to diversify mills into industrial applications, which in England

*Figure 1.2* WATER-POWERED FLOATING MILLS. French ms. illumination, thirteenth century. The Granger Collection, NYC – All rights reserved

and Italy most often took the form of the fulling of wool but in France was applied most productively to metallurgy. And even though the profitability of mills remained predominantly in the processing of grain, which limited the so-called "industrial revolution" in milling technology, often mills now saw a variety of uses at the same site, where adaptability became key to a mill's survival. In any case, industrial applications of milling technology did not really take off until the very end of the Middle Ages and on into the early modern period, when expanding markets created greater incentives to mechanize production in a number of industries.[66]

The last element to be harnessed during the high Middle Ages was blowing air, or wind. The windmill is in fact the one innovation that can truly be said to have originated in the medieval period, and it has been called "the most characteristic contribution of the age to the technology of exploiting natural power, and by far the most important."[67] It first appears mostly in England during the 1180s and 1190s, with a few other examples in northern France and Flanders.

However, there then followed a period of stagnation, evidently a trial period to try out the new technology, until another round of windmill construction burst forth in the second quarter of the thirteenth century. By 1300, it is estimated that 3000 windmills had been built in England alone. Typically, medieval windmills were constructed with four sails and with the millhouse set on a single, swivel post anchored to the ground so that the entire structure could be turned to best harvest whichever way the wind was

*Figure 1.3* CARRYING WHEAT TO A WINDMILL. "Obituaire de Notre Dame de Pres", after 1270, French School, Bibliotheque Municipale, Valenciennes, France / The Bridgeman Art Library

blowing. (The mill's balance point was at the front near the sails, rather than in the exact center of the millhouse.)

Contrary to popular perception, windmills did not so much replace watermills as supplement them, especially in flat, marshy areas such as the east of England that had little access to strong, high-flowing watercourses. (It may be no accident that windmills first appeared here, where the sight of sailing ships along the coast perhaps provided the inspiration for a solution to the lack of any other milling alternatives.) The one exception was the tide mill, which was steadily replaced in the high Middle Ages by the less vulnerable windmill. Even so, windmills were more expensive to maintain than watermills, with the costs of repairs typically eating up a third of revenues. This is why when patrons had a choice of natural power to harness, they usually went with watermills, which is reflected in the fact that they outnumber windmills by three to one at the height of medieval England's milling heyday.

However, windmills, especially towards the end of the Middle Ages, were to develop other applications besides grain and industrial milling, such as pumping out water in order to drain the ever lower lowlands of Holland and the ever deeper mines of Germany. This ensured windmills' presence in some areas of the Continent on into the early modern period.[68]

*Figure 1.4* WINDMILL BENCH END. English School, All Saints' Church, Thornham, Norfolk, UK. Photo © Neil Holmes / The Bridgeman Art Library

*Figure 1.5* MILLER AND HIS WINDMILL. Misericord carvings, French School, fifteenth century. Musée National du Moyen Âge et des Thermes de Cluny, Paris / Giraudon / The Bridgeman Art Library

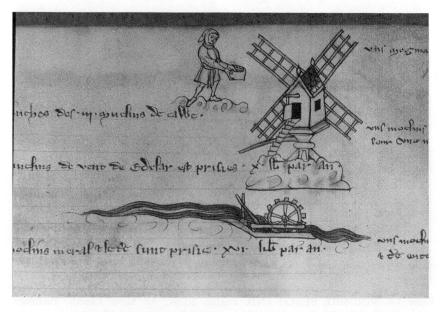

*Figure 1.6* WINDMILL & WATER WHEEL. French ms. illumination, c.1270. The Granger Collection, NYC – All rights reserved

## Collaboration, or exploitation?

Moving from the realm of the practical to the intellectual, there does seem to be some evidence that the warmer climes and technological advances of the high Middle Ages were reflected in a more optimistic outlook among medieval philosophers for man's role and place in nature. In the cultural "renaissance" of the twelfth century, there emerged a new-found "discovery of nature" and a neo-Platonic revival of the concept of the "world-soul" (*anima mundi*): Man was deemed to be fully a part of and one with nature at the same time that he had his own, unique nature (i.e., having both a corruptible body and an immortal soul), while God as the grand artisan of a universal creation or macrocosm was reflected in the human artisan who was able to shape and influence his immediate natural surroundings, or microcosm. These ideas found expression through the writings of such twelfth-century thinkers as Bernard Silvester, Honorius of Autun, William of Conches, and Alain de Lille.[69] For example, William of Conches in his *Philosophia Mundi* (Philosophy of the World) writes in chapter 15 that the *anima mundi*, or world soul, is said by some to be the holy spirit, but others identify it with "a natural vitality innate to things, by means of which some exist only, some live and feel, some live and feel and discern," which can be understood to correspond, respectively, to plants and other inanimate objects, animals, and finally man. Humans therefore have in them both a world soul and their "own [rational] soul," but this does not mean they have two souls, nor at the same time that the world soul is to be identified with the human soul. In Chapter 22, on the elements, William declares that the four elemental building blocks of nature, namely, earth, water, air, and fire, are all bound up indivisibly in the human body "just like a letter exists [only] in the composition of a word, but not on its own." William is therefore at pains to identify the elements with the human body and its humors before going on to discuss their respective properties of hot, cold, moist, and dry and their interrelationship with each other, in the manner of Isidore of Seville.[70] Bernard Silvester, in his *Cosmographia*, or *De Mundi Universitate* (the Cosmography, or Concerning the World Universe), introduces the allegorical figure of Natura, who in the first part of the work, the "megacosmus," appeals to Noys, or the Divine Mind, to restore order and harmony to a chaotic creation, which she does by stabilizing and balancing the four elements, which now complement and share each other's attributes instead of being in conflict. Meanwhile, Endelichia and Mundus are married together to form the World Soul. In a revealing passage towards the end of the megacosmus, Silvester refers to an "elementing Nature" (*Natura elementans*), in which "the elements come together with an elementing Nature [and] an elemented Nature with the elements." By this he seems to allude to a unity or oneness between the elements and the natural world that we saw also expressed by William of Conches. In the second part, the "microcosmus," Noys has Natura fashion humans in order to "bring consummation" to the cosmic design, in collaboration with Urania, or celestial existence, and Physis, or material existence.[71] Alain de Lille likewise develops the allegorical figure of Lady Nature is his prose work, *De Planctu Naturae* (The Complaint of Nature). After describing Nature's incomparable, virginal beauty as she descends from the heavens in her chariot, the author must then listen to a long diatribe from Nature about man's upending of the moral order, particularly in the rules of love. Due to its lustful and perverse disposition, the human race "commits monstrous acts in its union of genders," in which "some embrace only the masculine gender, some the feminine," while still others are of "heteroclite gender" in that they are "declined irregularly," being feminine through the winter and masculine in the summer. This seems to harken back to an eschatological agenda of the early Church fathers to fulfill the biblical text of "Be fertile and increase, fill the earth and master it" (Genesis 1:28), although Alain does refer in one passage to man as the mirror of the world and of nature itself, where the four elements are in "concordant discord," "single plurality," "dissonant consonance," and "dissenting agreement" that are at the foundation of both the human body and of all creation.[72] This idea is fully in accord with what William of Conches and Bernard Silvester also had to say on the collaborative relationship between nature, the elements, and man.

The thirteenth century was the age of the schoolmen and of the great systemization of philosophy, as embodied in such giants as Albert the Great and St. Thomas Aquinas. This trend applied no less to natural history, where it is best represented in the encyclopedic works of such authors as Alexander Neckam, Vincent of Beauvais, Bartholomaeus Anglicus (Bartholomew the Englishman), and Thomas of Cantimpré, who wrote, respectively, the *De Naturis Rerum* (On the Natures of Things), the *Speculum Maius* (The Great Mirror), the *De Proprietatibus Rerum* (On the Properties of Things), and the *Liber de Natura Rerum* (Book on the Nature of Things). Neckam, author of the earliest of these works, dating to the end of the twelfth century, describes the element of air as necessary for sustaining life, in that we cannot breathe without it, and compares its qualities to those of the soul. He also notes that it is the source of clouds, winds, and thunder, fills vacuums, and carries sounds, such as echoes. Water opens his second book, where it occurs in salty oceans and seas, rivers, springs, and wells. Neckam describes how water is spherical in nature, how it congeals into ice, how various rivers flow into the sea, and how the ocean moves back and forth in tides. Earth is the lowest element, at the center of the world, and sometimes moves in earthquakes caused by violent winds passing through "subterranean passages and caverns." Associated chapters discuss the diverse products of the earth, including coal, lime, metals, green plants and trees, and precious stones.[73]

Beauvais provides a standard introduction to the four elements in book two of the *Speculum Maius*, followed by explanations of the various manifestations of air, water, and earth in books four, five, and six along the lines laid down by the *Etymologies*, albeit now in more detailed form with reference to a host of authorities, the most important and often-quoted being Aristotle, Seneca, Pliny, Isidore of Seville, Avicenna, William of Conches, and sometimes the author himself. One focus that receives much more attention than in previous compilations is Beauvais' concern with the environmental impact of the elements upon human health, which of course anticipates similar concerns among plague treatise authors during the Black Death, beginning in the mid-fourteenth century. Thus, in chapter 113 of book four, Beauvais provides an extended discussion of "pestilential air," which he defines as air changed in its substance into putridity or its "very worst quality," which affects people everywhere and equally with diseases having "many very bad symptoms" present in "diverse bodies." Bad air can cause "disorientation, pain, excessive sweating, coldness in the extremities of the body, heat in the chest, dryness of the tongue, bad breath in the mouth, excessive anxieties, choleric vomiting, looseness [of the bowels], windiness, [and] very bad and diverse urine." The change to bad air generally has two causes, one of which is regional in nature in that it arises from a "dissolute fume" generated by putrefying fruits and vegetative matter and that comes out of lakes, crypts, ditches, and dead bodies; the other cause is temporal, in that it comes about through abnormal or unusual weather patterns during the seasons of the year, such as a hot and dry winter without any rain, or a rainy or cold and dry summer, or a hot and humid autumn. Such ideas obviously owe much to Hippocrates and Avicenna. Beauvais also notes how beasts can die "from a corruption of the humors or vital spirits" in the case of bad air caused by regional pollution, and how when a pestilence falls upon a land during a growing season, the fruits on trees can lose their color and dry up, whence those eating them "incur the worst kind of illnesses." Beauvais concludes this section on the air with a final chapter on how demons can reside in certain airy regions, particularly those that are the lowest and darkest, which anticipates another late medieval concern with how necromancers and witches can harness the power of the elements through their worship of the devil.[74]

In book five Beauvais is indebted to new information available from Arabic authors, such as the ninth-century Persian astrologer, Albumasar, that allows him to explain tides as connected with the movements of the moon, which was not explained by Isidore in the *Etymologies*. Concern with the health impacts of the elements continues with a couple of chapters on bathing, based on Rhazes and Haly Abbas, which can confer benefits such as humidifying and rejuvenating the body, removing accumulated filth, dissipating windiness, rarefying the humors, opening the pores, evacuating superfluities, mitigating pains, and restoring good digestion and good looks, especially when done after exercise and before a meal. However, it can also cause harm, draining one's bodily strength, heating up the heart, inducing fainting fits or nausea, and occasioning the descent of bad humors down the body, especially when done after eating. Basing himself on Avicenna, Beauvais lays out the eight "tastes" of water—sweet, bitter, sharp, salty, sour, briny, astringent, and oily—and their various operations and effects upon the body. He then goes on to recommend, once again quoting from Avicenna, which waters are best to drink, such as from spring-fed streams flowing over rocky beds, and warns against "pernicious and poisonous waters" to avoid, such as stagnant pond water and waters that have an overpowering odor and strange taste, that are unsettled, dense, and heavy, that are "quickly turned into stone," and those that have "an evil white scum" floating on the top and in which "evil foreign matter" is congealed. Bad waters can be corrected with evaporization and distillation and by boiling, but other remedies can be used specific to the kind of water being treated: Nitrous water is corrected with milk, aged wine, starch, and unripe cucumbers and other such vegetables thrown into it; crass, unsettled water with garlic, onions, leeks, and wine; bitter waters are corrected with sweet ones; and salty water with carob-wood and myrtle berries. Best of all is vinegar, which cures all bad waters without exception.[75]

In book six, Beauvais has several chapters on the phenomenon of earthquakes, with his information largely derived from Seneca, Pliny, and Aristotle. Chapter 30 discusses one of the more alarming environmental impacts of earthquakes, namely, the giving rise to pestilences, which anticipates the naturalistic explanations of the Black Death that were to be favored by such late medieval authors as Konrad of Megenberg. Earthquakes of course emitted noxious exhalations from deep within the bowels of the earth up into the air, producing an "eternal night" and generating "new kinds of diseases" that come about from the "many poisons that arise therein, spread not by [human] hand but spontaneously." The example quoted from Seneca is the earthquake that occurred at Pompeii in Campania in 62 C.E., which allegedly killed 600 local sheep and made birds fall down dead from the air, their bodies livid and throats swollen. In a later chapter, number 69, Beauvais, basing himself on the Roman architectural historian, Vitruvius, accurately describes how vapors arise from the earth heated up by the sun to eventually form clouds, and that by such humidity the earth can help eject humors from the human body in the form of sweat. This evaporation process, especially when winds from violent storms pass over rivers or stagnant pools or the sea, can sometimes make it rain frogs and fishes, which a century later observers took as an apocalyptic sign of the Black Death but here receives a purely naturalistic treatment.[76] In books seven and eight, Beauvais devotes himself to "terrestrial bodies" that dwell "partly in the bowels of the earth, partly on its surface," and that are products of the "earth's nature and fecundity and cultivation, also from its passions and vapors"—namely, minerals and stones. Here Beauvais takes particular note of the medicinal properties of certain minerals and stones, knowledge of which will be put to use by many physicians during the Black Death.[77]

After introducing the four elements and their properties in book 10 of his *On the Properties of Things*, Bartholomaeus Anglicus has separate sections on the air (book 11), water (book 13), and the earth (book 14). Basing himself on the work of the eleventh-century physician, Constantine the African, Anglicus describes the air as hot and moist, but even more importantly, because it is a light element, it is easily changed "into contrary qualities," namely, that it is infected by "corrupt and venomous" vapors from the earth and sea so that the mixture becomes a "pestilence vapor," by means of which diseases spread to men and beasts by virtue of the fact that all living things must breathe in the air and "no creature with a soul may live without" it. The quality of the air then affects the quality of the humors in the body, i.e., "troubled" air will created troubled humors and vital spirits, since the air is drawn directly to the heart and from the heart "to all the body." Therefore, air is the element that "most changes the body" since it "passes to the inner parts and to the [vital] spirits and is melded with their substance that gives life to the body." If the air is pure and clean then it preserves life, but if "corrupt and distempered, then air grieves the body most and corrupts it."[78] The rest of the book follows Isidore of Seville in describing the various phenomena of the air, such as winds, clouds, rainbows, dew, rain, hail, snow, thunder, and lightning.

Likewise, the books on water and earth are largely derivative of Isidore, Constantine, and the fourth-century *Hexameron* of Basil of Caesarea. Water, a cold and moist substance, is noted for nourishing growing things such as trees, herbs, grain, and grass, and for giving drink to beasts and men and "spirit and breath to fishes." Anglicus also follows Avicenna in describing the tastes of water and recommending rain and well water over melted snow. The earth, while being the "lowest" element, cold and dry and of little subtlety or simplicity, unlike the air, is nonetheless "steadfast and stable," supporting all life, and therefore was depicted as the "mother goddess" or *alma mater* in pagan times. Even so, there occur earthquakes due to "cold winds" or "vapors" battering against the hollow caverns of the earth, which Anglicus describes on the authority of Aristotle, whereas he quotes Constantine and the fifth-century Roman author, Ambrosius Macrobius, to point out that the fruitfulness of the earth can vary depending on location in relation to distance from the sun or blowing winds. Book 16 is devoted to "things that are engendered in the earth and in the veins thereof," namely, precious stones and metals.[79]

Although Cantimpré is indebted to Isidore of Seville, whom he quotes extensively, he does seem to have a new-found concern for how nature interacts with man. For example, in his 19th book, on the elements, Thomas praises the earth above all for its benevolent relationship with human beings:

> The earth is the element most accommodating to man. For just as the heavens encompass God or the angels, it welcomes us, man, as we are being born, and it always nourishes and sustains us once we have been brought forth into the world, enfolding us to her bosom like a smothering mother at our utmost need when we have been utterly spurned by the rest of the natural elements. This is the only element that is never angry towards man. Water drenches us in rain, pelts us with hail, drowns us in floods, while the air presses down upon us with clouds, lashes us with winds. But mother earth is always benign, mild, indulgent, always a handmaiden for the benefit of us mortals, it generates what has been harvested, it freely brings forth smells, tastes, colors, and juices or the rest of nature's goodness. And in good faith it pays back with interest the

credit that has been sown with seed. It produces for man those medical herbs, and, to be brief, it always yields what is necessary.[80]

By contrast, the waters, especially in the oceans, are mysteriously deep and moved by the moon, while the air is the realm of demons and gives rise to the winds, snow, hail, rain, thunder, and lightning. Perhaps worst of all, the air is also the source of corruptions, "from whence are generated infirmities and pestilences." There are also separate sections on the various manifestations or efflorescences of these elements in nature. For example, book 13 concerns the springs that are to be found throughout the world that "are drawn from the secret bowels of the earth"; books 14 and 15 treat of the precious stones and metals that likewise occur "in the bowels of the earth"; books 16 and 18 explain the seven regions or "humors" of the air—namely, dew, snow, hail, rain, honey, ladanum [resin], and manna—as well as meterological phenomena, such as thunder and lightning, falling stars, various kinds of winds, clouds, and rainbows. Much of this information is drawn from authorities such as Aristotle, Pliny, Isidore of Seville, and Matthaeus Platearius of Salerno (twelfth century).[81]

Finally, there is Albert the Great's commentary on the *Liber de Causis Proprietatum Elementorum* (Book on the Causes of the Properties of the Elements), erroneously ascribed to Aristotle. Perhaps one of the more interesting and original passages of this work is chapter 9, on floods, where Albert had to tread lightly, especially when discussing the causes of a universal flood, such as in the time of Noah, because the opinion that this was a natural event was a condemned one. Albert was careful to distinguish between universal and particular, or local, floods and to emphasize that *all* of the celestial and terrestrial causes coincide in the case of the former, which presumably includes the agency of God. However, Albert only considers natural causes of floods, which he states explicitly at the outset of this chapter. As physicians were later to do with explaining the causes of plague, Albert gives both higher and lower causes for universal floods: Among the celestial causes are conjunctions of the planets and the operations of the moon; the lower or terrestrial causes include vapors prevalent in the air that fall down as rain or vapors in the earth that "burst forth" into the waters and cast them out "from the depths of the abyss," much like a tidal wave. Particular floods arise from "fewer and weaker causes," such as winds created by an earthquake, a single vapor, or "some of the celestial powers".[82]

Undoubtedly, the great agricultural expansion of the high Middle Ages that went hand-in-hand with a burgeoning population fed into some of the encyclopedists' optimistic outlook, such as Cantimpré's paean to the earth, but, as we will see, this was eventually undone by the Black Death, when all three elements—not just the air—were seen as a potential source of the plague. But was this sense of collaboration with nature the predominant mentality and ethos among ordinary men and women of the high Middle Ages? It is probably impossible for us to ever recover what this mentality might have been, but some kind of clue—namely, as to what a learned cleric thought a peasant may have thought, if not what the peasant actually thought himself—may be gleaned from the *Vision of Gottschalk*, written in 1190 and part of the visionary literature that was usually associated with monastic protagonists. In this version, however, the hero is a peasant from the village of Großharrie in the Holstein region of Germany. As Gottschalk lies in a state of near-death from an illness over the space of five days, from the 20th of December until Christmas Eve, his soul is led by two angels on a tour of the medieval afterlife, encompassing hell, purgatory, and heaven, the last of which Gottschalk can only view from afar. Our interest lies near the beginning of this text, where Gottschalk is described as:

> a simple and upright man, poor in spirit and goods, a husbandman of a monastery—not himself a monk, but a farmer—a man with one wife who did not know any other woman, having with her a son and two daughters. He labors his whole life under various, long-standing burdens and infirmities. In times of health he does not twitter about, but rather he spends his days toiling to pull up beeches, oaks, and other trees by their roots, extending his arable fields by sowing his seeds and cultivating his fields so that he can eat his bread by the sweat of his brow.[83]

It is hardly credible that this rude rustic devoted much of his precious leisure time, if any there was, to happily contemplating his collaboration with nature. Instead, Gottschalk's only concern was to figure out how best to exploit the benefits that the elements, especially the earth, had to offer, and how best to avoid their ravages, so that he could feed himself and his family.

Yet our survey of the medieval thought-world during the Indian summer of man's relationship with the environment would not be complete without a detour to the perhaps unique exception of St. Francis of Assisi. As already mentioned in the Introduction, it was famously argued by the historian, Lynn White, Jr., that St. Francis embodied a radically alternative, indeed almost heretical, view of nature that stood in evident contrast to the normally exploitative attitude in Christianity—traceable back to the Old Testament—in which it was arrogantly assumed that all of creation was intended by God for man's exclusive use.[84] While studies of St. Francis' environmental outlook most often focus on his relationship with animals, he also did not exempt even the inanimate from his worldview that demanded mutual respect and interdependence between man and the rest of creation. For example, in his *Canticle of the Sun*, St. Francis at one point gives praise to all four of the elemental building blocks of nature:

> Be praised, my Lord, for Brother Wind,
> And for Air, for Cloud, and Clear, and all weather,
> By which you give your creatures nourishment.
> Be praised, my Lord, for Sister Water,

> She is very useful, and humble, and precious, and pure.
> Be praised, my Lord, for Brother Fire,
> By whom you light up the night.
> How handsome he is, how happy, how powerful and strong!
> Be praised, my Lord, for our Sister, Mother Earth.
> Who nourishes and governs us,
> And produces various fruits with many-colored flowers and herbs.[85]

In addition, the testimonial literature surrounding St. Francis attested to his miraculous control over the elements or else to how he drew edifying allegories from them.[86] According to one biography of the saint, his favorite element after fire was water, due to its powers of cleansing the soul during baptism.[87] From the perspective of a modern-day world historian, St. Francis' homage to the elements is almost Jain-like in its willingness to extend sentient awareness, if not an outright soul, to what we usually regard as non-feeling aspects of creation.

White's views have been criticized on a number of grounds, most of them objecting that there was a greater diversity of Christian ecological traditions, including ones that support the concept of man as a steward of nature instead of merely its exploiter, as well as objecting that plenty of other cultures besides Christianity have inculcated wanton destruction of the environment.[88] But the idea that St. Francis represented an exception to Christian norms of behavior towards the natural world may also be wrong. One study claims that St. Francis was fully a part of the ascetic monastic tradition that already evinced a deep appreciation for nature, going back to the Celtic wandering hermits and the Cistercian reform movement that gravitated toward wilderness sites. It was St. Francis who apparently reconciled a utilitarian, hierarchical, and ambivalent view of nature (going back to St. Augustine) with a respectful, chivalric, and almost mystical approach to the environment. Even a thirteenth-century establishment figure of the Church such as St. Thomas Aquinas, who is considered to be rather dismissive of non-human creatures due to their perceived lack of rationality, could share with St. Francis a recognition of the profound beauty to be found in nature. If some of Francis' ideas are rather unique and original, such as his almost Jain-like submissiveness or deference to creation—appealing to Brother Fire before undergoing a cauterization or allowing mice to disturb his meals and his sleep—nonetheless the humble simplicity and directness of his expressions were a part of the growing vernacular and mystical literature of the late Middle Ages that was becoming more popular and accessible to a wider, more literate reading audience.[89] Likewise, his emphasis on the potential for harmony between man and the natural world and an appreciation for nature for its own sake, as the "handiwork of God," was really no different from the message being pushed by other thirteenth-century poets who addressed Lady Natura, such Alain de Lille, Bernard Silvester, and the Occitan troubadours.[90]

Yet despite all this re-assessment of St. Francis' ecological legacy, there remains the nagging suspicion that in practical terms, it did not change medieval attitudes towards the environment all that much. While Francis' views may actually have been more acceptable to the Church than previously realized, appeals to one's better (green) nature are not likely to have moved men's minds so much as the imminent threat of ecological disaster. If this seems a rather pessimistic view of the motivations for the human capacity for change, along the lines of Hobbes or Machiavelli from the post-medieval age, then it is amply borne out, both by the cultural transformations that took place in the late Middle Ages and by our modern, present-day experience with global warming. Only ecological shocks on the scale of the Little Ice Age or the Black Death beginning in the fourteenth century can produce evidence of seismic shifts in perception of man's relationship with his environment, which are largely lacking from the thirteenth century. There is nothing like catastrophe for re-focusing attention back to the natural world.

## The Little Ice Age

Sharply colder and wetter weather succeeded the Medieval Warm Period during an era known to climatologists as the "Little Ice Age," extending from c.1300 to 1850. This is a time when it is estimated that average temperatures became 1 degree Celsius colder, when growing seasons in Europe shrank by as much as three weeks, when the altitude limit of crop cultivation and tree growth fell by some 200 meters, or 656 feet, as snow lines and glaciers advanced. The vineyards of England, of which there had been at least fifty during the Medieval Warm Period, began to decline into obsolescence (not all due to a worsening climate but also competition from Gascony), crop cultivation disappeared above Trondheim in Norway and became more difficult in the highlands of Scotland (or anywhere else that was 1000 feet above sea level), fish stocks moved away from cold North Atlantic waters, and advancing sea ice made contact with Greenland more treacherous, until the island was completely abandoned by the end of the fifteenth century. Moreover, the elements of wind and water became stormier, especially in the North Sea along the coasts of the Low Countries and Germany, where death tolls in the hundreds of thousands were reported in single incidences beginning in the mid-thirteenth century and extending into the sixteenth, as dozens of coastal parishes were simply swept away. This reminds us that the weather was not just becoming colder and wetter, on average, it was also becoming wilder and more unpredictable, swinging from one extreme to the next. In this way we should not think of the climate of the Little Ice Age as being uniformly bad, but as being more variable: The terribly rainy summers of the famine years during the second decade of the fourteenth century, for example, were sandwiched on either side by decades that were unaccountably warm and dry. Thus it was the unreliability of the elements, which had previously been more consistently mild during the Medieval Warm Period, that was perhaps most unsettling to those who lived through the Little Ice Age. Over the long term, populations could adapt to and even benefit from a worsening climate, such as by concentrating farming and housing away from unproductive uplands and coastland that were overprone to freezing and flooding, to stable areas where intensive development would be more reliably rewarding. It was in the short-term fluctuations of the weather, when a succession of bad harvests caused by the climate occurred on an almost yearly, unforgiving basis, as seems to have happened in 1315–22, 1363–71, 1408–19, 1437–40, and 1481–84, that the effects of climate change were most keenly felt.[91]

The causes of the Little Ice Age include a range of possibilities. Perhaps the leading contender is thermohaline circulation, or the Great Ocean Conveyer Belt, but other factors that may have also played a role include more extensive snow cover, which can reflect heat from the sun and thus make surface temperatures even colder, something known as the "albedo" effect; variations in the earth's orbit, known as "precession"; fluctuations in solar radiation, such as are indicated by sunspots; and increased volcanic activity that discharges solar-screening microparticles, or "dust veils," into the stratosphere.[92] Whatever the reason, how we know that the Little Ice Age occurred comes primarily from the evidence of proxy data, such as dendrochronology or isotope readings in ice cores, rather than from documentary, or qualitative, records.[93] Only the former can really track the dramatic downturn in temperatures and increase in wetness and trace it consistently over an extended period of time. While some scholars have attempted to tease out such patterns from chronicle accounts, such as by compiling decennial indexes of years in which the weather is commented upon as being generally mild or harsh and rainy or dry, these, while revealing some perceived extremes of temperature, do not clearly point to a Little Ice Age beginning in the fourteenth century, since such extremes were also noticed during the second half of the twelfth and during the thirteenth centuries.[94]

It is true that with the advent of the Little Ice Age, we start to acquire some extraordinary documents, such as the weather journal of William Merle, rector of Driby in Lincolnshire, covering the years 1337 to 1343, the first journal of its kind in history.[95] Merle's entries are, on the whole, terse and repetitive, merely noting days with varying degrees of frost, snow, rain, fog, wind, cold, heat, and occasionally storms, thunder, and hail. Slightly more dramatic were a local earthquake noted in March 1343 and a comet later that same year in September. Sometimes comparisons were made in weather from month to month or even year to year, such as when Merle noted a penetrating rain in the months of November and December of 1341, together with "a very great diversity in the temperature of the air" after September of that year, or that he had never known the North wind that blew in March of 1340 as being so warm, nor that the winter of that year to be as snowy as the previous year's. But, aside from a couple of notices of how the rain in summer was not heavy enough to "hinder the workers in the corn-fields," what we are missing is a perception of the role the weather may have played in the course of human events, as well as any sense of man's reciprocal responsibility for the behavior of the elements. For that, we have to turn to two environmental catastrophes that occurred on either side of Merle's weather journal and that thus went unrecorded by him: the Great Famine of 1315–22 and the Black Death of 1348–49.

## Earth, wind, and death

Perhaps the greatest climate-related event of the late Middle Ages was the extended period of wet and cold weather that triggered the Great Famine between 1315–22. Across Northern Europe—in England, Ireland, the Low Countries, Scandinavia, France, and Germany—contemporary chronicles testify to extremely rainy conditions beginning in 1315 with up to 100 days of continuous rain that often coincided with attempts to plant crops or harvest them. One English chronicle complained that so much rain had fallen that year that seeds rotted in the ground and "in many places the hay lay so long under water that it could neither be mown nor gathered." The wet weather continued in 1316, so that in England it was alleged that "there could not be found seven days of good weather," with the result that "little grain grew that year, nearly all of it having perished." In Germany the rain was said to have shadowed the entire agricultural cycle, falling when the seed was planted, when it sprouted, and finally when it came time for the harvest. According to an anonymous Austrian chronicler, 450 villages were washed away by an overflowing river in Saxony, and deluges similarly threatened populations in Austria, Poland, Hungary, and Meissen in Germany. Fields and valleys were inundated to the point that the water "destroyed hay and standing corn [grain]." (In some cases, flooding from excessive spring run-off was exacerbated by human impacts, such as soil erosion caused by arable farming and clear-cutting of forested watersheds.) Chroniclers in Belgium claimed that it rained nonstop from June 24 until August or harvest-time in 1316, such that "the inside of granaries stood nearly empty of flour." Storms and floods also buffeted coastal areas in eastern England, Normandy, and Flanders. In between the rain severe winters trapped ships in the ice, as happened when the Baltic Sea froze over in 1315–16 and 1321–22, while in 1317–18 England knew a winter that according to one poem was a thousand times stronger than any that came before. That same year in Romania the Moldava River froze over without interruption from November to March, becoming so solid that one could "walk over it daily, as if one's foot was passing over dry land."[96]

All this bad weather produced a famine whose mortalities ranged from 5–10 percent in Belgium and 10–18 percent in England, based on available statistics. While these death rates might be two to three times higher than normal, they were still not so high that medieval communities might not be able to fully recover their previous numbers within a generation or two, and there are indications that some did so, although overall population seems to have slightly declined in the decades after the famine and leading up to the Black Death (1348–49). Likewise, medieval agriculture entered a prolonged period of contraction after the famine, when during the worst years of 1315–16 harvest yields declined by as much as 40–45 percent in the South and 72–89 percent in the North of England. Arable cultivation gave way to pasture farming, which required less labor and hands-on management, and villages shrank or were deserted of inhabitants, trends that were greatly accelerated by the far graver demographic crisis of the Black Death. While reports of cannibalism, such as were recorded in Ireland, England, Poland, and the Baltic States of Livonia and Estonia, might be exaggerations retailed for effect, somewhat more believable is what the chronicler John de Beka told happened in Holland, where "many paupers gnawed on the raw carcasses of cattle like dogs, and at the uncooked grass of meadows like cows." The spectacle of his fellow human beings behaving like brute beasts is what seems to have disturbed Beka the most, rather than what they were actually eating.[97]

Did medieval people make any connection between themselves and the extraordinary events that were happening in the skies? As to be expected, divine chastisement played a big role in explanations for the Great Famine, building upon a theme going back to the Old Testament and other ancient attributions of natural disasters to the gods. But even though storms and other atmospheric catastrophes were assumed to come from on high, at the same time we have also seen from Agobard's testimony during the Carolingian era that humans were likewise believed to have the capacity to affect the weather. During the Great Famine, this belief manifested itself in the form of prayers and processions for a "suitable serenity of the air," such as those called for every Wednesday and Friday throughout the Southern province of England by Archbishop Walter Reynolds of Canterbury in July 1316. A hundred years later, in 1416, a similar call for prayers for the "serenity of the air" was issued by the current archbishop, Henry Chichele, who justified it on historical grounds by harking back to when the Emperor Theodosius had called down the Bora, or North Wind, to turn back the javelins of his enemies at the battle of the Frigid River in 394 C.E. During the Great Famine, similar prayers and processions are known to have occurred in France, Belgium, and Germany. Contemporaries' sincere belief that they had the power to change the weather during such times of crisis is attested by a chronicler from Malmesbury Abbey in England, who asserted that had it not been for the Church's intercessions, "we should have perished [from the famine] long ago." This is a pattern that will be repeated again, only this time on a much grander scale, a few decades later during the Black Death.[98]

The "Great Mortality" that first struck Europe in 1348 was undoubtedly the greatest natural catastrophe to occur during the Middle Ages, and perhaps throughout all of human history.[99] The explanation for this unprecedented and awesome event posed an enormous challenge to doctors and other academics who debated the causes of the plague, which for the most part were discussed in terms of man's relationship with his environment. Natural causes of disease entailed a corruption, or what we might today call a "pollution," of all three elements of air, water, and earth. However, in the eyes of most medieval physicians, it was the air that was of most concern, which was based not on empirical observation but on ancient and prevailing medical theory going back to the great Roman physician of the second century C.E., Galen: This held that the airy regions, being closest to the heavens, are most susceptible to celestial influence; that air is an element essential to life breathed in by every living creature; that air, especially in its corrupt form, can be transferred from one place to another by prevailing winds; and that air is intimately bound up with the other two elements, water and earth, by which it can be corrupted or in turn corrupt them.[100] Following an Aristotelian conception, air is also explained to be a substance that is "finer" or "more subtle" as well as being "more easily changed" and "more passible" than other elements, particularly the earth, and therefore it is more quickly and virulently corrupted; air also contains the two essential qualities for corruption, namely, heat and moisture, that are lacking in other elements.[101] The plague treatise penned by the medical faculty of the University of Paris for the king of France in October 1348 states authoritatively that "illnesses which proceed from a corruption of the air are more deadly, for bad air is more harmful than food and drink, because it penetrates quickly to the heart and lungs with its malice."[102] A number of later treatises from the fifteenth century continue to follow this line, some of them quoting the ninth century medical authority, Isaac Israeli ben Solomon, otherwise known to Europeans as "Isaac the Jew," in support.[103] In the words of one author, a "pestilence, according to doctors of medicine, is nothing else than a putridity of the air."[104] An anonymous treatise from south Germany from the late fourteenth century asserts that the notion of corrupt air as being "very often the cause of the pestilence" is something "on which all ancient [authors] are in agreement."[105] According to Hans Würcker, a doctor at Ulm writing in his native German tongue in 1450, of the three kinds of pestilence mentioned by Avicenna, that from the "poisonous, foul air" is the most dangerous as well as being the most common (as compared to "poisonous moistness" or "poisonous, foul water"), because "from it comes the quick death," in that the "air is drawn directly into the heart." Würcker also uses the German verb, "to pollute" (*verunreinigen*), when he writes that "the air gets polluted [*verunraint*] with bad smells and vapors and fogs and winds".[106] Other German treatises at this time assert that "all poison comes from the air," which runs directly to the heart and from there to the "emunctories," or places where boils form on the groin, armpits, or the neck.[107]

Plague doctors also devoted much attention to explaining exactly how a simple and pure element like the air could be corrupted and from what sources the corruption came. The standard explanation was first set during the Great Mortality of 1348, when a number of medical authorities, including the faculty of medicine at the University of Paris, Gentile da Foligno of Perugia, Jacme d'Agramont of Lérida, and the Moorish physician, Ibn Khatima from Almería, following lines laid down earlier by Avicenna, distinguished between two main causes of the plague, a higher and remote cause from the heavens or celestial bodies, and a particular or near cause arising here on earth.[108] Citing Albert the Great's *De Causis Proprietatis Elementorum*

(Concerning the Causes of the Properties of the Elements), the Paris masters declare that the plague was caused by a great conjunction, or lining up, of the planets Saturn, Jupiter, and Mars at 1:00 p.m. on March 20, 1345, a statement that is much repeated by other physicians, even down into the fifteenth century.[109] The planets then drew up many "evil vapors" from the water and earth which mixed with the air and "corrupted its substance" to the point that it was "hostile and repugnant to our nature."[110] Agramont's very definition of a pestilence (again traceable ultimately to Galen) is a "contra-natural" change in the air, either "in its qualities or in its substance," the latter of which he calls a "putrefaction" (as opposed to a mere "alteration") and which is the worst kind of change in that "it signifies something very repugnant and disproportionate in the air from which no spirit of life can rise." Therefore, it should be judged the true source of an epidemic or pestilence.[111] Khatima also makes this distinction, calling a "partial" corruption "the change of all or some characteristics without changing or spoiling the element [of air] as a whole," or in other words, "without changing its essence". A "total" change, on the other hand, is "caused by the corruption of the elementary components by means of decay so that the air is a wholly different mixture." In the former case, air is still air, but in the latter, air can no longer really be defined as air but is instead "closer to the rotten vapor than to absolute air, let alone good air." Khatima gives as examples rotten food "left in old pantries" and "wells in which animals died and where the stench remains locked up." In these cases, the air is so rotten and altered that it is "changed into a miasma which is dangerous for animals" and from which a human will die immediately if he breathes it in. Moreover, within such a miasma all light is extinguished and, historically, "thousands of people perished" in a single day.[112] Both Khatima and Foligno, following Avicenna, also explain that the air around us that we breathe in every day is never "simple and pure" air, but rather an adulterated mixture containing watery and earthly vapors, and that therefore air in this form can always putrefy, even though in its elemental state it does not.[113]

Near causes can come from various sources, including: rotting cadavers in cemeteries or left on the battlefield during wars; stagnant pools of water full of rotting matter such as in swamps, lakes, cesspits, and sewage or drainage ditches; rotting plants and animals; caverns and enclosed places in the earth whose vapors are released by earthquakes; stables full of the dung of beasts, etc., all of which give rise to bad vapors or "stenches" that can be carried from place to place by prevailing winds. Already there was a difference of opinion among these early plague doctors as to which cause was more important for explaining the origins of the plague. The Paris masters obviously favor the celestial cause, on the grounds that a universal pandemic of disease demands an equally "universal" source, i.e., from the heavens, whose influence can be felt everywhere; indeed, for them, even near causes can ultimately be traced to a higher one.[114] Ibn Khatima, who gives three causes of the plague to include changes in the air caused by "celestial bodies," changes of the air caused by unseasonable weather, and changes of the air from near causes, judges the first to be the true cause of the 1348–49 pandemic on the grounds not only that the disease was present in "most, if not all, countries" but also on that it seemed unaffected by the change of seasons, arriving at Almería at the beginning of June 1348 and lasting throughout the summer, fall, and winter up to the time he was writing, in early February 1349. However, Khatima is still wary of the celestial or astronomical cause on religious grounds, in that astrology or astronomy might pose an explanation of the plague that is too autonomous of God, who, according to the Prophetic tradition of Islam, is the direct source of the disease and admits of no other agency. Thus he says that "research" into the celestial cause "is still incomplete, because the knowledge of the laws of astronomy, as far as it goes, is not ascertained and we don't know whether they're based on facts."[115] This same ambivalence is to be found in the work of the fellow Muslim, Avicenna, who seems to favor near causes of disease in that he compares corruption of the air to that of a mostly earth-bound element, water, and thus Gentile da Foligno, who was famous for his commentaries on the entire *Canon* to the point that he was dubbed "the soul of Avicenna" (*anima Avicennae*), is likewise ambivalent towards the higher cause.[116] Jacme d'Agramont lumps all causes together in his discussion of a "universal pestilence" without seeming to favor one over the other, although he does state that unseasonable weather, while being "a change very dangerous to man and to all living things," is only a qualitative change in the air, not a substantial one, and he regards a corruption of the water or earth alone, no matter whether this be in their qualities or substance, to be incapable of causing a pestilence without causing a corresponding change in the air.[117]

The debate over the causes of the plague—whether these were celestial or terrestrial in nature and whether they originated in the air, water, or earth—was to continue throughout the rest of the Middle Ages, even though later plague doctors generally adhered to the basic principles lain down during the first pandemic, i.e., that there were two distinct categories of causes, higher and lower, and that plague usually resulted from a substantial change in the element involved.[118] Those preferring the astrological explanation included John of Burgundy, a physician from Liège, who wrote his treatise in 1365, and the papal physician based in Avignon, Raymond Chalin de Vinario, writing in 1382. Both doctors are well known for criticizing ancient authorities for their lack of experience in treating a disease like plague and for being allegedly "ignorant" of the true causes of "pestilential diseases" and therefore unable to cure them. In the case of Vinario, it seems clear that he mainly has Avicenna in mind, but Burgundy takes to task a whole host of revered names, including Galen, Dioscorides, Rhazes, John the Damascene, Geber (Jābir ibn Hayyān), Mesue (Ibn Masawaiyh), the eleventh-century Salernitan physician, Copho, Constantine the African, Serapion of Alexandria, Avicenna, and Algazel (al-Ghazali). From just reading these denunciations, one might easily be led into thinking that this represents a bold and progressive step forward for medicine, but it's not quite that simple.[119] Both Burgundy and Vinario were perhaps prescient for their own times, in that they pointed to the "wisdom" of astrology, a fashionable subject in fourteenth-century medical circles, as the most important source of knowledge about the plague; yet this stance hardly anticipates the modern scientific method. Burgundy, citing Hippocrates, asserts that the reading of the stars is "a science vital to the physician," and moreover it is one that he himself has proven by his own "practical experience," in that medicine "is of no benefit to the patient if it is given when the planets are contrary."[120] Vinario declares that a "true epidemic" comes from the higher, celestial cause, whereas a lower one can only give rise to diseases that are regional or local in scope, although sometimes the higher and lower cause can act concurrently. He further explains how each of the plague epidemics that have occurred thus far, including ones in 1348, 1361, and the present one in 1382, came about through conjunctions. Even though Vinario finds fault with some "eminent doctors of medicine," namely Avicenna, for touching upon celestial causes "only superficially," he does manage to find support for his position in the works of other Muslims, including Albumasar (the ninth-century astrologer) and Avenzoar (an eleventh-century physician), and in the work of Albumasar's fellow Persian, Massahalla (an eighth-century Jewish astrologer). He then gives in the next chapter a very detailed explanation of exactly how planetary influences can bring about epidemics.[121]

## Environmental causes of the plague

For more environmentally savvy explanations of the plague, we need to refer to treatises that champion the near cause. One of the earliest and more truly forward-thinking treatises to exclusively consider local sources of disease is by the fourteenth-century German scholar, Konrad of Megenberg, author of the *Buch der Natur* (Book of Nature), a natural history largely borrowed from Thomas of Cantimpré's *Liber de Natura Rerum* and completed in c.1349–50, after what the author tells us was a fifteen-year period of composition. In a separate treatise devoted to the causes of the plague, which Megenberg wrote in 1350 when he was at the papal court of Avignon seeking the patronage of Pope Clement VI, he rejects in short order higher causes of the plague, either from God or a conjunction of the planets. God cannot be the author of this plague, Megenberg says, because we see that "people have in no way amended themselves of any vice," nor that the disease has "struck down all mortal sinners," which would imply that God had sent down the plague for no purpose, "which is not to be admitted." This then leaves the plague as arising from a natural course of events, which nonetheless cannot be the remote cause of planetary conjunctions, because the celestial influence of the planets, especially Saturn, as they travel through the zodiac does not always correspond with the timing of the mortality. Making an observation that is confirmed by a modern study of the spread of plague based on chronicle accounts,[122] Megenberg states that the disease behaves erratically, "as if it moves with an accidental or involuntary motion," in that it "now moves towards the East and now in the opposite direction, and then changing direction leaps from the South to the

North and retraces its steps in places already visited," which he attributes to the propelling of plague vapors or fumes by "winds in the airy regions." Megenberg's preferred explanation is an earthly one, namely, "a corrupt and poisonous exhalation from the earth, which infected the air in various parts of the world and when breathed in by men suffocated them to the point of sudden extinction." His "proof" of his position, "grounded in the probable realm of natural science," is twofold. One is that air, when it "is shut up and imprisoned for a long time in the earth becomes so corrupted in some corner of the earth that it may be turned into a potent poison to the human constitution." This is demonstrated for Megenberg by the fact that wells, when they are first opened after being sealed and unused "for several years," typically suffocate with their fumes the first person to go into them, a proof that likely comes from Avicenna.[123] The second proof is that earthquakes are known to occur when vapors and fumes seeking to escape from "the bowels of the earth ... forcefully batter the sides of the earth ... they shake and move the earth, as is made clear by natural philosophy." Megenberg follows this up with eight arguments in support of his position, which include specific instances of earthquakes known to have occurred in the German-speaking provinces, along with other natural phenomena such as flooding, which according to another German treatise written around this time took place continuously "without interruption throughout the whole world in 1346," and he replies to six "doubts" raised against his position.[124]

That Megenberg is well in tune with the natural philosophy of his day is quite evident from the fact that, as we have already seen, earthquakes are discussed as a terrestrial phenomenon by all the thirteenth-century encyclopedists and natural historians, including Alexander Neckam, Vincent of Beauvais, Bartholomaeus Anglicus, Thomas of Cantimpré, and Albert the Great.[125] But while Megenberg is indebted to Cantimpré for explaining the physical mechanics of earthquakes, much of the rest of his discussion on earthquakes in book 2, chapter 33, of the *Buch der Natur* is clearly based on his own experience or knowledge; indeed, this is perhaps one of his more original contributions in the entire work.[126] Thus, we get the same references to actual, local earthquakes that Megenberg also cites in his plague treatise and which he must have either experienced himself or heard first-hand from others, such as the one that occurred in the town of Villach in Carinthia, Austria, on January 25, 1347, which Megenberg claims belched poisoned air that spread over 100 square miles along the Danube River and lasted for forty days and toppled churches, houses, and even two mountains. From seemingly the same store of his own personal information, Megenberg tells us that experiencing an earthquake is like being on a swaying ship; that as the air forces itself through "small cracks" and other gaps it can find, "the earth often hisses and whispers like a hundred thousand snakes or groans and bellows like horrible oxen"; that signs that one has happened include the sun turning red from the obscuring "earthly vapor"; and that "old women" tell "silly" wives tales about how earthquakes come about because the earth rests on the back of a huge fish with "its tail in its mouth," who shakes the earth whenever "it moves or turns." We even get the same proof that people die from earthquakes by referencing what happens to those who first enter long-abandoned wells, and Megenberg adds that this even happens in mining pits where "the same vapor has not been locked up in the earth for long," yet the miners who enter the pits "get dizzy in the head" and "fight like drunkards." Also coinciding with the plague treatise are arguments on how plague began in cities near oceans where the poisonous air was trapped in "veins" beneath the ocean which also made the water poisonous; how "burning and stinking fogs" were prevalent during the years of the plague; how the "poisonous vapor" released by earthquakes corrupted fruits, such as pears; how plague has lasted longer than constellations; and how the plague cannot be the work of God since sinners are still alive and well, such as the "Polish knights" who accompanied King Louis of Hungary's campaign against Naples in 1347–48 in order to avenge the murder of the king's younger brother, Duke Andrew of Calabria, and who "ate and drank early and lived fully, [and yet] nothing happened to them." Even though Megenberg rejects planetary conjunctions as an explanation of the plague of 1348, he nonetheless still admits the possibility that the swelling earthly vapors pushing against the walls of the caverns of the earth that cause earthquakes are themselves stirred "by the power of the stars," particularly the planets Mars and Jupiter. Here perhaps Megenberg is hedging his bets, since by rights a universal catastrophe requires a universal cause, and it may not have been clear to him yet that the Black Death was such a catastrophe, which would be beyond the scope of purely natural laws. If one admires Megenberg for describing natural phenomena that accord remarkably well with what has been documented by modern science, such as "outgassing," then one is puzzled by his comment, in both the *Buch der Natur* and his plague treatise, that from the vapor arising from earthquakes people can be turned to stone, which seems largely a biblical reference to Lot's wife in Genesis 19:26. Yet here Megenberg is also being "scientific," citing Avicenna and Albert the Great as his authorities, and noting that when it happens the stone into which people are turned is "mostly salt stone" and takes place in mountains "where salt is mined." The fact that sometimes "fire and flames issue forth from the earth" that burns "several villages and towns" and smells like sulpher points to the simultaneous eruptions of volcanoes.[127]

Another German commentator on the plague with an interest in natural history was a Master Bernard of Frankfurt, who composed a casebook during an outbreak of the disease in March 1382 but who at the same time referred to another work of his, entitled *De Natura*, in order to explain the causes of the epidemic. Like Megenberg, he gives earthquakes as a possible source of "universal vapors" (*nebulae*) spreading the disease, but he also discusses, on the basis of his own work (which perhaps was inspired by Cantimpré's *De Natura Rerum*), the "four universal fluxes" that "enter everywhere under the ground and issue forth from invisible places in the East, West, South, and North," and which ebb and flow like tides underground, propelling both air and water and cooling down the "lower regions." However, during a plague this underground water becomes infected and enclosed within "the bowels of the earth," thickening and stewing "like a porridge."[128]

Other plague doctors who seem to favor near causes of plague include the fifteenth-century physicians, Sigmund Albich of Prague, Blasius of Barcelona, Peter de Kottbus, and John of Saxony in Strasbourg. In his "Minor Collection" of 1406, Albich comments that "an astronomer places more faith in the motions of the stars and their laws, which is the higher cause," but that a physician places his faith in lower causes which, although being "less important, are [proven] to be true by demonstration," or syllogism. While medicine must pay homage to astrology, Albich seems content to leave that to the work of the ancients, such as the *Centilogium* and the *Almagest* of Ptolemy; physicians, by contrast, must observe what "is more true with regard to our senses, which has [but] one purpose, to decide what is the right thing to do here" when faced with a plague arising from a "lower origin," and about which a physician can do more, such as by prescribing a regimen against the local "corruption of the air."[129] Writing in the same year as Albich, Blasius of Barcelona is likewise dismissive of "presumptuous astrologers," who cannot be trusted with their prophecy, "poetic arguments," or other "recitations of proof," and who "grind down their life in [the study of] judgments devoid of foundation [and] dare to open their mouths [as if] they were most illustrious men of understanding." An admirer of Avicenna, like Foligno, Blasius appealed to the Persian's authority for not delving too deeply into the mysteries of the stars or the heavens, leaving the foreknowledge of all things to God. Thus it suffices to know that the "glandular pestilence" is ultimately "caused by the stars, but by what [causes] and how, we remain ignorant." Playing the ultimate skeptic, Blasius rejects that the plague arose either from the air or the "bowels of the earth," since there has been no wind, and "air that is still and calm is very pure," while he asserts that the symptoms of "glandular swellings" (*glandifera*) cannot be caused by "lightning flashes nor falling stars nor comets or the like," but from some other "efficient cause" which he fails to elucidate. If one tries to predict how long an epidemic will last on the basis of the impossibly esoteric calculations of astrology, he might have to wait 36,000 years to find the answer.[130]

Writing later in the century, Peter de Kottbus argues, like Raymond Chalin de Vinario, that plague can come from a higher or lower cause or both, but unlike his predecessor he is willing to admit that sometimes "a great and universal" pestilence can originate in a lower source "that is inferior in name alone, such as a local 'corruption of the air and water' or from "corrupt earthly vapors," as Avicenna says in the *Canon*.[131] John of Saxony, arguing on the basis of Galen, declares that "God and the heavens are not in and of themselves the cause of an epidemic," but rather, "God is a very remote cause," followed, in descending order of nearer causes, by the heavens, the air, bodily humors, "putrid air," and finally, "putrid vapor infused in the heart." Thus, for John of Saxony, the most important cause of plague takes place entirely within the human body and its immediate surroundings and has little to do with the distant celestial realm, which acts mainly as a "first cause" setting in motion "many other intermediate causes" that in the case of plague lie "between God and the epidemic." Proof of this for John of Saxony can be found in the fact that, while a universal first cause such as God should result in an epidemic that is "general and uniform," nonetheless, the pestilence that he was now experiencing was "deformed and particular," as evidenced by the many different kinds of plague symptoms that "are most clearly manifesting themselves in this [past] year in Strasbourg." These included "apostemes" or buboes on the armpits, groin, or "elsewhere"; other skin manifestations such as *antraces* or carbuncles and "erysipelas" or a rash of perhaps purpuric spots (called in modern medical parlance disseminated intravascular coagulation, or DIC), which may be indicative of septicemic plague; very high fever, apparently unaccompanied by other major symptoms and where the patients "die without any kind of pain" (again septicemic plague); and even cases where victims "die almost joking and laughing (*jocando et ridendo*)."[132]

Other schools of medical thinking were perhaps more ambivalent or divided about the true causes of plague. Two doctors practicing at the papal court at Avignon during the 1370s, John of Tornamira and John Jacobi, were rivals for the chancellorship of the medical school at the University of Montpellier, which seemed to determine them to take opposite sides on whether "a very great pestilence" could proceed from a "higher source" (Tornamira) or a lower one (Jacobi), with each one declaring that those who disagreed with them were "deceived."[133] Perhaps reflecting this division, Heinrich Ribbenitz, based at the University of Prague in the late fourteenth century, gives a confused and rambling explanation of the origins of the plague: At one point he declares that there are three kinds of disease outbreaks—a pestilence, an epidemic, and a "vintage"—which are caused respectively by a constellation of the planets, corrupt vapors "released from the earth," and "a putrid corruption of water." While this would seem to imply progressively less virulent occurrences of disease, based on Ribbenitz's earlier distinction between air corrupted either in its substance or in one of its qualities, in fact all of them seem to qualify for his fourth category of "corruption of something pure by mixing in with it something that is impure." Indeed, Ribbenitz warns his readers to be on their guard against stenches arising from "foul water," which was "especially a cause of this pestilence," such as subterranean caverns and wells and bodies of standing water that have become "corrupted and rotten," such as retting flax or sewage drains and gutters.[134]

Plague doctors also urged their readers to pay close attention to their natural environment in order to detect "signs" of a coming plague, which would allow one to prepare for it.[135] Many of these signs were atmospheric disturbances or phenomena, such as comets, shooting stars, and the like, or else "alarming" changes of color in the sky—such as a reddish, yellowy, greenish or bluish hue (the rainbow?), as well as dust or vapors obscuring the sun, an unnatural stillness to the air, fogs, mists, winds, thunder and lightning,

*Part I. Air, water, earth* 61

cloudy and unsettled weather, etc. This also could involve unseasonable weather, or climate that was inconsistent or counterintuitive to what was expected in any given season of the year, such as warm and rainy weather in the winter, or colder and drier weather in the spring and summer. Also predictive was changeable weather within the same day, such as that it looked like it was about to rain but didn't. (We have a saying in Vermont: If you don't like the weather, wait five minutes!) Above all, it was believed that plague prevailed during the autumn: Autumn-time was called "our enemy" and "the slayer of many men" because this was the season that was seen as being more changeable, humid and windy, and it was then that the summer's heat might linger and yet the sun, being lower in the sky, was not as effective in terms of burning up the air's moistures with its deflected rays. Other signs could include "cobwebs or powders flying through the air" when "putrid humors" are released from the earth; spoiled grain, fruit, meat, water, or fish as evidence of corruption in one of the elements integral to production of these comestibles; and the behavior of animals such as birds, reptiles, and vermin, who by instinct flee corruption in the air or ground or else signify a spontaneous putrefaction that gives rise to their very existence (see section on "Animals and disease" in Part III). Typically, the treatises cite authorities such as Hippocrates or Avicenna as their basis for making such statements, but sometimes they make observations on plague signs based evidently on their own experience. Therefore, the Paris masters claim to "speak from experience when we say that for some time now the seasons have not been regular," with the past winter less cold and very rainy, the spring windy, and the summer less hot than it ought to have been and "very unpredictable from day to day and hour to hour." And yet, we cannot always take such statements at face value, for a couple of German treatises from the fifteenth century repeat this passage almost word for word, as if it were their own.[136] Most puzzling is the rigid adherence to the orthodoxy that plague always came in the autumn, which we see extending from Gentile da Foligno in 1348 down to Primus of Görlitz in c.1464.[137] Yet the evidence from bishops' registers, wills, and other documents that allow us to track the plague month by month clearly shows that plague, even in the hotter Mediterranean countries such as Spain or Italy, usually peaked during the summer months of June, July, or August.[138] (Foligno in fact died on June 18, 1348, which was during a peak month for plague deaths in Perugia, according to testamentary evidence there.) It seems that late medieval doctors were not yet ready to leave the comfort zone of their reliance on authority, even when faced with contrary evidence before their very own eyes.

As part of their preservative regimens, doctors usually paid most attention to the first "non-natural," i.e., precautions with respect to the air, which was only natural as they saw its corruption as the most common cause of plague and the respiration of it as most essential to human health.[139] Sigmund Albich, for example, warns that a physician should "take great care to see to the rectification of the air," because many a patient who has died would not have "if the air had been rectified."[140] The first and most recommended piece of advice during a plague was to flee from wherever the air was "pestilential" or bad and seek out clean and clear air, but if this was impractical (as it was for the vast majority of medieval people, who had but one home to live in), then the alternative was to shut oneself in one's dwelling or chamber, especially when the weather was bad, closing the windows and doors or covering them with glass or waxed cloth, particularly those facing south and west (from whence blew winds carrying bad air), while keeping windows facing north and east open in order to ventilate the room, and fumigate the inside air with fires made from aromatic and dry, clean-burning woods, perhaps perfumed with a little incense or herbs, often made up in pill form called *trocisci*. (The benefits of enclosed or encapsulated air explained why nuns and prisoners often escaped pestilences, a question that was posed in academic treatments of the plague.)[141] If one lived in the forest, then trees should be cleared on the north side to allow the clean northern winds to blow in, but keep the house sheltered on the south (which is the exact opposite of advice now given for those wishing to install solar panels!). Otherwise, low-lying dwellings or caves are the preferred place of residence when the plague source is up high, but homes in the mountains are preferred when corrupt air originates from the valleys or terrestrial sources. (Again, one wonders how many had the luxury of choosing such locations.) One could also sprinkle the house with vinegar and fragrances and strew flower petals, leaves, and cut green boughs from aromatic trees, choosing "cool" aromatics (vinegar, roses, willow, water-lilies, sandalwood, grape vines) during summer and "hot" ones (aloe, amber) during winter. If one had to go out into the open air (as presumably most people did who had labor obligations such as peasants, merchants, or artisans), then one should wait until mid-day when the sun has had a chance to purify the air, and one should carry on his person smelling apples or fragrant herbs and perfumes (rue, sage, wormwood, camphor), wash his hands and face with vinegar (for the sour smell to ward off the bad air, not for the purposes of hygiene), keep in one's mouth a piece of fragrance (cedar, birthwort, gentian, zedoary), or else fortify the body with theriacs, pills, cordial electuaries, and appropriate foods and evacuations, such as blood-letting. Also advisable, but perhaps none too helpful, was to avoid breathing in too much of the suspect air by over-exercise or exertion, but one could also help in this regard by attending to his diet and evacuations, especially by bleeding. With regard to water and earth, doctors advised being choosy about ingesting anything that was of or in contact with these elements: Thus one should boil one's drinking water or seek out clean sources of it, such as clear wells or springs, rain water, or swift-flowing, rocky streams, and store it in earthen cisterns or wood-tarred barrels. One must also beware of fish and root vegetables, although almost any green-growing plant, animal, or body of water might be suspect at such a time if in contact with the corrupt air.[142] Once air corrupted by the plague entered the body, then the physician's job turned to a cure, because after being breathed in by the lungs or absorbed through the pores, the bad air or vapor was believed to immediately assault the heart, which was normally ventilated by clean air in healthy times, such that the vital spirits and blood in the heart were corrupted or else a putrefied moisture and unnatural heat generated around it, from which symptoms of the plague emerged.[143]

## Man-made pollution of the environment

Obviously, from an environmental point of view, it is the near causes and signs that are of most interest, but within this category, what are especially interesting are the plague sources identified as the products of human industry, since these clearly anticipate concerns about pollution in modern times. One of the most noxious of the man-made or small-scale industrial contributions to plague, as these were identified by medieval plague doctors, was the retting of flax or hemp, in which these plants were immersed in standing bodies of water in order to soften them up so as to separate their fibers from the stalk, which were then rendered into linen or hemp threads that were woven to make clothes, canvas, and rope. Tanning of leather also required steeping the animal hides in a water solution (often for a period of months or even years) in which oak bark was the main ingredient, but not before the skins had already been treated with lime and watered-down bird or dog excrement. Gentile da Foligno, in an uncredited nod to Avicenna, compares "the putrefaction of the air" to the "putrefaction of water in that there is a fetid softening," and he gives as an example "water with which wheat or flax is softened."[144] Raymond Chalin de Vinario gives as one of his lower causes of plague "the stenches of linen or hemp putrefying in water and of leather hides and pelts steeped in water," which was frequently to be found in the *viterbia* or leather-selling district of cities.[145] Heinrich Ribbenitz, writing in 1370, claims to have had personal experience of the dangerous effects of such water pollution, mentioning "fetid and infected water due to something put into the water" as a "special cause of the pestilence," and he gives as an example of such a cause "whenever flax or hemp is put into the water, which especially kills men, as I have seen [myself]."[146] A Lübeck doctor cites retting flax and hemp as capable of producing "a universal stench in our climate" that can give rise to an epidemic pestilence, such as one that began on July 20, 1410, in Göttingen, Germany, which lasted until the end of autumn and which was the worst seen in thirty years.[147] John of Saxony, likewise writing in the early fifteenth-century, was concerned about the filth flowing "continuously" into the common streets of Strasbourg, such as "effluent from birds' nests and kitchen waste," as well as pig dung owing to an abundance of the animals "feeding in the city," all "infecting the air during the whole time" of an epidemic of plague.[148] Also of concern were fish and meat markets, the dross or slag from smelting furnaces, sewage and drainage pipes or ditches and open latrines, stockpiling of manure, even the daubing of walls with lime.[149] Nor should we forget mining, since this was mentioned by Konrad of Megenberg as giving access to deadly vapors or fumes trapped within the earth.

It is easy to see how doctors' warnings translated into regulations and ordinances passed by urban communities to limit exposure to these health hazards, especially during a time of plague. During the Great Mortality of 1348, towns such as Florence and Venice set up temporary sanitary commissions in order to remove all putrid matter "from which might arise or be induced a corruption or infection of the air" and otherwise to "avoid the corruption of the environment." Even before the plague, beginning in the thirteenth-century, Florence and Bologna had regulations for cleaning streets and disposal of waste, which in Florence was especially directed at trades such as dyers and butchers who generated contaminated water and animal by-products whose "fetid odors" could give rise to corrupted air and "pestilential illnesses."[150] A complete set of ordinances designed to check the spread of infection during the Black Death survives from the commune of the city of Pistoia in Italy. A number of regulations, for example, concern the butchering of meat, evidently with the aim of preventing rotting or decomposing flesh adding to the stenches in the air or being sold for consumption to unsuspecting customers: Thus, sides of beef could no longer be hung up in the storehouse, horses could not be stabled in the shop, meat for sale over the counter was limited to what was slaughtered from a single animal, meat had to be sold on the same day it was slaughtered, all animals for slaughter had first to be inspected by appointed officials, and so on. The other main industry affected were the tanneries: Leather tanners could no longer cure their skins within the city walls in order "to avoid harm to men by stink and corruption," although this was later amended to only within certain districts of the city.[151] Another set of plague ordinances that survive and have been printed are those issued by the Belgian town of Diest from 1469, which were in turn adapted from the ordinances of the nearby capital of Leuven. It was at this time, during the mid-fifteenth century, that health boards started to become a permanent feature of the landscape of many an urban community in Europe. At Diest, a variety of workers and their businesses were affected by the plague ordinances in an effort to limit waste streams that might contribute to spreading the disease. Thus caretakers of the sick were not to dump anything into street gutters nor wash the clothes of their patients at the town bridge over the river Demer; barber surgeons could not set blood from their bloodlettings out in their windows or pour it out into the streets; no one could offer meat for sale from a house where a plague death occurred; clothes sellers had to certify that second-hand clothes for sale came from houses where there had been no deaths or illnesses due to plague; pigs could no longer run freely through the town streets; no refuse, urine, or "rotten matter" could be thrown out into the streets, and everyone had to clean the "sewer" or gutter in front of his dooryard twice a week "with clean water."[152]

Even in non-plague years authorities were concerned about pollution of air and water by medieval industries, although even then a connection was made between the stenches created by waste streams and disease. Butchers and tanners were the most common targets of urban regulations due to the waste streams of animal parts and contaminated water that they produced. Thus in 1361 and 1368, the city of London tried to ban its butchers from slaughtering within the city limits and dumping the entrails or offal of their slaughtered animals into the Thames; by a new proclamation of 1371, processing of meat was restricted to two districts of the city, at Stratford-le-Bow and Knightsbridge, which was being done, it was said, because of the "appalling abominations" and "stenches" that infected the air, as a result of which "sickness and other maladies have befallen residents and visitors to the city," and would continue to do so "unless some remedy could be devised." Despite the penalty of a year's imprisonment for violating the ordinance, it had little effect because the leading shambles in the city were exempt from its regulations.[153] Earlier, in 1357, concern about stenches emanating from the Thames and Fleet rivers in London prompted similar laws against dumping of household garbage and stable manure and called for waste to be cleaned from the streets by *rakyers* and transported outside the city on *dongebotes*.[154] In 1388 a statute of Parliament proclaimed that in London and other cities and towns throughout the realm, anyone who dumped "harmful matter, dung, offal, entrails and other ordure into ditches, rivers, waters, or other places should be responsible for having it completely cleaned up, removed, and carried away," on pain of a fine of £20. It was declared that this was in response to the fact that from such "filth" the air "is greatly corrupted and infected, and many illnesses and other intolerable diseases daily befall both the inhabitants and residents of the said cities, boroughs, and towns, and others visiting or passing through them."[155]

Other English towns such as York, Winchester, Exeter, Colchester, Bristol, and Leicester enacted regulations against their butchers in order to protect their waterways, even though the meat trade was practiced on a much smaller scale than in London. Leicester imprisoned any butcher found selling meat tainted "with any manner of sickness," while Exeter and Colchester assigned two meat wardens from the early fourteenth century to oversee the trade. In 1425, the brewers of Colchester complained against the tanners who were said to be "impairing" the water used for making ale. Waste streams from butchers, tanners, and potters were likewise legislated against in Paris throughout the second half of the fourteenth and fifteenth centuries in order to safeguard the Seine, which was becoming "infected and corrupted." The city of Namur in present-day Belgium enacted regulations during the fifteenth century that prohibited throwing refuse in the rivers Sambre and Meuse or in the defense ditches surrounding the city, and it hired a "varlet" to make sure the regulations were obeyed and to clean up any dumping that occurred. Namur sited its butchery and tannery industries further downstream than others that also required water, such as brewers and cloth-makers, which were in a more central location and thus evidently considered less polluting. (It is now thought that clothing dye may even have helped purify the water through its organic vegetable properties that settle out sediment.) Even the noise pollution from medieval forges could be the target of complaint. It is true that well before the Black Death pollution was a concern for medieval authorities: In 1307 England forbade burning of sea coal in limekilns as a result of complaints about coal fumes made in 1285 and 1288, and in 1310 London appointed a "Conduit Keeper" to preside over the city's water supply. On the Continent, Marseilles in 1253 ensured water from the river Jaret, diverted to irrigate gardens and supply tanners, did not flow back into the harbor; and Florence, beginning in 1322, regulated against piscicides used for fishing in the Arno, and in 1330 Barcelona controlled stenches said to be issuing from drains, sewers, and latrines built in the Jewish quarter. But it certainly seems that the Black Death added a new urgency and scope to such anti-pollution measures.[156]

Admittedly, none of this was very new. Even the efforts to control the inadvertent, man-made pollution that was the inevitable by-product of a large population living in an urban environment was addressed by Muslim authors centuries before. But what was new was the belief that humans could *intentionally* and *maliciously* pollute the air or water in order to cause harm to one's fellow men, namely by communicating to them a deadly disease, which dates from the very beginnings of the Black Death. One of the more remarkable treatises in this regard is the *Epistola et Regimen* (Letter and Regimen) by a Spanish doctor based at the University of Montpellier in southern France, Alfonso de Córdoba, who is thought to have been writing in 1348 or 1349, in the midst of the first outbreak of the plague. After considering two "natural" causes of the pestilence, namely, a planetary conjunction and an earthquake, Córdoba goes on to examine the case of an artificial, man-made cause, which he says is necessary in order to explain how the plague could have lasted so long and "spread throughout all the regions of Christendom," since this was beyond the scope of the other causes. In this last instance, the plague is deliberately instigated and spread by a human hand, which "proceeds out of a deep-seated malice through the most subtle artifice that can be invented by a profoundly wicked mind." Although Córdoba warns his readers to be on their guard against poisoned food and drink, "especially non-flowing water, because this can most easily be infected," the example he gives of how such a poisoning can be effected is a pollution of the air. He describes how a "certain formula" can be "well fermented" in a glass amphora, then thrown "forcefully" against some rocks opposite the city or town to be infected, after waiting for "a strong and steady wind" blowing towards the target. For Córdoba, this is "the worst kind of pestilence, because it cannot be averted except by those who are experienced." Presumably as one such experienced physician, Córdoba gives some remedies against this evil, even though "the wise counsel of doctors does not profit or help those in the grip of this most cruel and pernicious disease." These include fleeing the plague, a "theriac of Lemnian earth," which was a common remedy against poison by inducing vomiting, and Córdoba's special "pestilential pills," which "prohibit the infected air from penetrating the heart."[157]

Obviously, Córdoba's treatise meshes quite well with the accusations of well-poisoning that were leveled against Jews and others during the Black

Death, even though nowhere does Córdoba identify the Jews or anyone else as specifically behind this plot against "the [Christian] faithful who chiefly suffer from it."[158] One can speculate about whether Córdoba was influenced in the writing of his treatise by an independent poisoning accusation, or whether he himself contributed to the accusation's origins or its validation. On the one hand, the poisoning accusation had already been vetted in the spring of 1348 at Narbonne, Carcassonne, and Grasse on either side of Montpellier along France's Mediterranean coast: According to an official account of these trials, some "poor men and beggars of various nationalities" (but not Jews) were tortured into confessing that they put "powdered poisons" in the "water, houses, churches, and foodstuffs in order to kill people" and then were executed by being "pulled apart by red-hot iron pincers" before being quartered, having their hands cuts off, and finally burned.[159] Such gruesome displays of justice certainly would have garnered the attention of all who lived in the area. On the other hand, poisoning was a well-established topic in the medieval medical community, with its roots in ancient Greek and Arabic authors, and the plague seems to have given a fresh impetus among university physicians to revive and reformulate the poison concept as an explanation of the disease.[160] It is in fact hard to see how poisoning accusations during the Black Death could have appeared out of the blue without some impetus from knowledgeable experts who could advise on the subject. When a fifteenth-century Lübeck doctor, in the course of explaining how a pestilence can arise from bad food and water, states that "all physicians who are writing about poisons set down many foods in the place of poisons," or a Leipzig plague booklet published in 1490 gives a regimen for "what is good when the water runs poisonous from the well," is it any wonder that many would make the all-too-small leap from natural to artificial, intentional poisoning as a cause of plague?[161] As for why Córdoba chose to demonstrate the poisoning charge with the medium of the air as opposed to food or water, which was the usual basis of the actual accusations, this is easily enough explained by the fact that medieval physicians saw corruption of the air as the most potent danger during a plague. The Parisian chronicler, Jean de Venette, also seems to have associated the poisoning charge against the Jews as entailing a corruption of both air and water: In 1348, he reported that "men ascribed the pestilence to infected air or water," as the abundance of food at that time made its contamination less likely, and that "one result of this interpretation was that the infection, and the sudden death which it brought, were blamed on the Jews, who were said to have poisoned wells and rivers and corrupted the air."[162] It seems likely that Venette means by this that water, once allegedly poisoned by the Jews, in turn corrupted the air through rising vapors.

Córdoba was by no means the only medical authority to entertain the notion of polluting the environment with malice aforethought: In fact, a century later, in 1448, his entire example of polluting the air by brewing a poison in a glass amphora was quoted word for word by a Master Berchtold in order to demonstrate that plague is poisonously contagious.[163] Córodoba's contemporary colleague in Lérida, Jacme d'Agramont, considered among the universal causes of pestilence that the plague comes "from wicked men, children of the devil, who with venoms and diverse poisons corrupt the foodstuffs with evil skill and malevolent industry," although Agramont did not identify "such mortality" with the pestilence taking place in his own region. Agramont had obviously heard of the trials taking place in Narbonne and elsewhere, for he mentions that "in some regions near to ours there are now many deaths, as in Colliure [in the province of Rousillon on France's border with Spain], in Carcassonne, in Narbonne, in the barony of Montpellier, and in Avignon and the whole of Provence"; that Agramont was so well-informed was no accident, for the crown of Aragon controlled the barony of Montpellier until 1349, when James III of Mallorca sold it to King Philip VI of France.[164] In 1370 Heinrich Ribbeniz at the University of Prague continued to believe that human poisoning was one of "the chief causes of pestilences" because one of the "impressions on these lower parts" of a "maximum conjunction" of Saturn and Jupiter that came around every ninety years (its next advent was 1371–72) was that men were more prone to "sin against their gods," such as by poisoning. To prove his point, Ribbenitz cited the Jews of Milan, who were forbidden to go to a certain "mountain lying near the city" where there grew "an herb which is called *napellus* [monkshood], and it is the worst poison among all poisons, and it kills a man in an instant."[165]

Finally, a Lübick doctor in 1411 raised the specter of plague coming to Saxony from "some poisoned herbs, or from a poison invented by the wise Pol, as appears in his book where he instructs how to make such a confection." Who the "wise Pol" is who teaches how to devise poison is not clear, but the Lübeck author does not seem to identify potential culprits with the Jews, for he mentions that the last time they were "defamed of having poisoned Christians" was during "the first great pestilence" of 1348, as described by the famous surgeon, Gui de Chauliac.[166] Chauliac, in his *Great Surgery* of c.1363, had explored the accusation of many "uncertain" people that the Great Mortality could be attributed to "that the Jews had poisoned the world" or to "a deformity of the poor," such that "guards were posted in cities and towns" and forced anyone found with "powders or unguents" on their persons to "swallow them, fearing that these might be poisons." But he dismisses such rumors and asserts that the true cause of the plague is a twofold universal or "active" and particular or "passive" source, as had been expounded by the Paris masters in 1348.[167] Konrad of Megenberg, in his *De Mortalitate in Alamannia* (On the Mortality in Germany) of c. 1350, likewise dismisses the well-poisoning accusation against the Jews on the grounds that he had personally seen them die "in droves" from the plague in Vienna, where he was based in 1348 (a fact also mentioned in the *Buch der Natur*), and that even after the suspect wells and springs had been purified and sealed over and all the Jews killed or driven off, people still continued to succumb to the disease.[168] This logic was amply borne out in the case of Strasburg, where one of the largest Jewish pogroms occurred—2000 Jews burned alive over the course of three days—in February 1349, but where the plague came later that summer around the same time as a procession of the Flagellants.[169] It may have been experiences like this that put an end to persecutions of the Jews for alleged well poisoning, which do not seem to crop up again after 1348–50.[170]

## The poison thesis

But if scapegoating of some members of medieval society disappeared after the first, terrible onslaught of the Black Death, the idea of environmental poisoning as a cause of disease took a firm hold on medical thinking after 1348. Already by the early fourteenth century, a number of works on poison appeared, the most influential of which was the *De Venenis* (On Poisons) by the Paduan physician, Pietro d'Abano. Late medieval fascination with poison seems to have been inspired by a number of factors, including: the recovery in Europe of works by Aristotle, Galen, and Avicenna; concerns over the dispensing of a wider and more dangerous array of drugs by apothecaries; and the political intrigues at princely courts where vendetta poisoning was always a possibility. The subject of poison was now approached from the perspective of natural philosophy—namely, attempting to explain exactly what poison was and how it entered the human body and operated once there. This represented a major advance over classical treatments of poison, which often merely listed the varieties of specific poisons and their respective antidotes, or remedies. Abano's main contribution seems to have been one of emphasis: He adopted and expanded upon the notion put forward by the great eleventh- and twelfth-century Islamic philosophers, Avicenna and Averroes (Ibn Rushd, 1126–98), that the deadliest of poisons, such as *napellus* or monkshood (also known as "wolf's bane"), operated by their "specific form" or "total substance" (*tota species*) which is "contrary to the life of man," an idea that ultimately can be traced back to Galen. For example, Abano states at the outset of his work that "poison is the opposite of food for our body." This establishes that some kinds of poison are particularly deadly substances that act independently of Galenic notions of quantity or quality that hold true for other, less virulent poisons: Examples of the latter include whenever one ingests too much of something, so that it acts like a poison, or whenever an excess of a particular quality occurs in a substance, whether it be hot, cold, moist, or dry, that can make it a poison.[171] Abano explains that poisons that act through their specific form get their occult power from the stars, channeled as if through an inverted pyramid, but this leaves knowledge of the exact elemental make-up of such poisons rather vague. His main concern seems to be with how poisons can kill the body once inside it, which they do by going straight to the heart and attacking it and by converting whatever poison "touches in the human body to its own species of poison," thereby multiplying itself.[172] Such a concern, however, is more relevant to the history of medicine; what is particularly interesting for our purposes is how plague doctors expanded upon and applied Abano's ideas on poison to the environmental spread of disease, moving beyond the traditional conception of poisoning as simply the oral ingesting of animal, plant, and mineral substances. Specifically, they did this by conflating poison with air corrupt in its substance.[173]

Both ideas, that especially virulent poison acts through its "total substance," and that only air that has been corrupted in its very substance can cause pestilential disease, are ultimately indebted to Avicenna.[174] The next logical step was to connect and equate the two, but this only seems to have happened with the advent of the Black Death. It was in fact the plague doctors of the late Middle Ages who seem to have been the first to make an explicit connection between poison and disease, one that not even Avicenna made in the *Canon*.[175] Why they should have done so seems to have much to do with the extraordinary nature of the Black Death itself, a universal disease event unlike any other in human history. Poison, as a universal agent of death through its specific form, no matter the complexional imbalance or make-up of the individual, proved to be a particularly attractive way to make sense of so unprecedented a mortality. Especially during its first outbreak in 1348–49, plague seemed to obliterate all differences and to attack everyone, young and old, strong or weak, indiscriminately. Just as physicians searched for a universal cause of a substantial corruption of the air that was spreading the plague, poison could serve on a microcosmic level as a universal cause of the disease taking hold and spreading within and among human bodies.[176] Poison thus easily assimilated with the miasmatic theory of corrupt air, since both involved concepts of substantial change. Doctors also tried to assimilate the poison thesis with the humoral theory, though I think with rather less success.[177]

One of the first and strongest advocates of the poison thesis as an explanation of plague was Gentile de Foligno, who, as a fellow Paduan, was perhaps particularly open to the ideas of Abano. Even so, Foligno chose the more iconoclastic of the two options as to how to conceive of poison, namely, that poison acts through its specific form or whole substance, rather than through one of its qualities (both of which could have been justified by Abano's work). In his *Long Consilium*, Foligno explains that plague is caused in the body by "a certain poisonous matter that is generated around the heart and lungs, whose stamp is not from an excess of its primary qualities by degrees, but due to its property of being poisonous." This poisonous quality of plague also made it highly contagious, as the "poisonous vapors" were breathed in and out by victims. Foligno is also heavily indebted to Abano, whom he quotes directly, in his answer to the seventeenth and last of the "doubts" or talking points on the plague that he raised with his students and that forms the last chapter of his casebook—namely, "how is infected air drawn in by the heart?" Foligno replies that this happens when the "pestilential poison or poisonous air" is drawn in by the lungs and then through the arteries to the heart, where it multiplies and turns everything it touches into poison, thus corrupting the vital spirits and causing the heart to stop beating.[178] Abano's influence, and in turn that of Avicenna and Averroes, is also self-evident in treatises that

Part I. Air, water, earth    71

define poison or poisoned air as being "by its very nature the opposite of life" and having the quality of attacking or destroying the human constitution "with all its power" or "with full force" and so "brings death to man."[179] Sigmund Albich in 1406 compared poison's ripple effect on the vital spirits of the body to "like when a stone has been thrown into water, one circle generating another," and he advised that a cure for plague had to be treated "much like the cure for a [deliberately] poisoned man."[180] Going beyond poison's effects on humans, a Master Bernhard of Frankfurt in a treatise from 1381 described poison as something whose "make-up is unlike that of anything else," in that it could "cool down and prohibit growth in vegetables and sentient things, [and] in whatever they come into contact with."[181]

Foligno's championing of the poison thesis with respect to the plague did not go unchallenged at the time. A colleague at the University of Naples, Giovanni della Penna, wrote a plague casebook in 1348 specifically in order to refute Foligno's opinion that plague was caused by poisonous matter generated in human bodies, which Penna declared to be "impossible." Rather, Penna believed that plague was caused by the hot and dry choleric humor when "placed in the path of a burning heat," so that it easily "boils over," say, when in contact with the breathed-in corrupt air, even though "it induces effects that are near enough and similar to those of pharmaceutical poisons."[182] Despite these objections, other plague treatise authors contemporary with Foligno and Penna seem to have bought into the poison thesis: The Paris masters referred to "poisonous vapors" among the near causes of plague; Jacme d'Agramont described how corruptions of plants and animals "acquire the property of poisoning" to those who eat them; Konrad of Megenberg talked of a "poisonous exhalation from the earth" released by earthquakes that had been trapped for so long in the ground it was "turned into a potent poison to the human constitution"; and an anonymous medical practitioner from Montpellier writing in 1349 explained how the retrograde aspect of Mars reduced the air to "its windy and poisonous nature," so that then a "poisonous moisture" abounded in plants and animals which, once eaten by humans, created a windy and poisonous moisture or matter in man, which the brain expelled through the optic nerves of the eyes so that plague could be contagious by look alone.[183]

After the initial outbreak of the Black Death in 1348–49, it is striking how ubiquitous references to poison become in the plague tract literature of the late Middle Ages. Of the some 300-odd plague treatises written between 1348 and c.1500 that I have personally examined, I would be hard pressed to find one in which the term "poison" or "poisonousness" does not occur in some form.[184] Thus, in the same way that poison was believed to manifest itself in the body as "poisonous matter," "poisonous humors," "poisonous superfluities," "poisonous blood," or "poisonous symptoms" such as plague boils, plague also manifested itself in the environment as "poisonous vapors," "poisonous air," a "poisonous fume," "poisonous moisture," "poisonous muck" (*fimi venenosi*), "poisonous, foul water," "poisonous animals" with their "poisonous breath," and so on.[185] And just as poison acting through its specific form posed a challenge to the humoral theory of disease causation within the body, so poisonous air in the environment posed a challenge to the complexional or qualitative explanation of miasma, i.e., why air should become corrupt in its substance. Apparently, the issue was a matter of some debate within the medieval medical community: In c.1464, a German physician and professor of medicine at the University of Paris, Primus of Görlitz, wrote a treatise (clearly intended for an academic audience) in which he explained that putrefied air was a "universal" cause of plague but that this was not due to a corruption of one of its "primary qualities"; rather, it was entirely the result of a "poisonous quality". Unfortunately, Görlitz refused to say anything more "on this controversy for the present for the sake of brevity," since his position was apparently "an especially difficult [point] to maintain," one that was "very much at variance with the consensus of other statements that have gone before" and that touched upon "great and profound questions." Nonetheless, the university professor fully expected that he would be impelled by his colleagues to pursue this subtle "matter of disputation" in the future, "by way of lucidly demonstrating and maintaining what I have said."[186]

What is clear is that the poison thesis reformulated the miasmatic theory of a disease's spread in a new way, one that had momentous implications not only for the history of medicine but also for the history of human attitudes towards the environment. For it fundamentally changed how medieval people viewed the environmental impacts of corrupt air or water: These were now seen as not only much more pernicious and threatening to society's well-being, but also as much more amenable to human as well as natural causation, since poisoning was easily understood as both a man-made as well as a natural phenomenon. A "poisonous air" or vapor, after all, sounds much more sinister than simply a "bad air" or a "corrupt air." It also apparently was too autonomous of God's ultimate causation for the comfort zones of Islamic and Jewish doctors, such as Ibn Khatima and Rabbi Isaac ben Todros, a physician based at Avignon who wrote a plague treatise in Hebrew called the *Well of Life* in c.1376. These authors omit references to poison altogether and instead explain the causes or impacts of corrupt air in terms of a "decay."[187] Poison likewise put medieval doctors more on the spot with regard to their curative regimens, since they no longer could explain failures as due to the patient's complexional indisposition to being cured.[188] It also implied different approaches in treatment, using drug therapies and antidotes as well as sweating regimens in place of or alongside more invasive procedures like blood-letting and lancing of bubes, since remedies were needed that could safely extract poison without driving it further into the body and that could attack it in its "whole substance," i.e., by incorporating poisonous ingredients in their recipes. Especially if the plague poison was perceived as acting through its specific form, one whose elemental make-up was ultimately unknown, then only a universal remedy like theriac, that also acted by means of its specific form, or herbal compounds and precious stones that got their occult virtues directly from the stars, could be effective against the disease.[189]

*Part I. Air, water, earth* 73

During the fifteenth century and on into the sixteenth, many physicians, especially those like Conrad van der Weyden and Antonio Guaineri who wrote tracts that combined the dual topics of plague and poison, became interested in moving beyond poison as simply an analogy for disease causation and exploring it on its own terms as an autonomous, foreign, and incredibly potent (albeit natural) substance that was "fundamentally harmful" once it invaded the human body. Poison also became crucial to new theories of disease contagion, such as were famously formulated by the sixteenth-century physician, Girolamo Fracastoro.[190] All this ensured that the concept of poisoning the environment had a persistent and pervasive place in medieval understandings of disease and related catastrophes, one that remained alive and well down into the early modern period.

Although Jews may have been let off the hook after 1348–50 owing to the unsustainability of the charges against them, new scapegoats for the poison accusation emerged by the fifteenth century, namely, the alleged witches who became the target of the great witch-hunt and witchcraft trials that extended well into the modern era. Poison, due to its occult properties like other agents of natural magic, was an especially apropos medium for witchcraft, especially love magic that trod a fine line between harmful and beneficent practices.[191] Poison was also explained at this time as an agent of the devil's power.[192] Some of the charges against witches read very much like what Alfonso de Córdoba wrote or what was accused against alleged poisoners during the Black Death a century before. Thus, one of the earliest treatises written on witchcraft, the *Errores Gazariorum* (Errors of the Cathars) of 1437, told of a "synagogue" or witches' assembly at which members made up powders from "the internal parts of children mixed with poisonous animals" that were then "scattered through the air by a member of that society on a cloudy day." All touched by the powder "either die or suffer serious and lingering illness," and this was said to be "the reason why in some villages of a region there is great mortality, and in other areas there is much bad weather."[193] In 1460 in the Lyonnais region of France, an unspecified number of witches were tried on thirty counts of diabolism and sorcery, among which was that they received "powders made up by demons through [wicked] artifice, which they secretly sprinkled into food or drink [and] caused diverse and grave infirmities, which indeed frequently inflicted deadly and very long lasting illnesses." Thus, accusations of poisoning the air and water in order to communicate disease once again make their appearance in the witch trials towards the end of the Middle Ages.

## Weather magic

Equally influential for the history of the witch-hunt were the accusations of weather magic that form their own genre of *maleficia* within the records of witchcraft.[194] Two witchcraft treatises from the 1430s, the *Formicarius* (Ant-Colony) of Johannes Nider and the *Errores Gazariorum*, recount confessions obtained under torture from alleged witches that they were able to call up at

*Figure 1.7* WITCHES BREWING UP A HAILSTORM. German Woodcut, 1489. The Granger Collection, NYC – All rights reserved.

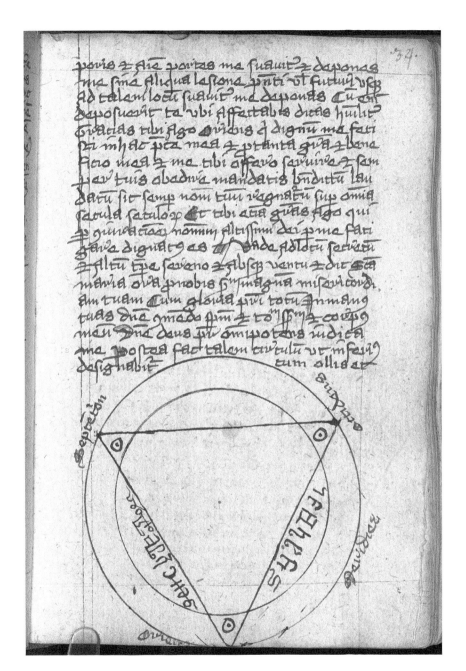

*Figure 1.8* AN INCANTATION TO OBTAIN A FLYING THRONE, conjured by invoking 33 demons of the air, from a fifteenth-century necromancer's manual – BSB Clm 849. © Bayerische Staatsbibliothek

will "immense hailstorms and poison winds with lightning" or else bring down "storms and bad weather" at demons' command in order to "break up ice" on top of a mountain.[195]

In the middle decades of the fifteenth century, a cluster of weather-related trials took place in Switzerland and its border regions. At Metz in the Lorraine region of eastern France, unseasonably cold weather that included a killing frost in the spring of 1456 that ruined the best wine-grape harvest anticipated in forty years led to the trial of a dozen men and women on a charge of sorcery, whose alleged leader was burnt at the stake after confessing that he and his companions caused the frost by throwing a "mixture made by the art of the devil" into a nearby spring. In 1481 and 1488, more trials on a charge of weather magic took place in Metz after unusually heavy rains and extended stormy weather once again led to failures in the grape harvest, when a total of thirty-eight accused witches, all but four of them women, were burned.[196] A well-worn charge was that witches were able to fly or travel through the air to their Sabbaths: A necromancer's manual from a Munich manuscript from the fifteenth century gives details on how such magical transport was to be done, by conjuring demons to produce a flying horse, a flying throne, or a magical ship. In the case of the flying throne, the necromancer was to call upon a total of thirty-three demons "who are potentates of the air," including "Boreal," "Aurora," and "Zephirim," and upon whom he was also to call if in his journey he felt threatened by "storms or serpents or birds or other terrible things".

The ship, of course, required conjuring eight "spirit dwellers of the water, who seek to endanger the ships of sailors," while the flying horse was conjured from three demons bound by the four elements as well as the sky, sun, moon, and stars.[197] Closely related were legitimate exorcisms under the auspices of the Church that tried to conjure away "clouds and hailstorms or tempests."[198] If all this seems highly reminiscent of the "storm-makers" and "cloud-sailors" investigated by Bishop Agobard in the ninth century, the difference now is that such weather-witches are seen to be in league with the devil, which makes them a far more dangerous enemy of the Church, and indeed of all of medieval society.

In sum, by the end the Middle Ages, largely in response to the plague, society became aware of two main culprits of polluters: those who poisoned the air, water, or earth through industrial processes, and those who did so under an evil, demonic influence. Not only did the poison thesis change the way doctors approached plague medicine, it changed the way late medieval people viewed their environment, as something that could fall under the sinister influence of man and that in turn could profoundly affect his health and well-being. This, I believe, is approaching a modern view of the pollution of the air, water, and earth. By the time we reach the Romantic era of the late eighteenth and nineteenth centuries, it became the opinion of many who had grown disenchanted with the Industrial Revolution—the "dark, Satanic mills" of William Blake's poem—that the two kinds of polluters, industry and the devil, were in fact one and the same.

# Part II

# Forest

To many minds, the forest is synonymous with the untamed, untrammeled wilderness, where the writ of human civilization does not run. It is an idea that runs deep in the human psyche, going back to the very first story on written record, *The Epic of Gilgamesh*. Dating to the third millennium B.C.E. from the region of Mesopotamia in present-day Iraq, *The Epic of Gilgamesh* tells the story of a legendary but real king of Uruk, Gilgamesh, who goes on a series of adventures with his friend, Enkidu, before the latter dies and Gilgamesh embarks on a quest for immortality in order to avoid his friend's fate. On one of their adventures, they journey with fifty men bearing axes to the "Land of the Living," which is also identified in the sources as the "Land of the cut-down Erin-trees," suggesting that already in Neolithic times there was considerable exploitation of the forest, especially as Gilgamesh and Enkidu must cross seven mountain ranges before reaching trees of timber-bearing size. That Gilgamesh and his companions are there for timber to be used in building projects back in Uruk, which are described as going on day and night at the beginning of the poem, is indicated by the fact that they dress the logs of their branches before laying them "at the foot of the mountain." Traditionally, the forest denuded by Gilgamesh is identified with the Cedars of Lebanon (which also feature in the Hebrew Old Testament) that still grow today in isolated pockets to the west of Iraq, either in the Mount Hermon range of the modern country of Lebanon or in the Amanus Mountains of north Syria. However, alternative theories place the Land of the Living to the east, in the land of Dilmun in present-day Bahrain or in the Zagros Mountains of south-central Iran. Likewise, while there is general agreement that the Erin tree was a fragrant, light-colored wood suitable for building timbers, the specific species with which it is to be identified are variously named as cedar, pine, juniper, or cypress. The fearsome "monster" Humbaba, who dwells in the forest and has been appointed by the gods as its guardian, is often equated with the chief deity of Elam, a rival civilization to Mesopotamia's that lay just to the east in south-west Iran.[1] After Gilgamesh and Enkidu use a sword and an axe to slay Humbaba, despite his pleas for mercy and offer to cut down the trees himself, the Erin trees "shiver" for at least two leagues in lament for their "watcher" and the intruders next turn their weapons on the forest itself, felling trees and clearing their roots "as far as the banks of Euphrates." Herein lies an allegory for man's innate fear of the forest and of all the wild secrets it contains, for which he seeks remedy by laying waste to the sheltering boughs of the trees. *The Epic of Gilgamesh* seems to establish a precedent for measuring the progress of civilization in terms of how much of the forest can be converted into arable land.[2] However, under the later empire of the Assyrians, an inscription from the reign of Tiglath-Pileser I (c.1115–1077) describes how "groves" or plantations of pine, algum, and fruit trees imported from conquered territories as tribute were planted back in Mesopotamia, which apparently had never before been grown in the region.

## Pre-Christian tree cults

At what many consider to be the height of ancient culture, during the Greek and Roman periods, the forest gained some respect as sacred groves were set aside for protection in both popular mythology and under the law. Some groves, along with springs and mountaintops, were considered to be the original temples or homes of the gods and therefore were held to be sacrosanct, while individual trees were believed to be inhabited by the dryads or nymphs, who cursed the man who cut them down. Pliny the Elder (23–79 C.E.), in his *Natural History*, talks of how "trees formed the first temples of the gods, and even at the present day, the country people, preserving in all their simplicity their ancient rites, consecrate the finest among their tress to some divinity." Thus, as in the Greek pantheon, certain tree species were associated with specific gods (such as the oak to Zeus or Jupiter), and the majesty of some groves inspired no less awe than temple idols decorated with gold and ivory. Pliny then goes on to describe the various products from and uses of trees for which we should be grateful.[3] In order to preserve their trees, some Greek and Roman communities went so far as to pass local ordinances that protected with severe penalties sacred groves from felling, cultivation, or pasturing of animals within their precincts. This seems to have resulted in the survival of some old growth forests, but one has to believe that other trees outside the sacred area did not fare so well.[4]

Greek and Roman culture also contributed the Orphic and Pythagorean schools of philosophy, which held that all living creatures, plants included, possessed souls that were perpetually reincarnated within a cyclical universe: This naturally precluded any harm being done to a fellow creature on the grounds that one could potentially be harming a former fellow human being. This approach is very much akin to the ancient religion of Jainism founded by Vardhamana Mahavira in sixth-century India, and it is close to the modern outlook of an integrated biological system whereby what is done to the environment will rebound back onto the perpetrator. While the philosophy was mainly directed at interactions with animals, it also implied that one should not cut down trees or larger plants for sustenance, instead living off the fruit that could be plucked without harm, such as apples or grapes.[5]

On a more scientific level, the systematic study and classification of plants, the origins of botany, were undertaken by Aristotle's student, Theophrastus (c.371–287 B.C.E.), and his efforts received some stimulus from the flora and fauna discovered by another of Aristotle's students, Alexander the Great, during his campaigns to the East. Relying mostly on second-hand reports rather than direct observation, Theophrastus was nonetheless one of the first to study plants on their own terms, i.e., as organisms that sought to propagate their own species through the dispersal of seeds, rather than simply as food for men and animals. He classified trees based on their appearance, habits, and properties, observing not only their bark and leaves but also their root systems below ground, and he observed the effects of environmental factors upon plant growth, including climate change. Theophrastus also noted that certain types of trees were adapted to their particular habitat, such as whether they grew up in the mountains or on the plains, and he delineated the various species of trees that grew in each of the known regions of the world. In addition, he was a pioneer in distinguishing trees from shrubs. One of his oddities, from a modern point of view, was his attempt to distinguish male and female species of tree, generally characterizing the softer, easier to work wood as female, the harder and more knottier kind as male.[6] Roman authors, on the other hand, were more practical than theoretical, being mainly concerned with tree management on their estates. For example, the farming author, Columella (4–70 C.E.), describes the coppicing of trees, whereby some species such as oak and chestnut are cyclically cut down to their stumps in order to promote regrowth, in his *De Re Rustica* (On Rustic Matters), while in his short treatise, *De Arboribus* (On Trees), he addresses tree cultivation and maintenance, to include pruning, grafting, and protection from insects.[7]

Cults to woodland gods and goddesses centered on sacred groves likewise existed in Celtic and Germanic cultures during the Roman period and beyond. Strabo (64 B.C.E.–24 C.E.) attested to oak sanctuaries in Galatia in Asia Minor, while Cassius Dio (c.150–235 C.E.), Tacitus (56–117 C.E.), and Lucan (39–65 C.E.) claimed that human sacrifices were performed in sacred forests in East Anglia dedicated to the goddess Andraste, on the island of Anglesey off the coast of Wales, and at Marseilles in Gaul. Grove goddesses were also worshipped at the spring sanctuaries of Buxton and Bath in England, at Grenoble (dedicated to the "Nemetiales") and in the forest of the Ardennes in France (site of the cult to Arduinna), and near Speyer in Germany (which served as the capital for the Nemetes tribe devoted to the goddess Nemetona). As with mountaintop deities (which often doubled as forest gods), local cults were usually paired with Romans gods: Thus, Nemetona was coupled with Mars, and the Roman god Silvanus, described as the "god of wild nature and the woodland," was worshipped at a number of local shrines to gods associated "with vegetation, hunting, the forest and fertility," such as Vosegus of the Vosges mountains in France, Sinquates of the woods near Gérouville in Belgium, and Callirius, the "god of the Hazel Wood," at Colchester in England.[8]

Trees were venerated, both individually and as a whole forest unit, by the Celts because they were seen as a link or bridge between the earth and sky, between the under- and upper-worlds, between the chthonic and celestial realms, as symbolized by the roots going into the ground and the trunk and branches reaching up to the heavens. With their cyclical rhythms following the seasons, trees were also symbols of death and resurrection. The oak tree was a particularly important object of veneration: The druids of Gaul, according to Pliny, climbed a sacred oak tree on the fifth day of the moon and cut off a mistletoe growth with a gilded sickle, which they believed cured all barrenness and was "an antidote for all poisons" when taken in drink.[9] Tacitus testifies to the role that the druids played in native Briton resistance to the Roman conquest, rallying their own side and intimidating the opposition with their "dreadful imprecations." In retaliation, the Roman troops destroyed sacred groves where, it was alleged, the druids "deemed it indeed a duty to cover their altars with the blood of captives and to consult their deities through human entrails."[10] Archaeological excavations have uncovered stone columns or pillars decorated with oak leaves and acorns that were dedicated to the Celtic version of the Roman sky-god, Jupiter, and which were designed to imitate sacred trees. Other trees besides oak were also paired with specific Celtic gods or tribes, such as the beech tree with the Gallo-Roman god Fagus and the yew and the elm with, respectively, the Eburones and Lemovices tribes of Gaul. Physical remains of Celtic tree cults likewise include what seem to be sculpted images of woodland divinities, such as the bas relief of a bearded, toga-wearing god holding a billhook in his left hand and a pine cone in his right from Aix-en-Provence in southern France, or a female statuette figure from Sens dressed in a tunic and whose head, neck, and shoulders are enfolded in the foliage of a pine tree, perhaps symbolizing the anthropomorphic attributes awarded to trees by the Celts.

Forests were also venerated as healing sanctuaries, such as the *Forêt d'Halatte* near Senlis in Gaul, where sculpted images again survive, this time of the bodily parts and of pilgrims who hoped to be cured by the resident deity.[11] With the formal adoption of Christianity as the state religion of the Roman Empire by the end of the fourth century, and its subsequent spread through forceful conversions and missionary work, many of the sacred groves and cultic centers of tree veneration in the pagan world came under attack. The Emperor Theodosius (379–95) issued a decree in 392 proscribing pagan sacrificial rites and temple worship, including fillet or ribbon offerings tied around the trunks of trees as a form of prayer to the resident woodland deity. In the next century, the Councils of Carthage (401) and Arles (452) forbade the worship of trees and ordered that all such objects of veneration be destroyed. In the first half of the sixth century, Bishop Caesarius of Arles preached against worshippers of trees and springs, while in the latter half of the century, Archbishop Martin of Braga in Portugal railed against those who "offer sacrifices on mountaintops and in leafy glades" and Pope Gregory I (590–604) urged the destruction of "tree cults" in Gaul and that those who

*Figure 2.1* GREEN MAN, an example of anthropomorphic representation of perhaps a woodland deity. Roof boss, English School, Norwich Cathedral, Norfolk, England. Photo © Neil Holmes / The Bridgeman Art Library

"reverence trees" in Spain be punished as an example to others. During the seventh century, St. Amand was active in Flanders contending against pagans who "worshipped trees and woods as their God," while in 660 the Council of Nantes ordained that those trees "that the common folk reverence and hold in such veneration be dug up by the roots and burned." In 681 and 693, the Council of Toledo warned that the "venerators of stones" and "those honoring the sacred sites of fountains and trees," among other idol worshippers, "should know that they who are seen to be [thus] sacrificing to the devil freely expose themselves to death."[12] But whether this be a figurative death of the soul or rather death at the hands of a lynch mob is not entirely clear.

Perhaps the most famous example of tree destruction in the name of Christianity was that carried out by St. Martin of Tours during the fourth century. It is worth quoting the full account of the incident, as told in the *Life of St. Martin* by Sulpicius Severus:

> Again, in a certain village he [Martin] had demolished a very ancient temple and was proceeding to cut down a pine tree that was close to the shrine, when the priest of the place and all his pagan following came up to stop him. These same people had been quiet enough, at Our Lord's command, while the temple was being thrown down but they were not prepared to see the tree felled. He painstakingly explained to them that there was nothing sacred about a tree trunk and that they had much

better be followers of God he himself served. As for the tree, it ought to be cut down because it was dedicated to a demon.

Then one of them, more audacious than the rest, said to him, "If you have confidence in the God you say you worship, stand where the tree will fall, and we will cut it down ourselves; and if your Lord, as you call Him, is with you, you will not be harmed."

Martin, with dauntless trust in God, undertook to do this. Thereupon all the assembled pagans agreed to the bargain, reckoning the loss of their tree a small matter if, in its downfall, it crushed the enemy of their religion. And as the pine leaned to one side, so that there was no doubt on which side it would fall when cut through, Martin was bound and made to stand on the spot chosen by the rustics, where they were all quite sure that the tree would come down. Then they began to cut down the tree themselves with great joy and delight. A wondering crowd stood at a little distance.

Gradually the pine began nodding and a disastrous fall seemed imminent. Standing at a distance, the monks grew pale; and, so frightened were they as the danger drew near, that they lost all hope and courage, and could only await the death of Martin. He, however, waited undaunted, relying on the Lord. The tottering pine had already given a crack, it was actually falling, it was just coming down on him, when he lifted his hand and met it with the sign of salvation.

At that—you would have thought it had been whipped like a top—the tree plunged in another direction, almost crushing some rustics who had ensconced themselves in a safe place. Then indeed a shout went up to heaven as the pagans gasped at the miracle, the monks wept for joy, and all with one accord acclaimed the name of Christ; and you may be sure that on that day salvation came to that region. Indeed, there was hardly anyone in that vast multitude of pagans who did not ask for the imposition of hands, abandoning his heathenish errors and making profession of faith in the Lord Jesus.[13]

There are several salient points to be observed from this account. For one, it is remarkable how the pagan natives of what is likely to have been western Gaul, where the Celtic druidic religion was still prevalent, came to the defense of a tree, even in preference to a human construction, a "very ancient" temple, which they were content to see "thrown down." At the same time, the pagans seemed willing to sacrifice the tree if it meant the end of their persecutions, perhaps in order to save yet more "sacred" trees? Nonetheless, the way that Martin narrowly averted death is puzzling, for if the woodsmen knew their work, they would have known exactly where the tree would fall and how to deliver the axe blows to get it there. Perhaps in the end they were unwilling to take the life of a human being in this bizarre kind of trial by ordeal pitting the Christian versus the pagan gods. On the other hand, as every experienced logger knows, falling timber can sometimes be deceptive and land in the most unexpected places. The very real dangers inherent in felling trees, even when loggers went out in teams, are witnessed by the old Anglo-Saxon, Lombard, Visigothic, Welsh, and Norwegian laws, which provided redress in case a man was killed or accidentally pinned by a tree so large that his comrade was unable to free him and so had to leave him in the woods to die.[14] How many actually gave up their pagan faith as a result of this ritual is debatable, but the loss of their beloved tree to little or no result must have inevitably dampened morale.

Other famous examples of such "dendroclasm" in the West come from Germany during the eighth century under the Carolingians. In Willibald's *Life of St. Boniface*, it is told how the "apostle of the Gernmans" came to the town of Geismar in central Germany, where "he attempted to cut down a certain oak of extraordinary size called in the old tongue of the pagans the "Oak of Jupiter." As Martin had discovered in France, the great crowd of pagans who had gathered to watch Boniface's sacrilege were, at least initially, very protective of their tree, "bitterly cursing in their hearts the enemy of the gods." But like Martin, Boniface also performed an alleged miracle, when at the first superficial notch of his axe the great tree suddenly crashed to the ground in four equal parts, which duly overawed the heathens into belief, and a Christian oratory dedicated to St. Peter was then built on the spot from the fallen wood. Then in 772, Charlemagne had the mighty *Irminsul* or "world tree" of the Saxons pulled down near what is now Obermarsberg during one of his brutal Saxon campaigns. This cult has obvious similarities with the *Yggdrasil* or ash tree that was at the center of the religious typography of Norse myths during the Viking age. In the East, similar assaults on sacred trees, such as the cypresses of Syria, were made by Christian orthodox monks under the late Roman and Byzantine Empires during the fifth and sixth centuries.[15]

Are we then to infer that Christianity was inveterately hostile to the nature worship of trees? Should we juxtapose a Christian attitude that was destructive of the sylvan environment with a pagan one that supposedly nurtured and guarded its woods? The consensus among scholars of Celtic druidic culture seems to be no, that Christianity cannot be blamed for early clearances of the primeval forest. This is indeed a case where, in a sense opposite to the old familiar proverb, we *are* seeing too many woods in places where only single trees were felled, for one tree does not a whole forest make. There are instances, after all, where Christians did champion sacred groves or at least sought to convert them to the new religion. One such example from the fifth century, as told in *The Golden Legend*, is that of St. Germain d'Auxerre, who in his youth threatened to kill his predecessor as bishop, St. Amator, because he had cut down and burned his favorite pine tree on which he had suspended the heads of the wild beasts he had killed in the chase.[16] Some sacred groves were adopted by Christian monastic communities and successfully preserved for posterity, such as the holm-oaks of Montelucco above Spoleto in Italy, or a cedar-grove in Lebanon under the auspices of the Maronite Church.[17] At the same time, the Celts proved no less capable than the Greeks in denuding the countryside of forest cover outside their sacred groves of protected trees.[18] Exploitation of the woodland resource, it seems, had a non-ecumenical basis in the early Middle Ages.

## Surviving wildwood at the start of the Middle Ages

There is obviously much debate and uncertainty about the state and extent of the so-called primeval forest, or wildwood, as this existed and survived in Europe after the end of the last Ice Age around 11,000 B.C.E. and down through the succeeding Mesolithic (c.11,000–7,000 B.C.E.), Neolithic (c.7000–3,000 B.C.E.), Bronze (c.3000–1200 B.C.E.), and Iron Ages (c.1200 B.C.E.– 400 C.E.). Most studies of especially the earlier eras have relied on pollen analyses left behind by the trees themselves in ancient bogs, which nonetheless can still be distorted by the fact that some tree species, such as oak and birch, leave behind more pollen to be recorded than others, such as lime and ash, which may thus be under-represented in the real or actual woodland mix as this existed on the ground. In Britain, where much of the best evidence has survived and been analyzed, it is estimated that in c.4500 B.C.E. birch and pine predominated in the highlands of Scotland; oak, hazel and elm in the Scottish lowlands, northern and south-west England, Wales and Ireland; and lime in the midlands, East Anglia, and southern England.[19]

Man-made destruction of the wildwood probably first began in the Mesolithic Age but on a rather small scale, with little clearings perhaps for housing and animal grazing made with the help of intentionally set forest fires, although this is difficult to do in the climate of Britain (except on pine trees). Tree clearing picked up in the Neolithic Age with the advent of agriculture, which seems to have progressively advanced into the forest uplands from river valleys beginning around 6500 B.C.E. in Europe and perhaps 4000 B.C.E. in Britain, when the carbonized remains of cultivars such as wheat and barley begin to emerge. This is the time when a catastrophic "Elm Decline" occurred throughout Britain, which likely was caused by introduction of an elm disease by Neolithic settlers, although young elm shoots are also a favorite food for grazing animals and even for humans. Whereas previous clearings in the Mesolithic Age were believed to be only temporary, with the forest quickly regenerating due to the lack of sustained maintenance of the newly opened vegetation, the Neolithic Age saw the first permanent leveling of localized areas in England, including the East Anglian Breckland, the southern chalklands, the Somerset Levels in the south-west, and along the coast of the Lake District in the north-west. The clearings are thought to have been accomplished through a combination of "slash and burn" agriculture and stone axes and other tools, modern replicas of which have proved to be remarkably effective, even when compared to chainsaws. Modern experiments have also demonstrated that slash and burn clearings are much more conducive to the introduction by man of new plant cultivars than simple clearings with axes, in which the old undergrowth that had existed before the clearing quickly regenerates. It is also during the Neolithic Age that we have the first evidence of woodland management, with the archaeological uncovering of wooden walkways built to give footing across the boggy peat of fenlands, such as the Sweet Track in Somerset, which was made with a combination of planks cut from oak timber laid across an understructure of poles made from ash, lime, elm, oak, alder, hazel, and holly wood. Since the poles, which were also used to make wattle hurdles by interweaving the rods into a fabric-like design, are remarkably uniform in diameter, it is believed that they could only have been produced through the deliberate growing of the desired-sized underwood by means of coppicing. These techniques were to be continued down into the succeeding Bronze and Iron Ages.[20]

With the advent of metal technologies, particularly in the harder medium of iron, ancient societies were able to pick up the pace of forest clearing, mainly for the purpose of creating more arable and pasture land. During the Bronze Age, large, permanent clearings were created in both lowland and upland areas in Britain, with the lime tree species going into decline mainly due to human exploitation rather than climate change, as evidenced by the succession of agricultural pollen and the dying out of some wild animals, such as aurochs, that depended on forest habitat. The invention of the hand saw as a wood cutting tool in addition to the axe probably greatly facilitated tree-clearing operations at this time. Forest clearances became even more extensive during the Iron Age, when open areas were created "nearly everywhere in Britain," such that by the Roman era (c.500 B.C.E.–500 C.E.), the pattern of the modern open landscape and human settlement was said to have been largely in place, with only the north-west of England and Scotland retaining their forests intact down into the early Middle Ages. Meanwhile, new tree species such as ash, elder, and maple that were more light-dependent began to take advantage of the open space created.[21]

The Romans increased demands upon the woodland not only through agriculture, but also in order to service urban industries that relied on wood fuel, such as cremations, public baths, brick-making, glass manufacture, salt production, iron and lead smelting, and so on; in addition, large-scale construction projects and ship-building required equally large-scale timber resources.[22] Yet at the same time, the Romans also brought to the far-flung areas of their empire such as Britain extensive knowledge of woodland management, such as coppicing, which they had learnt in Italy and which could be used to continuously service these industries without making great demands upon the virgin forest. Nonetheless, the Romans settled and exploited Britain more extensively than has traditionally been thought. This is borne out by detailed studies such as of the village of Hanbury nestled in the forest of Feckenham in Worcestershire, where pottery scatters demonstrate that there may have been more widespread settlement and cultivation than in later times, when medieval woodsmen were clearing trees that had grown up over Roman cornfields.[23] A conservative estimate is that half the forest cover of Britain was already gone by 500 B.C.E., almost half a millennium before the Romans even arrived. Indeed, the late Iron Age and Roman eras have been called Britain's "most active period of wildwood destruction in history."[24]

On the Continent, there are plenty of literary references to the clearing of Mediterranean forests by ancient authors. A host of Roman writers such as Cicero, Lucretius, Strabo, and Tertullian celebrated man's creation of a "second nature" within the natural world that had been created by the gods: This was a landscape of man's own imagining, whereby the woods were forced to retreat at the hands of the agriculturalists, the creators of the *latifundia* or great estates. Other causes given by the ancients for deforestation included natural and man-made forest fires and floods, but above all it seems that the demand for timber for various uses mentioned above is what placed the most pressure on the forests. A few ancient voices were raised in protest to this deforestation: Perhaps the most famous is Plato's rant in the *Critias* on how the denuding of the Attican countryside outside Athens led to rampant soil erosion, which he strikingly compared to "the skeleton of a body emaciated by disease." However, some historians believe this to be literary hyperbole in which Plato harkens back to a mythical golden age. In the fifth century C.E., Sidonius Apollinaris mildly protested at the cutting down of the forest on both sides of the Apennine Mountains of Italy, which apparently had been going on "for far too long a time."[25] But modern studies based on archaeological evidence suggest that vegetation degradation and resulting soil erosion had already been taking place throughout Italy well before Roman times during the Neolithic period, so that the idea of a primeval wilderness continuing for long stretches without interruption and untouched by the hand of human influence existed "virtually nowhere" on the peninsula.[26] Elsewhere, there is likewise some question as to how much of what the ancients describe as wooded actually consisted of trees: Even the great boreal coniferous forest that is believed to have prevailed across northern Europe may have contained more of a mixture of open grassland or groves interspersed with actual tree stands. Tacitus' reference to the "horrid woods" (*silva horrida*) of Germany, such as the Teutoburg Forest where three Roman legions were wiped out in 9 C.E., may actually be a reference to "thorny woods," where open scrubland with dense undergrowths and thickets would have been just as conducive to ambushes as surrounding old-growth woodland.[27]

Whether all this forest clearance can be called an "achievement," as is described in some histories of the ancient woodland, is a matter that is open to debate, though of course by questioning what seemed so eminently practical and necessary to contemporaries would be imposing our modern sensibilities onto the past.[28] Of one thing we can be certain, however: If this "achievement" is truly what some historians say it was,[29] then this implies that much of the responsibility for the permanent clearing of the forest, and for the landscape being what it is today, actually pre-dates the Middle Ages. Even while clearing continued apace into the Middle Ages, it was left up to our medieval forbears to try to manage what was left and hopefully keep the wilderness from disappearing altogether.

## The early medieval woodland

The French historian of the early medieval forest, Charles Higounet, produced a map in the 1960s, which has been much reproduced since, that purports to show the distribution of the forest cover in Europe on the eve of the so-called "great clearances" (*les grands défrichements*) between 1000 and 1300. Based upon references to some 142 separate woodlands during the high Middle Ages, the map shows great swaths of nearly unbroken forest from southern France eastwards across nearly the whole of the German Empire and central Europe, with more isolated islands of woodland in northern France, Italy, Spain, and England. The neat and tidy appearance of the map, however, is deceptive, for even Higounet himself admitted that his map was only a bare "outline" and that his article was merely a "starting point" for future research that was "constrained by the mediocrity of the documentation" available for what happened to the European forest cover between the fifth and eleventh centuries.[30] A main criticism in recent years of such overly pat generalizations about the early medieval forest concerns the very definition of the contemporary term, *forestum*, or of its many equivalents (i.e., *silva*, *wald*, *bosch*, *gualdus*, etc.). A "forest" in the Middle Ages meant something quite different from how we think of it today: It mainly existed for our medieval forbears as a legal entity, rather than as an economic or ecological one. Therefore, a forest could actually include treeless fields of arable or pasture if these came within the purview of property rights and laws that defined it as such, according to the customs and traditions of the locality. More often, however, a medieval forest reflected the complex intersection between the wild, untamed woods and the cultivated open space of the village: A place where man left his mark on nature and was an integral part of it, actively managing the forest for his own benefit and, perhaps incidentally, for the benefit of woodland creatures and even of the trees themselves.[31] This is a practice that, of course, continues down to the present day (where the currently favored term is "sustainable forestry"), but one could argue that it had its most formative experience during the Middle Ages.

A good example of the above argument is the fact that in much of Europe, including lands ruled by the Lombards, Franks, and Anglo-Saxons, contemporary laws make it clear that forest was mainly understood to be land set aside by royalty as a hunting reserve, where the woods served as browse and habitat for game, rather than for the sake of the trees themselves.[32] As any experienced forester knows, managing a woodlot for the sake of wildlife habitat is quite different from managing it for timber or other wood product resources. Modern forest management plans often call for small clear-cuts in order to create the woody underbrush that is favored as browse for deer or as cover for game birds and other woodland species.[33] In this case, a forest that contains open land is actually a good thing, provided that it is being allowed to grow back into forest, or at least woody undergrowth. A forest's definition at the time of the Middle Ages was therefore extremely fluid, highly dependent on the specific temporal or geographical circumstances: While the term occurs in plenty of surviving documents from the early medieval period that became the basis for maps like Higounet's, only a close examination of the legal context for each occurrence (something that apparently has yet to be done) can establish how much tree cover in Europe there really was. Many areas shaded as forest on Higounet's map may have contained no trees at all, while those that are blank may hide unknown tracts of woods.[34] We probably will never know the true nature and extent of the forest in the early Middle Ages absent the detailed surveys and information on woodland management that had to wait to emerge until later centuries.

Where we are able to examine forests in some detail from the early Middle Ages, as has been done for the Weald in south-east England, the Ardennes in north-east France, and the Odenwald in south-west Germany, we find great regional variation in terms of how each forest was exploited by man and how far it was allowed to remain in its natural state. The Weald, one of the largest blocks of uncleared woodland in Anglo-Saxon England, was nonetheless from the eighth century heavily exploited as woodland pasture, mainly for pigs, a fact that is known from royal charters granted to ecclesiastical institutions for the right of pannage. The dens and pasture that pigs carved out of the Weald were then probably the launching point for human settlements, which begin to emerge by the tenth century.

In the Ardennes, the woods were managed much less intensively than the Weald during the early Middle Ages, but there was also far less unbroken forest, the area being more like a patchwork of dense woodland interspersed with open land: The former was kept mainly as a hunting reserve for royalty, while the latter was managed much like any other farming settlement, perhaps using slash and burn cultivation techniques. The two seemed to coexist side-by-side quite harmoniously, according to Carolingian records of twenty-five royal fiscal estates in the Ardennes. The Odenwald, a largely unexploited forest granted out piecemeal by the Carolingians to various monasteries, is perhaps the most typical example of pristine woods assaulted by systematic human colonization and large-scale clearing in order to make way for village settlements, which seem to emerge by the ninth or tenth centuries. In each of these woods there was a different starting point of conditions on the ground and a different pattern of ensuing development. If a generalization has to be made, it is that there was probably more continuity than previously thought between Roman settlement and exploitation of the European forest and that under the Anglo-Saxons and Carolingians. In northern France, for example, it is estimated based on archaeological excavations that 50 percent of Roman villa sites continued to be occupied down into the early Middle Ages, precluding any notion of widespread rural desertion. In England, archaeological

Part II. Forest 89

*Figure 2.2* HARVESTING ACORNS FOR THE PIGS in the month of November, from the "Breviarium Grimani", c.1515, Flemish School, Biblioteca Marciana, Venice, Italy. Giraudon / The Bridgeman Art Library

evidence likewise indicates that the country was densely settled in Roman times and that these sites were continuously occupied down into the Anglo-Saxon period, with new villages formed on marginal lands only when older settlements ran out of room. At the same time, theories of a demographic collapse in the wake of the fall of the Roman Empire, particularly after the advent of the First Pandemic of plague beginning in Constantinople in 542, seem to be overblown. All this then implies that there was less recovery of the forest during the supposed interruption to civilization created by the fall of the Roman Empire in the fifth century and that there was less overall woodland throughout Europe during the early medieval period to bequeath to the high Middle Ages.[35]

Despite the regional diversity in management of the early medieval forest, as alluded to above, there was some standardization of policy under the Carolingians, particularly Charlemagne (reigned 768–814). In his *Capitulare de Villis* (Capitulary concerning the Royal Domains), issued in c. 800, the emperor ordered that "our woods and forests be well kept; and where there is a place to be cleared, they should clear it and not allow the woodland pasture to increase; and where there ought to be woods, they should not allow them to be cut down or damaged excessively."[36] This same decree also provided that "the wild beasts within our forests be well kept" and provided for the regulation of pannage. Such conservation policies of the forest seem to have had a long tradition in Charlemagne's realm, building on earlier precedents among the Germanic kingdoms that succeeded the Roman Empire: For instance, the sixth-century law code of King Gundobad of the Burgundians stipulated that any subject, Burgundian or Roman, who cleared part of the common forest had to compensate for it with another tract of comparable size, while the laws of the Salian Franks, Lombards, Visigoths, and Bavarians from the sixth to the eighth centuries decreed penalties for burning, cutting down, or digging up valuable trees, including oak, beech, and various fruit, nut, and olive trees, which apparently some did out of "envy" for the other's orchard.[37] References to excessive cutting in the royal forest, which was to be curtailed by Charlemagne's *forestarii* or specially appointed foresters, are usually taken to mean excessive coppicing, or the taking of too much undergrowth from tree stumps. Coppicing was also a big industry among the Anglo-Saxons, where the wood harvested was used for fuel as well as fencing and wattle-work in buildings.[38] At the same time, King Ine of Wessex (reigned 688–726) imposed heavy fines for the destruction of trees, making a distinction between whether the tree was felled by the axe or by fire; in the latter case, the fine was double, since the wood was no longer usable, or in the phraseology of the law code, "fire is a thief," whereas "the axe is an informer." This distinction was lost, however, by the next century during the reign of King Alfred (871–99).[39] On the other hand, it could also be royal policy to encourage assarts, or clearances, of the forest, since then presumably there would be more agricultural produce to tax. This is precisely the strategy that Charlemagne adopts in the *Capitulare Aquisgranense* (Capitulary of Aachen) of 802–3, in which he orders that in each village, "wherever they find good men, they should be urged to clear the woods, so that our *servitium* be ameliorated."[40]

Yet a far more powerful engine of forest clearances, if such was royal policy, was to grant a part of the woods to monasteries, as we have seen was done in the Odenwald. The quintessential example here is the monastery of Fulda, founded in 743–44 by the hermit Sturm in the forest of *Bochonia* in Saxony. According to Eigil's *Life of St. Sturm*, written in the early ninth century, the hermit was given a mission by his patron, Bishop Boniface, to found a new monastery in the wooded "wilderness" of Saxony after he had already spent the earlier part of his life at Hersfeld in "a wild and uninhabited spot," where he and his companions "could see nothing except earth and sky and enormous trees." Eventually Sturm found the exact place he was looking for to found his new monastery, at a site called *Eihloha* by the river Fulda, which he reached only after wandering for many days through the "frightful wilderness," each night cutting down small trees to make an enclosure for his donkey against the wild beasts of the forest. Once he was granted the land by Carloman, mayor of the palace of Austrasia (741–47), Sturm together with seven companions set about clearing the site of trees "as far as they were able," but they were greatly helped in this task when Boniface arrived with "a great throng of men," who were able "to cut down the woods and clear the undergrowth" within a week. Yet even this iconic tale of a virgin forest in eastern Germany falling before the axe of Christian civilization is not so cut and dried as it might seem. Within the story, our skepticism should be alerted by the fact that Carloman has to command the nobles of Grabfeld to surrender any claims they may have to the site and that there was some resistance to Sturm taking possession before the mayor finally granted his charter. Our suspicions are only confirmed by modern archaeological excavations that have revealed a seventh-century Merovingian villa and church lying underneath the monastery foundations, which indicates that this was not quite the "desert" or "solitude" it is made out to be in the *Life*, especially since some of the woods had to be cleared for agriculture during the earlier period in order to support such a site. Even here, in a place apparently far from Roman civilization, there already was a pattern of settlement predating the Carolingian efflorescence. We therefore have to be wary of taking at face value any portrayal of "wilderness" in hagiographical literature such as this, since the very wildness and unspoiled nature of the woods may well be exaggerated as part of a "pious topos," whereby the forest is synonymous with the forces of the devil that the hero has to contend with and, in the end, conquer or tame.[41]

Perhaps our most reliable measure of the state of the forest at the end of the early Middle Ages is the Domesday Book, compiled by William the Conqueror of England in 1086. This lists approximately 7,800 individual woods, with the largest concentration of forest seeming to come in the midlands and south-east of the country. (The Weald is once again the biggest contiguous block of woodland in England.) The historian of the British countryside, Oliver Rackham, estimates that all this adds up to about four million acres of woods, or 15 percent of the total land area of England. Rackham further argues that a comparison of Domesday with Anglo-Saxon charters recording perambulations of land boundaries and with Anglo-Saxon place-names signifying the presence or absence of woods indicate that there was much continuity across the ninth, tenth, and eleventh centuries, and that Domesday Book was pretty accurate, leaving out little of the forest that may have escaped the surveyors. If we also accept Rackham's estimate of England as being 50 percent wooded in 500 B.C.E. and a third wooded in 586 C.E., then this represents a remarkable feat of clearance even before the high Middle Ages, with the bulk of it coming in Roman times, since Anglo-Saxon evidence does not generally record large-scale disappearances of woods, such as the existence of a forest which by Domesday is completely gone. If we then compare the Domesday survey with that of later centuries, such as estimates that England was 10 percent wooded in 1350, we would conclude with Rackham that "by 1086 about nine-tenths of the process of clearing the land of trees had been accomplished," with the high Middle Ages left to do only the one-tenth remainder, or five percent of the land area of England (about 1.6 million acres).[42] It should be noted, however, that other historians dispute the accuracy of Domesday or argue for a continuous state of flux in Anglo-Saxon settlement patterns. Nevertheless, the discrepancies persuasively point to *more* human settlements than those revealed in Domesday rather than less, largely as the result of fragmentation of ancient estates that went unrecorded in 1086, so that many villages were subsumed under the name of a single manor held responsible for taxation purposes. Moreover, the heavy tax dues that are recorded from Anglo-Saxon times indicate that already agricultural estates were being intensively worked, such that "the rural resources of England were almost as fully exploited in the seventh and eight centuries as they were in the eleventh."[43] Consequently, Domesday in this scenario would still *overestimate* the amount of pristine woodland that existed by the end of the early Middle Ages rather than underestimate it. On the Continent, standard assumptions are that forest cover went from four-fifths of the total land mass in 500 C.E. to 50 percent by 1300.[44]

## An era of "great clearances"?

The Anglo-Saxons, even though they could not match the "achievement" of the Romans in terms of clearing and in fact may have continuously managed farmland left behind instead of letting it fall into disuse after the Romans abandoned Britain, still managed to clear at a greater rate than the Normans or even down into modern times up to the twentieth century, grubbing out an average of 32 acres of woodland every day for some 500 years (as opposed to 17.5 acres per day during the high Middle Ages).[45] After the Domesday Survey of 1086, the new settlement pattern across England appears very mixed: In East Anglia and the south, most counties report *lower* rates of settlement after the Norman Conquest than before, since these were already highly developed in Anglo-Saxon times; in the south-east of the country, by contrast, there was much new activity, largely at the expense of the vast Weald (nearly half a million acres of woodland cleared in the Sussex Weald alone). The North also saw intensive colonization during the high Middle Ages, but much of it may have moved into waste that had been created during the Norman Conquest, and it may have contained a much higher percentage of pasture rather than arable compared to the rest of the country. It also seems that by the fourteenth century, some land in the north was already reverting to forest even before the Black

Death due to the marginal quality of the soils. The Midlands present a varied, fragmented patchwork, because most of the assarting was undertaken spontaneously by peasant tenants, with no clear pattern or organized agenda imposed by landlords, since in the East Midlands clearing was not a requirement of tenure. Limitations imposed by royal forest law may also have slowed development in some places. Overall, most assarting seem to have been concentrated in the West Midlands, the south-east, and the south-west, but with additional clearances in Berkshire, Hampshire, Suffolk, Yorkshire, and Derbyshire.[46]

If Rackham is correct in his assessment that a comparatively small amount of clearing ensued in England between c.1000 and 1300, at a time when it is generally hypothesized that population throughout Europe more than doubled, then this implies that what forest remained at the time of the Conquest was intensively but effectively managed (especially during the reign of Henry III in the thirteenth century) in order to meet the voracious needs of humans without making overly destructive incursions upon the woodland. An alternative or concurrent theory is that some of the woodland being cleared in the thirteenth century was actually secondary growth, having been grain fields in Roman times and then having reverted back to woodland during the Anglo-Saxon period, so that this represents in fact no change to the original, primeval forest.[47] Yet this cannot explain all of the slow place of deforestation in Britain, especially if Anglo-Saxons were also maintaining continuity on some Roman farms. If we then opt for the conservation scenario as the primary explanation, this would be quite at odds with the *grands défrichements* or "great clearances" postulated for the Continent during the high Middle Ages.[48] What is behind the discrepancy?

It is the influential French historian, Georges Duby, who seems largely behind the *grands défrichements* thesis, which he first put forward in 1962 in *L'Économie rurale et la vie des campagnes dans l'Occident medieval*, translated as the *Rural Economy and Country Life in the Medieval West*.[49] Duby's ideas of half a century ago seem hardly dislodged in current writings about the medieval forest.[50] Claiming to speak on behalf of "historians of all countries, but especially Frenchmen and Germans," Duby declared that "the age of medieval rural prosperity is the age of land reclamation," with "the culminating period of reclamation" coming during the twelfth century, when the Cistercian monasteries with their ready pool of lay brethren labor reached the height of expansion of their movement, deliberately seeking out the "desert" wastes of forest "wilderness" so extolled by St. Bernard of Clarivaux.[51] For Duby, the "new system of land settlement" during the high Middle Ages, particularly that of "dispersed habitation," whereby the "intervening spaces" between existing hamlets and villages were cultivated and developed by isolated homesteads, constituted "an extremely important change in man's attitude to and his relationship with nature," one in which permanent rather than temporary enclosures of the landscape were now the norm.[52] But even Duby was forced to admit that the documentation upon which he relied was very fragmentary and incomplete, such that "this great movement of expanding exploitation which changed the face of the medieval world and upon which so much else depended remains very little known."[53] Research since Duby has shown that the fabled Cistercian "reclamation" of waste or "wilderness" areas, such as by clearing forests or draining swamps, is largely a myth. Land-grant charters to Cistercian foundations in France during the twelfth century demonstrate that most land was already occupied and cultivated long before the white monks moved in, and rarely were they given carte-blanche to cut down trees and drain marshes; if anything, Cistercians were more inclined "to preserve forests than to cut them down," since woods were part of the diversified economy upon which the monasteries depended and thus "were not enemies to be overcome but useful resources to be managed." Moreover, what woods remained in cultivated areas that the Cistercians inherited were all the more valuable for maintaining the semblance of "horrible places of vast solitude" that was part of the original ethos of the mother house foundation at Cîteaux. Even when Cistercians were granted usage rights to woodland, they were often restricted to gathering windfall or cutting certain tree species (oaks were generally off limits) or trees of a specified size. Far from being the land-clearing pioneers of legend, Cistercians were not eager to acquire marginal land, and what reclamation that did take place was usually carried out by tenants.[54]

If much of the tree clearing during the high Middle Ages is thus hypothesized to have come at the hands of small peasant farmers, for the most part making illegal "assarts"—the term used in the records to refer to the grubbing out of woodland to make way for arable or cultivated land—then this hypothesis can hardly be tested, since the assarts would mostly go unrecorded in the official documents. However, a glimpse may be had of such assarts in the English forest eyres that tried offenses against the crown's rights in the royal forest, which were recorded on rolls whenever the court was in session at particular locales, usually at very irregular and long-separated intervals. Many of these assarts, though technically illegal, were tacitly allowed by the crown or by private landowners due to the lucrative fines that could be levied for them or the valuable crops that could be sown. For example, at Feckenham Forest in Worcestershire, which comprised 185 square miles, three rolls of assarts recorded between 1238 and 1248 list a total of 219 separate assarts and purprestures, or enclosures, for a total acreage of approximately 638, or an average of about 3 acres per assarter. (Since no dates are given for the assarts, it would be hazardous to attempt an estimate of the average acreage assarted per year.) While a few of the parcels can be quite large, up to 58 acres in size, and some of the assarters are well known, such as the prior and bishop of Worcester, one's general impression is that the vast majority of the unknown offenders were making assarts of roughly an acre or a fraction thereof, indicative of peasant farmers on a small scale. Indeed, one of the more commonly sized and also the smallest of clearances seems to have been "one curtilage," listed no less than thirty-five times, which is usually defined as the area of land immediately surrounding a dwelling and which obviously was the bare minimum of clearing needed to erect a house.

*Figure 2.3* FELLING TREES in the month of March, from a Book of Hours by Simon Bening, c.1540. British Library, London, UK / © British Library Board. All Rights Reserved / The Bridgeman Art Library

Many of these assarts mention crops being sown on the vacant land, particularly of oats; however, on nearly every assart one of the crops listed was a "crop of wood," indicating a managed forest that was being coppiced for wood production and thus not entirely cleared. Also revealing are the cases of foresters accused by inquisition juries sitting during the 1240s of taking money or bribes in order to allow assarts and purprestures, showing us how many clearances may have taken place in spite of the law.[55] By way of comparison, 60 assarts and purprestures totaling 134 acres, or an average of just over 2 acres per assarter, were returned in the New Forest in Hampshire, which measured 150 square miles, between 1216 and 1244, and a rent roll of assarts compiled in 1250–51 at Sherwood Forest in Nottinghamshire, which comprised 156 square miles, returned 106 assarts totaling 168 acres, or an average of approximately 1.5 acres per assarter. As at Feckenham, the New Forest records point to a variety of assarts, including not only crops but also pasture as well as "thicket" (*rifflettum*) and alder, indicating coppicing for periodic undergrowth.[56] Within the county of Staffordshire, the forests of Cannock and Kinver had different rates of assarting in the nine years in between the 1262 and 1271 eyres, at 1.6 acres and one acre respectively, while the Forest of Dean in Gloucestershire likewise records an average of one acre per assarter or less at the eyres of 1258 and 1270.[57] All this proves that not everywhere throughout the entire royal forest was the woodland assarted with the same degree of intensity and hence, environmental impact.

One of the difficulties faced by the "great clearances" thesis is that much of the local evidence that can build up a picture piece by piece of the inexorable retreat of the forest comes from England, which supposedly was the exception to the great clearances rule. For example, one of the best case studies for how a medieval village was carved out of the surrounding woodland comes from Hanbury, nestled in the royal forest of Feckenham. In the course of the thirteenth century, it is estimated that 1000 acres were cleared during the "great expansion" of the village; as in the case of other Feckenham assarts recorded in the eyre rolls that encroached on the northern and western fringes of the forest, those at Hanbury were largely the work of many individual tenants cumulatively nibbling away at the forest in small parcels on their own initiative, rather than the clearances being organized by great lords or monasteries.[58] Perhaps land colonization was more pervasive in eastern Germany and central Europe, where since the days of Charlemagne new settlements went hand-in-hand with political and religious policy directed against native pagans and which was continued into the high and late Middle Ages by the military order of the Teutonic Knights.[59] However, one should note that in many of the twelfth-century charters that Duby reproduces to show how both lay and ecclesiastical lords organized clearing of the "waste" or "wilderness," the men hired to head up these efforts were Hollanders or denizens of the Low Countries, whom we saw in Part I were especially known for their expertise in draining marshland, rather than cutting down forests.[60] The consensus seems to be that the absence of "unequivocal evidence" precludes a "definitive answer" as to exactly how much forest was cleared during the high Middle Ages. Based on the information we *do* have, best estimates are that by the end of the medieval period, the forest cover of central Europe was reduced from roughly 70 percent to less than half, while France went from being over half forested to a quarter covered.[61] But now let us turn to the comparatively well documented case of the royal forest of medieval England, and how the king and his forest officers tried to manage it in the face of population pressures upon the woodland resource.

## A brief history of the royal forest of England

In spite of its storied unpopularity, anecdotally reported in contemporary chronicles and more popularly in the legends of Robin Hood, the royal forest system of medieval England is also acknowledged to have slowed the pace of deforestation before the axe and the plough.[62] From a modern perspective, for those who champion the woodland, one roots for the "cruel tyranny" of the crown and its forest law rather than the "people's hero," Robin Hood! But of course, it is too easy to make judgments on past environmental policies and behaviors with hindsight, for who can say how posterity will judge us and our response in the face of the current phenomenon of global warming? One cannot fail of being accused of anachronistic arrogance unless one makes allowances for the contemporary dilemmas facing our medieval ancestors, who desperately needed to expand agricultural production to feed growing populations already subsisting at the edge of starvation through marginally productive farming techniques. Even so, there are lessons to be learned from an examination of the management and implementation of the royal forest of England, even if it constitutes the exception to the rest of Europe, which is probably due to the extraordinary application of central authority in the wake of the Conquest of 1066, in contrast to more regional power bases that were able to survive on the Continent. We might even find that medieval Englishmen—even those "oppressively" subjected to the forest law—had a far greater appreciation for their forest resources than is too often assumed.

The royal forest system of England—consisting collectively of some seventy separate forest districts by the thirteenth century—originated, of course, in the aftermath of the conquest of the country in 1066 by William, duke of Normandy, who applied Norman forest laws to his new island kingdom with an apparent ruthlessness that was decried by contemporary chroniclers. However, despite the Conqueror's reputation for depopulating the countryside (largely confined to the North in response to persistent rebellions there) and despite the hatred that the new forest law was widely reported to have aroused in the English populace, the Domesday Book evidence of 1086 seems to indicate that royal forests were mainly established in areas that were not well cultivated due to poor soils or unfavorable terrain and that therefore were always thinly settled. Hence royal forests were claimed on the clay soils in the south-west and the Midlands, on sandy or gravelly land in south-central

England, on barren moorlands and uplands in the south-west and the North, and on marshland in Lincolnshire. More densely populated and agriculturally productive land in the south-east and East Anglia tended to be much less "burdened" with the forest law (the notable exceptions being royal forests at Ramsey, Somerset, Huntingdonshire and almost the entire county of Essex). Aside from being sparsely populated and well wooded, royal forests also tended to have a high concentration of royal estates. However, it is true that in some cases, the area of the royal forest was extended by devaluing land within the reserve and thereby inducing owners to exchange their private holdings for ones just outside the forest, or sometimes by evicting them outright. A good example is the New Forest created in Hampshire, about which much is known because it received its own separate section in the Domesday survey. Here, hundreds of families were forcibly moved from the area, aside from those induced to exchange their holdings. The afforestation process was somewhat reversed beginning in the thirteenth century, but even so, the royal forest covered a quarter of the land mass of England at this time (with actual woodland probably more like between 10–15 percent), albeit at its height during the twelfth century the forest would have been even greater than this.[63]

We've already noted above how the royal forest could include land that was not actually forested, since one of its main purposes, albeit not the only one, was as a royal hunting preserve; nonetheless, woodland was largely targeted for protection as cover for game. While their Anglo-Saxon predecessors did enjoy the hunt and asserted their rights to the chase within their own private estates like that of any other landowner, the Norman kings went beyond this by claiming exclusive hunting rights on any or all lands of their subjects. This meant that, even outside the royal demesne, private landowners lost a certain degree of control over their own land, such as being forbidden to cut down trees that could serve as habitat for wild animals, aside from being forbidden from hunting the animals themselves. (Eventually they even had to maintain officials at their own expense to enforce the king's writ within their own forests.) Undoubtedly this was what was largely behind much of the supposed resentment of the forest law and push for disafforestment in later centuries. At the same time, however, many forest communities retained common-rights to the woods, also known as "estovers"—such as *housebote*, *fencebote* or *hedgebote*, *firebote*, *ploughbote* etc., or in other words, the right to gather wood for building houses, making fences, for firewood, and for making agricultural equipment—and from the time of Henry I charters were granted to ecclesiastical communities for multiple use rights to their woods.[64] These rights were highly cherished, as evidenced by a petition in the time of King Edward I (1272–1307) by the men of Easingwold and Huby in the forest of Pickering, who complained that the justice of the forest, Sir Robert de Clifford, prevented them from having their usual rights of *housebote* and *hedgebote*, so that whereas formerly they were entitled to thirty oaks a year for *housebote*, they were now reduced to between ten and fifteen oaks.[65] At the forest eyre of 1334 in the forest of Lancaster, no less than two towns and two monasteries pleaded before the court for their customary rights by charter in the forest, which included timber for repairing houses and fisheries, firewood, and blackthorn to enclose gardens and pastures.[66] The Norman approach to the forest seems to have had its origins in royal policy under the Carolingian rulers, which was then extended by the Norman dukes within their own duchy even after the demise of Charlemagne's empire down into the tenth and eleventh centuries. There was also some continuity with Anglo-Saxon forest administration in England, where "woodwards," "wardens," or "ministers of the forest" became the foresters in the new regime with the same duty of guarding the woods and supervising their harvest, although the number and variety of officials burgeoned under Norman forest law.[67]

Already by 1086, the time of the Domesday survey, William the Conqueror was enforcing the forest law beyond the limits of the royal demesne, extending the jurisdiction of the royal forest even into the estates of great landowners like the earl of Shrewsbury and the bishop of Winchester. These policies were continued under William's son and successor, William Rufus (1087–1100), despite his promises during his turbulent succession to restore to his subjects "their woods and their chases," i.e, to surrender royal rights to vert and venison. Nonetheless, later in his reign foresters appointed by the crown were evidently enforcing forest law, since a royal writ ordering them to cease and desist in Ramsey Abbey wood makes an exception for "beasts and assarts." At his coronation Henry I (1100–135) likewise issued a Charter of Liberties promising relief from forest law, yet he too maintained the reach of the royal forest and even added to it, creating new forests in several counties, while his laws enacted fines for cutting wood both within and without the royal forest. It is during his reign that we have the first record—in a pipe roll dated 1130—of forest eyres being held, presided over by a number of royal justices and even by the king himself on occasion. A fully fledged administrative structure for the royal forest seems to have emerged at this time, with foresters charged with purview over illegal clearing and cutting of wood, whether for cultivation, firewood, or building timbers, even when taking place on the landowner's own property. However, at the same time many exemptions were being granted by charter, particularly to monasteries, and the fact that fixed payments were assigned for violations, such that foresters were compensated out of this income, indicates that assarts were being committed on a regular basis. In addition, the fines levied on foresters testify to their sometimes lax enforcement. During the succeeding reign of King Stephen (1135–54), which was wracked by civil war, exemptions to the forest law multiplied exponentially to the point that the law was said to have collapsed or even ceased to exist, as each side sought to win over supporters by such means.[68]

The Angevin period of royal forest law inaugurated by King Henry II (1154–89) marked a revival and reorganization of the royal forest, such that it reached the greatest extent it was ever to achieve in its history and acquired its main features of administration that was to last through the succeeding centuries. Perhaps Henry's single most important act in this regard came towards the end of his reign when he enacted the Assize of Woodstock, sometimes also known as the Assize of the Forest, in 1184. Among its new provisions was that landowners within the royal forest now had to appoint their own foresters or woodwards responsible for upholding the forest law on the king's behalf within their boundaries, and that a jury of twelve knights was charged in each county with returning offences to the venison and the vert before the forest eyres. In addition, private landowners within the forest had to give securities to the king against harming the woods on their own land and were not to sell or give away wood but could gather it for their own use. Anyone caught trespassing against the vert in the royal forest was to be attached by the foresters and his transgressions written down on rolls to be presented before the forest justices for judgment. The assize also provided that all assarts, purprestures, and waste within the royal forest be regularly inspected and recorded; that all summoned by the chief forester to hear pleas against them must appear before the forest eyre; and that any royal forester who allowed for destruction in the king's woods without cause was to answer for it with "his own body."[69]

Despite promises to punish offenses with "full justice as exacted by Henry I," by far most transgressions were met with fines or, at worst, confiscation and exile, since this proved far more lucrative as a source of income to the royal treasury than capital punishment. Nonetheless, the number of people recorded in the forest courts as fleeing arrest and thereby willing to become fugitives or outlaws, incurring confiscation of all their possessions, testifies to the fear with which the forest law was regarded. Indeed, the assize was given teeth with the appointment of four justices of the forest—two knights and two churchmen—assigned to each of four administrative units into which the whole of the royal forest was divided, along with two members of the royal household assigned as keepers of the venison and vert in each quarter with authority over all foresters. Each of these four commissions of justices seem to have held their eyres simultaneously, giving the impression of a universal and coordinated system of forest law outside of which no one could escape. All justices, keepers, and foresters were furthermore required to swear an oath that they would uphold the assize, and indications are that, despite occasional fines for lapses in duty, forest officials could also prove intensely loyal to the crown, risking popular opprobrium and much worse in fulfillment of their charges. Indeed, the main complaint against the foresters under King John was that in a sense they were enforcing the forest law *too* well, by exacting so many or highly exorbitant fines on the crown's behalf.[70]

Henry II had a clear interest and devotion to the forest law: Well before his major reorganization of 1184, Henry personally undertook a royal visitation of the forest in 1175, when he meted out justice rather impartially by imposing substantial fines upon both supporters and enemies of the crown alike. Clearly, the Assize of Woodstock was part of Henry's drive to impose a uniform system of justice throughout the realm that started with the famous Assize of Clarendon of 1166, which first introduced the jury system to England and applied it to persons regardless of status, including churchmen who claimed exemption under their own law and courts. However, forest law was also claimed to be unique as lying outside the system of the common law, being completely dependent "on the arbitrary legislation of the king, so that what is done in accordance with forest law is not called 'just' without qualification, but 'just' according to forest law." This was according to Richard FitzNigel, the king's treasurer, in his *Dialogue Concerning the Exchequer*. Nevertheless, forest law did parallel the common law in many ways, such as by adopting a similar judicial machinery, including the eyre and later the inquisition systems during the thirteenth and fourteenth centuries, while forest courts, especially in the New Forest, also heard an array of criminal accusations even if they had nothing to do with the vert or venison, such as rape, theft, or murder, much as common law courts did. FitzNigel also claimed that the most prevalent offense against the vert was that of "waste," which he famously defined as the woods being cut so severely "that a man, standing on the half-buried stump of an oak or other tree, can see five other trees cut down round about him." Waste mainly seems to have been done in order to satisfy demands for building timber or to bring more land under cultivation; but other encroachments upon the forest recorded by the pipe rolls include the building or excavation of houses, mills, salt pans, and mines.[71]

Despite the impression of strict enforcement of the forest law under the Angevins, the evidence on this is mixed because of the many exemptions—which pick up notably during the reigns of Richard and John—that are recorded for those able to pay for the privilege, in which once again monasteries feature prominently. These included rights of assart—whose penalties were, again, so regular as to indicate a long-standing policy—as well as rights of common pasture and rights of free chase of game; in the last case the land passed out of the royal forest entirely and was administered by a private individual, albeit on the same terms as under the forest law. On the other hand, the substantial sums that could be raised by amercements levied in the forest courts, such as the total in excess of £12,000 recorded for 1175, when Henry II presided over the forest eyre in person, attest to a widespread and impressive enforcement that could be achieved of the forest law. The proceedings of 1175 may have been especially severe, but other years under the Angevins average in the thousands of pounds, and the prospect of being haled before the forest eyre and fined must have daunted and forestalled many would-be loggers and assarters. The universal hatred in which some chief foresters were held, such as Alan de Neville under Henry II or Hugh de Neville under King John, also attests in a kind of backhanded way to their effectiveness.[72]

Forest policy changed little under Richard I (1189–99), except that towards the end of the reign the Forest Assize of 1198 provided for the rather harsh penalty of mutilation of one's eyes and testicles (most likely resulting in death) for offenses against the venison, whereas capital punishment had been reserved only for the third offence under the Assize of Woodstock, while offenses against the vert still only resulted in fines. But it was the turbulent reign of King John (1199–1216) that saw the most important change in the forest law since the Assize of Woodstock, with no less than three clauses in the Magna Carta of 1215 devoted to the royal forest. Clause 44 provided that no one living outside the forest need appear before the forest eyre unless actually charged with an offense or standing surety for someone else. Clauses 47 and 48 revoked all forests created in the reign of John and provided for an enquiry by twelve knights into "all evil customs" relating to forests and foresters. These clauses seem to have been triggered by the forest eyre of 1212, which was widely considered the most oppressive since 1175, exacting in excess of £4000. The real significance of Magna Carta, it seems, was in reversing the idea of the royal forest as being outside the bounds of the common law and subject solely to the king: John was forced to declare that, "We hold our forests and beasts not to our own use only but also to that of our loyal subjects." As with the whole general intent of the Great Charter, the forest law was now seen as answerable to the constitutional will of the entire realm, a crucial precedent that was to be a source of conflict for future generations.[73]

With respect to the forest, Magna Carta was superseded in 1217 by the Forest Charter enacted during the minority of Henry III (1216–72). In terms of clauses that related specifically to the vert rather than to the venison, this provided for the disafforestation of all forests created by Henry II, Richard, and John; exempted any private forest owners from fines for purpesture, waste, and assarts since the reign of Henry II; allowed free men to pasture their beasts and make a "mill, fishpond, dam, marlpit, ditch, or arable land" within their own woods; and in general granted "to all persons the liberties of the forest and free customs they previously had within forests and without." The main issue that remained a source of contention between the king and his subjects was the real extent of the royal forest. A perambulation of 1225 claimed to disafforest "all new forests throughout England" in line with the charter, but after Henry III came of age in 1227, he began to dispute the perambulation, claiming that it had erred in disafforesting what Henry II had merely claimed back for the crown after the chaotic civil war of Stephen's reign. Henry III also took the position that rights of pannage and common pasture in the forest, which could potentially reduce the amount of wood cover for beasts, should revert back to those under his father before Magna Carta.[74]

The issue of the extent of the royal forest and perambulations of it, with a view to disafforesting as much as the barons could get away with, continued to reverberate throughout the rest of Henry III's reign and on into those of his son and grandson, Edward I (1272–1307) and Edward II (1307–27). Like Magna Carta, the Forest Charter of 1217 became a touchstone of all later disputes between the king and his subjects over the forest, being re-confirmed in 1267, 1297, and 1300 as a condition of continued support for the crown. The last confirmation was accompanied by a new perambulation of the royal forest, more thorough than that undertaken at the beginning of the reign in 1278–79; nonetheless, it was soon nullified by Edward I in his Ordinance of the Forest of 1306, which tried to sweeten this setback to disafforestation by promising to reform abuses of forest officials by which the English people were claimed to be "miserably oppressed, impoverished and troubled with many wrongs, being everywhere molested." The Ordinance went on to state that sometimes foresters indicted and arrested people on their own authority, without a proper inquest by a jury, out of "hatred" or a desire to "extort money from someone." From an environmental standpoint, however, the foresters seem to have done their job well, for petitions alleged that they interfered with clearing of trees even after land had been disafforested.[75]

Abuses by forest officials were again targeted in the general Ordinances of 1311 under Edward II, who had to concede to yet another perambulation of the royal forest and the prospect of still more disafforestation in 1316. As in 1306, prospects for the royal forest improved when the king found himself back in the political ascendency over his subjects in 1322. Once more petitions reveal that foresters were interfering with clearing and cultivation of forest land in line with a royal writ directing that forest boundaries be restored to what they had been in 1217, thus nullifying all perambulations and associated disafforesting that had taken place since the Forest Charter. While we may admire this particular aspect of Edward II's tyranny, the victory for the forest was short-lived because disgust with Edward's reign reached the point that the realm acquiesced in his murder and deposition at the hands of his own wife, Isabella, in 1327. In that year, a new statute was enacted as part of the inauguration of the reign of Edward III (1327–77), which included among its provisions yet another confirmation of the Forest Charter of 1217 and a restoration of all perambulations undertaken in the reign of Edward I, which ended up disafforesting some areas that the charter itself had confirmed as forest. Despite such conundrums, the statute is seen as representing a significant "turning point," mostly for the worse, in the fortunes of the royal forest, since subsequently crown policy throughout the rest of the late Middle Ages surrendered the initiative over forest policy to Parliament, while enforcement of the forest law was allowed to lapse, with the forest eyre ceasing altogether after 1368. This is taken to mark a "decline" of the royal forest in the sense that it no longer took center stage as a political issue between the king and his subjects, even though the forest itself may have made something of a comeback by benefitting from the massive human depopulation caused by the Black Death.[76]

The thirteenth century has been called the "height" of the royal forest system, when the English crown achieved perhaps its greatest effectiveness in administering justice through the forest eyre and the greatest reach and size of its forest bureaucracy. All this is very much in accord with the growth of central authority and monarchical power as a whole in England and elsewhere in Europe during the high Middle Ages. Under the fully-elaborated forest system, the royal forest was administered by a hierarchy of officials headed by two chief justices of the forest and under them the wardens (also called stewards or bailiffs in the records) who, sometimes acting through deputies, were in charge of each of the individual forests throughout the land; but it was the foresters (described either as "riding foresters" on horseback or "walking foresters") and woodwards on private estates, along with their grooms or pages (often listed as *garciones* or "boys"), who did most of the actual leg-work on the ground enforcing the forest law among the populace. Much like our game wardens of today, they were expected to patrol the forest, keeping a lookout for trespassers, or else respond to reports of trespass which the common folk were required by law to communicate to the foresters whenever they knew of them. In addition, there were unpaid officers selected from among the local landowning gentry in the forest who assisted the foresters in their work: The verderers kept their own rolls of offenders against the forest law which they were supposed to present at the forest eyre and in general assisted with the apprehension or "attachment" of such offenders, while the regarders held the "regard" that was supposed to take place every three years preparatory to the forest eyre in which they surveyed the condition of the forest and reported any damage or "waste" to the tree cover that may have occurred. The agisters were responsible for collecting pannage and other rents for right of pasturing beasts in the forest.[77]

The main event in the administration of forest law was the forest eyre that traveled about the country presided over by the chief justice of the forest and other itinerant justices, with all other forest officials and all landowners in the forest supposedly in attendance. This court heard major offenses against the vert and venison, which, along with assarts and wastes, were recorded separately in the court rolls. According to sixty-six "chapters of inquiry" for the eyre as set down in the mid-thirteenth century, the itinerant justices wanted to know, among many other things, about: any fallen-down trees in the forest, whether these had been cut down with or without the forester's consent or blown down by the wind; who stole, sold, or gave away wood from the royal forest, whether by day or night, including whole trees, branches, or underbrush, and what was their value; who took windfall wood and whether they had debarked trees in order to cause them to fall down; any waste or destruction of the vert, including by burning in order to create pasture for animals; any assarts and purprestures, or "encroachments," made since the last eyre (which were to be viewed by the regarders), the total number of acres in each assart, and what rent was paid per acre; any enclosures or parks made in the forest or any woods disafforested; any houses, hays or fences, and ditches made to the harm of the forest and its wild beasts, which were to be thrown down; whether the agisters had pastured animals "faithfully" so as not to overburden the forest or be grazing during the improper season, and if they had collected their pannage dues fully (in particular who kept goats in the forest and how many); who were notorious "malefactors of the vert," how much of the wood they wasted or made to deteriorate, and to what extent the foresters had colluded in this waste; who had customary license to extract firewood, timber, charcoal, or to collect dry wood and thorns, and how much this contributed to the deterioration of the forest; how many charcoal-pits, charcoal-maker huts, marl and peat pits there were in the forest and what

Part II. Forest    105

their damage or detriment was to the woods; and if any forester took a bribe for relaxing an attachment for a vert offense, with any forester allowing for destruction of the forest without just cause to be imprisoned at the will of the justices.[78] However, wardens also held their own local forest courts or "attachment courts," which they were supposed to convene every 40 days, at which minor crimes, such as the cutting down of wood or trees valued at four pence or less, were presented and fined. In addition to this there were general inquisitions into the state of the forest and special inquisitions called ad hoc in response to a specific offense known to have occurred in the forest. Paralleling developments in the common law, inquisitions gradually replaced the eyres in the administration of forest law. The "swanimote" court held three times a year convened agisters to account for pasturage of beasts in the forest.[79]

## The evidence of the eyre rolls

Particularly for the forest eyre, some records survive that give us a glimpse into the workings of the administrative machinery of the forest system as well as into the human impact upon the forest itself. For example, at a forest eyre held in the 1240s for Feckenham Forest in Worcestershire, the inquest jury reported how, in addition to their other extortions, all the foresters committed or allowed so much waste of the forest "that the quantity is impossible for them to estimate." This same eyre reported an impressive number of "houses" built to the detriment of the forest, totaling at least 228 such structures, not including a windmill and a forge. Nearly all of these buildings seem to have been designed as barns or stables, for from them was said to issue a total of 1451 animals, comprising mostly cows or oxen, horses, sheep, and pigs. It is not clear how many of these buildings were made at the expense of clearing the forest: In thirty-nine entries the houses are mentioned as being erected within already existing assarts or purprestures, and the presence of 260 "draught animals" certainly testifies to cultivated open fields lying nearby. In three instances only is it specifically stated that the structure was built within the tree cover (*tectum*) of the forest. Indeed, the main threat that these buildings may have been perceived as posing to the forest was in housing beasts that ate and trampled upon browse and new growth, which both deprived wild deer of food and prevented regeneration of trees; the 110 goats listed would have been particularly destructive in this regard.[80]

In two later eyres for Feckenham Forest held at Worcester in 1270 and 1280, a total of 433 vert offenses were tallied with fines amounting to approximately £32, which was roughly equivalent to the number of vert offenses recorded in the Forest of Dean in Gloucestershire in 1282.[81] Normally the exact nature of the vert offense is not specified, but in one case at Feckenham, a William Samuel of Kings Coughton was fined 40 pence for two oaks. Taking the standard value of an oak at this time at 5 pence, Samuel was therefore fined four times the value of his original theft, and if we take this rate as at all typical, then the total number of trees represented by the vert offenses at

Feckenham would amount to 384. It was also uncovered in these eyres that there was a hard core of twenty-eight "common destroyers" and charcoal burners of the vert residing in the forest of Feckenham; evidently, in order to hide their crime many of these men went into the forest to do their cutting at night, as in the case of the wood of William FitzWarin at Hadzor, which was wasted by half a dozen men or more. As in the 1240s, some of the worst offenders could also be the very foresters assigned to guard the forest. In 1270, for instance, it was discovered that two riding foresters under Robert Streche, keeper of the bailiwick of Alta Foresta until his death in 1262, made "great waste" in Walkwood and Astwood, while under Streche's successor, Walter Mareschall, the riding and foot foresters gave away and sold a total of 120 oaks and 180 lime trees and had "two carts almost continuously on the go throughout that time to remove wood from the king's wood and from the other woods in the bailiwick, to sell in local markets all around, by which carts great damage was done to the large timber and branches in the said woods." In addition, the foresters allowed other "wrongdoers" from neighboring towns to enter the bailiwick with carts "to carry off wood for timber, so that the bailiwick was destroyed by these wrongdoers both as regards oaks, pollards, and limes and as regards branches." It was further stated that "many of these wrongdoers were accustomed gradually to undercut oaks, pollards, and limes in the bailiwick, which afterwards fell down in a slight wind." In this way Mareschall "had the profit from 47 oaks," which he had sold in order to be made into charcoal, and another "35 oaks were found undercut in this way, which will soon fall." The result of all these depredations was that the forest in the bailiwick was "almost completely destroyed" and could in fact now be called wasted. Meanwhile, in the king's hay (enclosure) of Lickey, the keeper, Richard de Monte Viron, and his foresters had not only turned a blind eye to "many vert offenders" but themselves felled sixty oaks and eighty pollards and tried to hide their waste by uprooting the stumps and covering them over with soil in a cultivated assart; in a rare case of expressing their indignation at the harm being done to the forest, the regarders also said that in addition the keeper and foresters had cut many branches off of trees, a practice they described as "most vile" (*vilissima*). A decade later much the same was charged of Adam de la Bould, steward of the forest, who besides felling trees and lopping off branches to be sold or given away as gifts, he had permitted and pardoned men guilty of wasting their woods, even when apprehended by his own foresters. (A total of nine new wastes of woods was recorded in this eyre.) Both Mareschall and Monte Viron were deprived of their offices and they and their foresters sent to prison for their offenses, although Mareschall was able to make bail until his court date.[82]

Eyres in other counties are no less revealing. At the pleas for the New Forest held in c. 1257, 123 vert offenses were recorded whose fines totaled approximately £7. Here again, justices were making offenders pay four times the value of their original take, based on the entry that fined Hugh de

Brehull a half mark for four oaks, which gives us a total of about eighty-four trees represented. Somewhat unusual, however, is the fact that three men were hanged for their vert offenses, one of which was committed at night, which would be an unusually harsh penalty at this point in forest law: Perhaps the foresters took the law into their own hands or else their offenses were considered so numerous in this regard that it merited the death penalty? At the New Forest pleas of 1276 and 1280, a number of "malefactors" or "habitual offenders" of the vert were identified, but some of them included the foresters themselves, such as four who were convicted by jury of making numerous gifts and sales of wood to the destruction of the forest, including to charcoal and potash makers, and who had also not presented their many attachments for vert offenses but had presumably taken bribes for their own profit, for which they were imprisoned and their offices forfeit. At the New Forest pleas held in Southampton on July 9, 1330, a total of thirty men considered "habitual offenders" or "common malefactors" of the vert, including seven men described as "malefactors and destroyers of oaks and beeches without number," were merely fined after being apprehended. This inquest is also remarkable for the sheer number of trees and other wood reported as illegally cut down, totaling no less than 516 mature oaks and beeches (including four pollards), 310 small oaks and beeches, 230 saplings, 44 cartloads of oak, beech, and other vert wood, 44 cartloads of branches, 8 cartloads of thorn, 17 hollies, and an unspecified amount of bark stripped from oak trees. In a number of instances the chief keeper or his deputy and the foresters themselves were said to be responsible for this destruction, the most egregious cases being the 200 oaks and beeches that two foresters allowed to be cut down in their bailiwick of Frythham and the more than 100 saplings that another forester in Lyndhurst cut down for his own profit. However, in the case of 47 oaks that were charged to the keeper, John de Chaucombe, it was proven that these were legitimately harvested by order of the king's writ for the purpose of repair and enclosure of the manor and park at Lyndhurst. Sometimes the forest was sacrificed for industrial uses, such as the 40–50 men allowed to carry out cartloads of wood from Burynges destined for the saltworks at Lemyngton, or the eight charcoal-makers in four different bailiwicks allowed to operate their kilns within the forest.[83]

At the 1271 eyre for the forest of Cannock in Staffordshire, the number of vert offenses were less than half of what they had been at the 1262 eyre, 44 versus 114, but the total fines collected at each eyre were virtually the same, at around £9, representing approximately 36 trees felled in each instance. The 1271 eyre for Cannock also uncovered rampant corruption among the foresters, who took for themselves or allowed to be felled by others a total of 188 oaks, 46 birches, and many alders, holly saplings, and oak branches. In Alrewas Hay, which was described as wasted of underwood, William Cardon, the forester there, moved more than £15 worth of wood and timber with a two-horse cart for delivery to Lichfield and elsewhere over the course of two and a half years, while in Hopwas Hay, also described as wasted of underwood, the forester, William of Drakenage, operated a cart daily for a whole year in which he moved a total of £6 worth of wood and timber. In Gailey Hay, Henry Botte of Pillaton Hall attempted to hide his illegal sale of seven oaks and 46 birches and alders while forester there by covering "the resulting stumps with clods and moss" and by fixing thorns over them in a vain attempt to prevent these from being seen by the regarders. These men were delivering most of their earnings to Hugh of Eynsham, riding forester for the whole forest, who also kept under him three men and a woman burning birches, limes, and other trees for at least a year in Alrewas Hay in order to make ashes to sell to cloth-dyers. It was also at this eyre that some new wastes of woods were discovered, including Hopwas Wood, owned by Lord Philip Marmion (a frequent poacher of deer), where three charcoal burners were allowed to operate "to the great destruction in the wood"; Walsall Wood, owned by William de Morteyn, which suffered during the Second Barons' War of 1264–67, so that "almost nothing remains"; and the woods of Prestwood and Pelsall, owned by Theodosius de Camilla, dean of Wolverhampton, where instead of dead and dry wood being sold from the "older and drier oaks," as originally contained in the king's grant, the "best and finest" oaks were felled "to the great damage and detriment" of the forest, although this was said to have been done by Camilla's bailiffs without his knowledge. However, such illegal wastes and felling of wood could be easily matched or exceeded by the legitimate sales and gifts authorized by the crown: At the 1262 eyre, a total of 403 oaks from five hays were recorded as sold or gifted by King Henry III over a twelve-year period from 1249 to 1261.[84]

At the Rutland eyre in 1269, a long list of extortions and abuses were brought against the hereditary warden of the forest, Peter de Neville, who traced his family's custody of the office by inheritance back to the time of King Henry I. Nevertheless, the verderers, regarders, and jury of the eyre charged Peter and his foresters with felling an astounding 7000 oaks and other trees, whose worth totaled £350, during his time in office, which was said to have gone towards timber for Peter's houses and fuel for his limekilns and charcoal pits, as well as being sold for profit. Also damaging to the forest was Peter's appropriation of "small vert," such as thorns and hazelwood, which normally was used to enclose cleared land in order to keep out beasts and allow the woods to grow again: A good example of this was Peter's failure to enclose the park of Ridlington after a sale of trees there, so that "very many animals" were pastured who "ate the shoots of the stumps of the oaks which had been sold and of the underwood which had been felled." Even worse, Peter "caused a great part of those stumps to be uprooted and made into charcoal, so that it can never grow again." On the other hand, in the case of Stokewood, which had been cleared of thorns and underwood for the use of the king's brother, Richard, earl of Cornwall, Peter kept the wood "in defence" beyond the three years specified for regeneration, in defiance of the town's common right of herbage, and was now levying illegal fines for escape of domestic animals into the wood. The majority of complaints, however, centered on Peter's extortions of fines for his own use from both guilty and innocent trespassers of the vert and venison in the forest, which he persuaded his victims to part with by imprisoning them in his own private gaol at Allexton in Leicestershire, instead of in the official prison at Oakham Castle in Rutland, where prisoners were forced to sit in chains in a pit full of water lying at the bottom of the gaol. After being convicted of these crimes Peter was removed from office, but it was subsequently restored to him later that same year upon sureties for a substantial fine, whereupon Peter seems to have continued as before until he was finally outlawed in 1274.[85]

At the eyre held for Sherwood Forest in Nottinghamshire in 1287, the justices found that the Forest Charter of 1217 had not been well kept since the last eyre of 1262–63: The justices had to remind the verderers that they were bound to hear pleas of minor trespasses against the vert and venison every 40 days in the attachment court and present major pleas to the justices in a single roll rather than one for every verderer in his bailiwick. All those who committed trespasses against the vert, wherever these were committed—whether in the king's demesne woods or outside them—and of whatever degree—whether felling whole trees, lopping off branches, or cutting undergrowth such as saplings, hazels, and thorns—were to be arrested and imprisoned, being allowed to find bail only if they dwelled within the forest and had not committed a third offense. The implication here, of course, is that the foresters and other officers were being too lenient and not doing their job of enforcing the forest law. Nevertheless, the detailed records of this eyre specify a total of 610 oaks, 57 young oaks or oak saplings, 31 stumps, and an unspecified amount of timber, branches, dry wood or windfall wood, and heather in at least twelve cartloads taken in 52 separate vert offenses, five of which took place at night. However, over half the oaks, 383, were listed as taken by the abbot of Rufford from his own wood between 1262–66 for building work, for which he was quit because he produced charters proving his right to such timber. All this does not include the 635 oaks that Kings Henry III and Edward I gave away to various men and provided for repairs to the castle of Nottingham and the pond dam at Clipston.[86] A later eyre of 1334 listed 809 offenders against the vert, representing a comparable number of trees felled since in most cases the offense was the taking of a single oak fined usually at 2–3 times its value, for a total of £64 13s. 4d. Here, 128 offences were committed at night, over a quarter of the total.[87] Finally, also in 1334, the similarly detailed record of the eyre held for the forest of Pickering in North Yorkshire counted 190 mostly "green" oaks, 1 "young oak," 45 oak saplings, 2 green oak stumps, 21 "oak shingles," 40 horseloads of green oak, 9 wagonloads of alder, and an unspecified amount of oak trees, saplings, branches, as well as holly and thorns in 95 separate offenses. In one case involving a professional woodcutter, the offenders were apprehended as they were carting away the oak in a wagon drawn by four oxen at night, while in the case of Adam Prudhomme, who "appropriated several oaks," he was caught as he was attempting to sell boards made from his timber to "different men residing in the forest." The fact that a couple of other offenders were carpenters reveals a similar motive behind their theft. An unexplained carelessness on the part of John de Shafaldon, woodward, was behind six oaks torn up by the roots in Deepdale Springs. In a separate indictment, one of the foresters, Sir William de Percehay, was accused by the jury of overstepping his claim to "cablish" or windfall wood by cutting off "huge branches" of oak partly severed and hanging down to the ground and digging up "huge roots" of windblown oaks, all of which he could not possibly have done just with his billhook, as allowed by the claim. When Percehay came and denied the accusation, the verderers and regarders confirmed that the forester had taken "three cartloads composed of huge branches, stems, and roots, in contravention of what he was allowed." Although Percehay was imprisoned and deprived of office, he was able to purchase it back for a fine of 13 shillings and 4 pence.[88]

To sum up, how effective was royal forest law in protecting the greenwood? When vigorously enforced, as it seems to have been during much of the thirteenth century, the time of greatest demographic and agricultural expansion (perhaps explaining the frequent conflicts between crown and subject), forest law is asserted to have been effective in preventing "landowners and farmers in areas covered by it from agricultural improvements that involved clearing the land and extending the cultivated area."[89] Recently, it has been argued that from its very beginnings in Anglo-Norman times, the royal forest was managed by the crown not just as a recreational hunting preserve and home for wild game, but rather as a multi-use, renewable resource, in which its productive capacity for timber, fuel, and food for domestic animals was very much valued and strategically dispensed to grateful subjects (particularly monastic establishments) in ways that find parallels with modern sustainable forestry practices.[90] Basically, the forest was viewed as producing a crop that needed to be periodically harvested just like grain growing on arable land, except that it had the great advantage of automatically replenishing itself. Therefore, as Oliver Rackham has rightly pointed out, just because a wood was cut or even cleared does not mean that it disappeared; quite the contrary, this cyclical process could happen dozens of times throughout a wood's history.[91] To permanently clear a wood, by the laborious procedure of uprooting the stumps and root ball structures, might indeed be taken as a sign of desperation or wrongheaded carelessness, since the sowing of arable crops would soon exhaust the soil.[92] This is why "old assarts" listed in the forest eyre rolls are frequently recorded as lying fallow or used as pasture or even producing a "wood crop," and why "stumps" are valued in vert offenses at a rate just as equal to, if not greater, than whole trees, because they are still living things capable of producing wood products at an even denser rater than their unfelled cousins. It is also why in so many of the records, whether of the New Forest or the forests of Feckenham, Sherwood, or Pickering, when a wood is described as wasted it is often a wood "wasted of old" that has evidently grown back sufficiently to be now "wasted anew."[93]

## Managing the king's woods

The techniques of woodland management were fairly straightforward, falling into the categories of coppicing, suckering, and pollarding, whereby even after a tree was cut down, so long as its root system remained intact, it continued to send forth pole-like tree growths either from the stump or "stool" (in the case of coppices) or at ground level (the "clones" of suckers that emerged directly from the root system), or anywhere from 6 to 15 feet above ground (pollards), in which case the trunk was referred to as a "bolling." The last method was used especially in woodland pasture, so that deer or cattle might not eat the emerging shoots and thereby destroy the future wood crop. (These days, one can still see pollarding in practice along tree-lined streets at the level of power or telephone lines.) It was also rather more inconvenient or even dangerous to harvest, which is perhaps why William le Gale fell to his death climbing an oak in Brambelwode in the New Forest, which the court judged simply as a "misadventure."[94] By the time of the Domesday survey of 1086, nearly the whole of the medieval English forest exhibited evidence of human management of some kind, whether as primary or secondary woodland; the only virgin forest or wildwood left seems to have been in the forest of Dean in Gloucestershire. The coppiced "underwood" (*boscus* in Latin) was usually harvested on short rotations, called "fells," averaging 4–8 years during the Middle Ages, although longer rotations of 15–20 years appear during the late medieval period as demand seems to have fallen in the wake of the Black Death, and the resulting wood products were generally used for fences, wattle-work, and for fuel, especially charcoal-making. A continuous series of records from the mid-fourteenth century onwards at Hardwick Wood in Cambridgeshire shows underwood being felled at a rate of between 6–8 acres in the 1360s and '70s and then steadily declining to 2 acres or less by the late fifteenth century.[95] Underwood allowed to go too long without cutting was said to "spoil," as happened in the Black Prince's parks of Shotewi and Rusti in 1347, where it was recommended that trees now be cut every year to make up the difference.[96] Timber trees, also known as "maidens" or "standards" (*silva* in Latin), were typically used for beams or boards in building work and were cut on much longer cycles, requiring more far-sighted techniques of woodland management. The evidence of such timber frames as survive from the Middle Ages indicate that these were usually harvested between 20–70 years, although ages of 100 years and longer are also common, and ranged in size from 4.5 to 36.5 inches in diameter. In a typical medieval wood, the taller canopies of timber trees naturally existed side-by-side with the lower growth of the more regularly coppiced underwood (designated as "coppice-with-standards"), which of course required different rotational cutting cycles, although sometimes a whole wood might also be cut down all at once.[97]

Thus, in 1256 King Henry III ordered the sale of both "old oaks" and underwood from seven bailiwicks in the Forest of Dean in order to realize a profit of just over 1000 marks (about £666).[98] Even when felling timber,

*Figure 2.4* COPPICED TREES in the foreground, with oak standards left standing in the cleared area behind, from a Book of Hours by Simon Bening, c.1540. British Library, London, UK / © British Library Board. All Rights Reserved / The Bridgeman Art Library

*Figure 2.5* THE "BIG BELLY" OAK in the Savernake Forest of Wiltshire, England, a nearly 1000 year old tree that is thought to have taken root during the time of William the Conqueror. © Jim Champion http://www.geograph.org.uk/photo/153419

nothing was allowed to go to waste, for the whole tree was used. This can readily be seen from the 50 oaks felled in the king's park of Guildford, Surrey, for his building works there, when the guardian of the king's works, Ranulph of Combreiton, sold their "shrouds" or topmost branches not used for timber for the tidy sum of 17 shillings.[99] Such "lops," "tops," and "crops," as well as bark and chips, were a much prized perquisite of both commoners and foresters, for which one can sometimes see them jostling in the forest eyre rolls.[100] While oak was the typical timber tree of medieval England, plenty of other species show up in the managed woodlot, including ash, maple, birch, sallow, hazel, alder, elm, lime, hawthorn, and so on, with the major exception of pine, which is truly killed when cut down. Coppicing greatly prolonged the life of a tree. An ash can live more than 100 years beyond its normal lifespan when coppiced, while a hazel can more than triple its age through the process, which is indeed why some of the trees from the Middle Ages survive today.[101]

A wood managed very much along the same lines as it was in medieval times can still be seen in the Bradfield Woods of Suffolk, which dates back in its coppiced form to at least the mid-thirteenth century, when it was managed by the Abbey of Bury St. Edmunds.[102]

There is no doubt that sometimes impressive numbers of trees were recorded as felled in the eyre rolls, whether by command of the king or taken furtively by his subjects. Yet despite this and the sometimes equally impressive amounts of acreage listed as assarted or wasted, one can't assume that the end result in all cases was the permanent destruction of the forest.

Vert offenses such as are recorded in the eyre rolls can be interpreted in two ways, either as evidence of numerous evasions of forest law or rather as evidence of its vigorous enforcement. On paper, there were plenty of safeguards built into the system to ensure adequate protection of the forest. A general stock-taking

*Figure 2.6* MODERN COPPICING in Bradfield Wood © Robin Chittenden/Frank Lane Picture Agency/Corbis

was conducted for all justices and wardens, both upon their entering and leaving office, which was done either by holding an inquest into the current state of the woods in order to establish a baseline from which to measure future performance or by the official rendering an account for past management of charges.[103] In addition to the usual machinery of the regard and the eyre in terms of ferreting out damages and abuses after the fact, foresters and verderers were supposed to preview any proposed harvests or assarts to see whether these could be done without causing long-term, permanent harm to the vert, and similar policies were apparently practiced in Normandy as well. If approved, the cutting of trees was to be done wherever it would be "least injurious" or "least harmful" and cause the "least damage" to the forest.[104] In one instance, in 1281, King Edward I, after commanding an inspection by the warden of the Forest of Dean, Ralph de Sandwich, in collaboration with local foresters and verderers, decided to allow the abbey of St. Peter's in Gloucester to fell the whole of Hope Mansel Wood on a rotational system over the course of four years because the forest cover was so old and dense that it was "less fit for the beasts to feed in than the trees which would grow anew."[105] Local manorial woods also operated along the same lines, greatly restricting tenants' rights to fell trees without supervision.[106] Even woods that were cleared were clearly intended to grow back again, as indicated by the policy of fencing them in to protect young shoots from overbrowsing by game or cattle. For example, in the late thirteenth century the bailiffs of the forest of Pickering were taken to task for agisting cattle in Scalby Hay, even though "the wood is growing again in a most wonderful manner, and there are countless shoots an ell high [45 inches] and more which the cattle will nibble and waste if they are not removed."[107] In some cases, the process of regrowth was helped along by scattering oak acorns or transplanting "imps" or young saplings nurtured in special enclosures called "impyards."[108] Any private landowner who allowed woods to be wasted within the royal forest was liable to have these confiscated and could only get them back by paying a fine at each eyre until the woods had grown back through coppice regeneration.[109] Enough fines and penalties were levied for lapses in duty to show that these safeguards were taken seriously. As one combs through the records, it is impressed upon one how serious it was to be caught in breach of forest law. The result was either substantial fines (albeit these were often waived for the poor); confiscation of land, carts, horses, and other goods and means of making a livelihood; or else imprisonment, exile, and even death. The long reach of forest law did not end even upon death, for one reads of heirs being held to account for the sins of their fathers.

On the other hand, it is quite evident that the main concern of crown policy was to exploit the forest law for its revenue-raising potential and not necessarily for the sake of the best interests of the woods themselves.[110] The fabled long intervals in between the sittings of the eyre, with defendants having died in the interim or lain so long in prison that they were automatically released as having already served their sentence, made it a very inconsistent instrument of forest law.[111] Local attachment courts presided over by the foresters and verderers, whose rolls were to be presented before the eyre when it did meet, were designed to fill in the gaps and complement the supra-regional courts, but not enough of their records survive to get a really good picture of how well they functioned. One warden's roll that does survive, that of Henry Sturmy for Savernake Forest in Wiltshire covering the years 1296–1305, shows that like the eyre, the local attachment courts were held on an irregular basis, between 1 and 4 times a year, but also that vert violations were pretty low, running between 1 and 6 oaks a year.[112] The attachment rolls for the forests of Quernmore and Wyresdale in Lancashire between 1290–1307 record a total of 24 offenses against the vert, mostly for cutting green oaks or oak branches, but also some alder, hazel, holly, and hawthorn. The worst offense was that of Simon the Forger who cut down 100 green oaks, but this was unusual; other offenders included carpenters, fullers, and blacksmiths, three of whom thought they could build additions for the earl's castle at Lancaster using a green and two dry oaks cut down in their employer's forest of Quernmore, for which they were promptly caught.[113] Manorial courts, which had local jurisdiction over the woods of the manor, also reveal few felling offenses; mainly defendants were hailed for theft of existing wood supplies already harvested and stored in cartloads or bundles.[114] But when one reads in the eyre rolls of sometimes hundreds or even thousands of trees felled at a time for human uses, it is impossible to know exactly what impact this made upon the living forest. Much of what was harvested, at least for fuel, could be dead or dying, which is why a distinction is made in the records between green and dry wood. In the early fourteenth century, John Dalton, constable and keeper of the castle and forest of Pickering, sold "dry oaks and trees for the most part dead, and trees which day by day were going back" in Scalby Hay, and out of the proceeds he purchased a "fresh store of trees," presumably for replanting.[115] In 1299–1300, the New Forest courts heard a total of 67 offenses of dry wood, for which the crown collected £2 and 6 shillings.[116] A great storm that struck England in December 1222 produced a boon of windfallen wood throughout the royal forest in 49 separate woods, which must have kept everyone in fuel and perhaps some timber for a time without having to make fresh harvests of the greenwood.[117]

The sordid tales of foresters' rampant corruption can be entertaining and impressive, as we have seen above, but at the same time the very reason that we know about them is because they eventually *were* brought to account, even if not always in a satisfactory manner, despite sometimes strenuous efforts at evading detection, such as the forester who attempted to hide the stumps of 600 oaks with turf.[118] We also have to remember that there must have been many cases in which forests were well run and the forest administration ran rather smoothly but which did not make it into the records perhaps for this very reason, that their history was unremarkable. An exception is the testimony of an inquest jury sworn at Corby in Northamptonshire on October 2, 1253 concerning the forest of Rockingham, where the jurors stated that the vert and venison were "well preserved" and that the foresters were "faithful," knowing "nothing else of them."[119] A view of Cannock forest in Staffordshire which was ordered by King Henry III in 1235 as part of a general survey of the entire royal forest system found 16 of the woods or hays to be "well kept in respect of the vert," or of the "oak and underwood," while 14 were "much wasted" either of old or recently; in two other cases the woods were much wasted with respect to oaks but the underwood was well kept. This marginally positive balance sheet was in spite of the fact that the local forester, Hugh de Loges, was so corrupt that he was eventually removed by 1246. In the king's hay of Gailey, for example, "many of the best oaks were recently felled" in order to erect a building for Hugh at Rodbaston, which he claimed was his right. Nonetheless, a wood could still recover from such treatment, as was the case in the hay of Teddesley, which was reported as "well kept in respect of oaks and underwood" and where "a reasonable number of beasts were seen," despite the fact that "many stumps of oaks were found" which had been felled by royal order to help the town of Stafford rebuild its walls.[120]

On the other hand, courts were suspicious when juries had no offenses to report at their inquests, as the twelve jurors discovered who were each fined 40 pence or else half a mark (6 shillings and 8 pence) at the eyre for the New Forest held in 1280, because they "presented nothing in their verdict of the new pleas of the crown which have arisen since the last forest eyre."[121] Sometimes foresters were accused unjustly, as was Hugh of Goldingham, steward of Rockingham Forest, who stood accused of "many and great trespasses to the venison, vert, pannage, and other things in the forest" but was found not guilty by the verderers, regarders, and other jury members at the eyre held in 1255; instead, it was judged that the townships that had brought the charge, which included Geddington, Brigstock, Stanion, and Oakley, had accused Goldingham "falsely and out of hatred."[122] In November 1276 an inquest into the New Forest held at Romsey in Hampshire heard how at least 28 jurors had been suborned to bring false accusations against foresters of offenses against the vert and venison and in one instance of the capital crime of murder; ironically, it was fellow foresters and verderers who seemed to be behind this miscarriage of justice, including a verderer who was described as "the falsest man in the county."[123] At the Sherwood Forest eyre held at Nottingham in 1287, three successive wardens of the forest, which included the constable of Nottingham Castle, were accused of taking 600 oaks for their own use from Beskwood Hay, but they were all acquitted after it was determined that the oaks went for the legitimate repair and maintenance of the castle works, including its mill, fish-weir, fishpond, and charcoal pits.[124] (One of the accused was no less a personage than Walter Giffard, archbishop of York.) During the early fourteenth century, John Dalton, constable and keeper of the castle and forest of Pickering, was accused of appropriating 300 oaks in Scalby Hay and at least 132 oaks elsewhere in the forest, but he was able to account for all but five oaks: 110 oaks for fortification, repair, and fuel for Pickering Castle; 71 oaks as gifts from the earl of Lancaster to various men; 28 oaks for a house built for Edmund Crauncester; and the rest as sale proceeds for replanting the forest. Likewise Dalton's successor, John Kilvington, was charged with appropriating at least 623 oaks and branches from the forest, of which he claimed a total of 443 oaks and 50 branches for repair of Scarborough Castle, but was only able to produce the king's writ dated 1335 that accounted for 163 oaks at the exchequer, and for the rest he had to pay a fine of £20.[125]

It is not to be denied that forest officials were unpopular. At the eyre for Rockingham Forest in 1251, one of the verderers, Richard of Aldwinkle, reported a conversation he held on September 13, 1250, with one William the Spenser, in which Spenser, after refusing to return the courtesy of Aldwinkle's greetings, informed him that, "I would rather go to my plough than serve in such an office as yours."[126] Time after time one comes across the at least passive resistance of towns fined for failing to show up for the eyre, although in many instances an unrealistic criterion of "full" attendance was applied simply as an excuse to levy a fine and raise cash.[127] Rather more active resistance was shown by Geoffrey of Sudborough, chaplain to the church of St. Peter in Aldwinkle, Northamptonshire, who shot and slew a forester, Stephen of Moulton, with a barbed arrow as he "stood on a certain oak," almost certainly one that had recently been felled or coppiced, in a private wood around 1248, or the brothers Richard and Rocelin de Lyndhurst, who murdered a forester by the name of Wasmerus at Guldenelak in the New Forest on May 15, 1256, and who had been harbored both before and after the crime by the abbots of Beaulieu and Tychefeud and the priors of St. Mary's and St. Denys in Southampton. One of the brothers, Rocelin, was apprehended and sentenced to be hanged, along with a harboring relative, Ralph de Lyndhurst, while the abbot of Tychefeud paid a fine of £10 after being detained in prison.[128] Rather more harmless but just as revealing is the case of Robert le Baud, master of the hospital of Huntingdon, who appeared before the eyre in 1255 and "alleged that the lord king ought not to have attachments either of vert or of venison, and that neither the foresters nor the verderers ought to make any attachment thereof in the town of Huntingdon." However, the court was not amused by Baud's outburst and sent him off to gaol and fined him half a mark because "by his chatter the court was disturbed and the business of the lord king hindered."[129]

Sometimes forest officials were impeded by other royal officers who sided with offenders, such as when the foresters and verderers of the forest of Huntingdon tried to dismantle the houses of Vincent of Stanley which "had been raised to the nuisance of the forest" but who were forcibly driven back by the bailiffs of the sheriff of Huntingdon, Philip of Stanton, just as they were laying "their hands on the said houses to unroof and pull them down." When the verderers and foresters approached the sheriff to apprise him of their commission, Stanton replied "that they had no order thereof from him and disavowed their deed entirely." When Stanton was summoned to the eyre in 1255 to answer for his obstruction, he freely admitted the fact; clearly, his sympathies were with a man whose family or animals were about to be rendered homeless.[130] In a sense, the system whereby foresters were reimbursed out of the perquisites of their office almost ensured abuses, but at the same time, the crown was well aware that it needed to keep a vigilant eye on its foresters, else it might be in danger of losing the peaceful obedience of its subjects as well as the cover of its trees. The case of the renegade warden of the Forest of Rutland, Peter de Neville, who went beyond the petty extortions and corruptions of some foresters to engage in forcible imprisonment of his victims in a noisome gaol must have stretched the patience of forest dwellers to the breaking point, but just as troubling were the deeds of Norman Sampson, riding forester of Huntingdon, who tortured a man "upon a harrow and pained him sorely" until his victim gave him twelve pence on the spot "that he might be released from the said pains," and later handed over a further five shillings. A harrow was a tined farming implement, very much like a bed of nails, that was drawn by horses over ploughed fields in order to smooth them out prior to planting. One can imagine how much pain Sampson caused by placing his victim upon it, if the harrow was turned upside down with the tines facing upwards.[131]

## Disafforestment and the rise of private woodland

Englishmen's drive for disafforestment arose almost from the very beginning of the forest law, yet this does not necessarily mean that they had the explicit goal of denuding the forest as soon as they could for their own benefit. Common folk could recognize just as well as the crown that the woods were a valuable renewable resource which it was in their best interests to preserve for future generations, and they could be just as indignant at the harm they perceived being done to their forest. At a perambulation of the forest of Somerset in 1278–79, the inhabitants submitted a list of grievances against the Forest Charter, among which was that the woods were still not disafforested as they ought to have been, but they also complained, among many other things, of being burdened with too many riding and walking foresters who fine them excessively and illegally for wastes, purprestures, and assarts, and that foresters, in brewing their scotale which they force people to buy, "fell trees for their fuel in the woods of the good people without leave, to wit, oaks, maples, hazels, thorns, felling the best first, whereby the good people feel themselves aggrieved on account of the destruction of their woods."[132] A wide variety of peasant laborers and craftsmen depended upon the forest to make their living, including woodcutters, carpenters, coopers, sawyers, wheelwrights, makers of various wooden tools and goods, charcoal-burners, potash and lime makers, iron-forgers, glass manufacturers, potters, coal miners, tanners, dyers, rope-makers, and licensed hunters and fowlers.[133] The records of the forest of Lancaster demonstrate that when control of the forest passed into private hands with the granting of the earldom to Henry III's son, Edmund Crouchback, in 1267, along with the right to hold forest pleas independent of the crown from 1285, the vert could be just as well protected and forest law just as well enforced as under royal agents.[134]

Another snapshot of a private forest is provided by the registers for the bishops of Hereford, whose diocese included the Forest of Dean and who managed woods as part of their episcopal manors. A good series of these registers extending from the late thirteenth century to the very end of the Middle Ages allows us to see how the bishops managed their forest over the whole of the very time coinciding with the decline of royal forest law. While the bishops of Hereford were zealous in their assertion of independent forest rights, as evidenced by their preservation of royal charters to that effect going back to the time of Henry I, they were every bit as willing as the crown to prosecute trespasses against their woods and ensure that foresters were performing their expected duties, with threat of excommunication added to the enforcement toolkit that the bishops had at their disposal.[135] Private woods managed by other ecclesiastical institutions, especially monasteries burdened with debts and seeking to raise ready cash, were not so well managed. The prior of Huntingdon and the abbot of Peterborough were cited during visitations by the bishop of Lincoln in the 1430s and '40s for felling "copses" of woods and pocketing money from the sales for their own profit or that of their friends. In the case of Huntingdon, it was said that the prior destroyed the coppice stumps even "when they were sprouting again," while the abbot of Peterborough was charged with making a "great sale" of copses and "thick trees" whose like had not been seen for the past hundred years, and that he caused woods to be felled "that were not fit for cutting." At both Peterborough and Bardney Abbey, it was complained that woods that were felled were not enclosed so that the stumps as they were sending forth new shoots were "spoiled" or "wholly destroyed by the beasts which enter therein."[136]

Perhaps what Englishmen wanted most from disafforestment was control over their own forests, which they believed they could maintain just as well, if not better, than the king and his meddling agents.[137] Nevertheless, it seems a truism that when the royal forest system failed or ceased to function, the forest suffered. This happened, of course, during times of civil unrest, such as Stephen's reign or the Second Barons' War in the time of Henry III. In the reign of Edward II (1307–27), one of the leaders of the baronial opposition, Thomas, earl of Lancaster, was said to have allowed "many strange things" to be done by his bailiffs, foresters, and verderers in the forest of Pickering, including having them make purprestures and forfeitures "whenever they could against the king, often rebelling against him and making others who are the king's tenants rebel."[138] This was undoubtedly a politically motivated accusation, however, as the records of the forest of Lancaster actually show a decline in vert and venison offenses during Thomas's stewardship, whereas disorder and plundering broke out in the northern forests only upon the earl's execution in 1322, carried out in the main by the earl's supporters in the forest as a kind of protest against the crown.[139] The decline in the enforcement of forest law during the late Middle Ages, marked by the demise of the forest eyre after 1368, also seems to have seen a corresponding increase in disturbances, judging from the increase in hardened malefactors of the vert, foresters among them, reported when the eyre did meet or in later inquisitions that replaced the eyre up to 1377.[140] Disafforestment also accelerated after 1327, the year of the statute that re-affirmed both the Forest Charter of 1217 and the peramublations of 1300, which signaled a new willingness on the part of the crown to surrender the initiative over forest policy to the commons as part of the negotiations that opened the reign of Edward III. By 1334 it is estimated that the area of the royal forests had declined by a third from a century earlier.[141] Even when there was an announcement of reafforestment in 1347, Edward III made it clear that it was not his intent to hinder private landowners in the forest "from felling the woods and selling or otherwise making profit of the same," which contradicted the enforcement efforts that had been undertaken by his own forest officers.[142] In 1383 a statute enacted under Edward's grandson, Richard II (1377–99), declared that any forest official making a wrongful arrest was liable to pay a double indemnity to the injured party. Thus, even though the administrative system of the royal forest continued basically unchanged at this time, with the sole exception that the office of ranger first makes its appearance around the middle of the fourteenth century, the purview and effectiveness of these men were now severely curtailed. For example, the regard completely disappears from forest records after 1387.[143] Ironically, it was at this very time when the royal forest ceased to be an issue of contention between the crown and its subjects that we see the emergence in popularity of the legend of Robin Hood, which perhaps harks back to an earlier time in the imagined oppression of the forest law.

A major factor in the decline of the royal forest system was the crown's growing reliance on the instrument of taxation for its revenue, for which it needed regular parliaments rather than eyres, and correspondingly less dependence on income from the forest.[144] Forests were still valuable, of course, for their capacity to produce timber and fuel, which was only to grow in importance with the emerging shipbuilding and metallurgical industries down into the modern period, and thus there was still the incentive to manage the woodland. But without the supervision of the forest law, there is a question of whether good management practices could be maintained. The forest of Lancaster, for example, showed good productive capacity of oak timbers throughout the fourteenth century, but it appears to have entered terminal decline during the fifteenth century due to overexploitation and neglect.[145] The Forest of Dean, which suffered the loss of just a few thousand acres from the time of Domesday to the end of the thirteenth century, when it comprised about 30,000 acres in all, then was reduced to 15,000 to 20,000 acres, a loss of a third to a half, by the mid-sixteenth century.[146] A revival of sorts of royal forest law emerged at the very end of the Middle Ages during the reign of Henry VII (1485–1509), founder of the Tudor dynasty, when the forest eyre was resurrected in some counties in 1488 after a hiatus of well over 100 years, after which it was never to be held again. This temporary victory for forest law was bittersweet, however, as the records show widespread disregard for the law in the interim and trees felled on a scale larger than ever before, perhaps as a legacy of the widespread disorder during the War of the Roses.[147] A list of what was despoiled from the forest of Pickering during the reign up to 1502 counted a grand total of 816 oaks, 90 stumps, 89 wagon-loads of wood, and 124 other items of timber used to make various products such as forks and stiles; in addition several hays and woods had been found to be wasted and destroyed due to over-pasturage of cattle or over-selling of wood from there by the king's own forest officers. As a result, a flurry of orders were sent out by the crown to halt the "cutting down and fellyng of our woodes" and remedy the "lak of good officers and good oversight of the same."[148] However, the fact that landowners within the royal forests of England and Wales were encouraged by an act of Parliament in 1482 to enclose their woods for seven years after each coppice felling in order to allow for regrowth against the grazing of deer and cattle shows that much of the initiative for protecting the remaining forests had passed to private hands.[149]

But while royal forest law, when enforced effectively, might have prevented the worst abuses, it is claimed that in the end it was not so successful "in stabilizing the landscape where it operated."[150] This is because there were allegedly too many exceptions allowed for assarts or other encroachments for a fee, which the crown was all too greedy to obtain as income. Such fines may have been an annoyance to potential assarters but evidently did not deter them altogether from their activities. However, this is not entirely true, as there could be some notable exceptions to this rule. Especially in the vicinity of royal hunting lodges, the crown had an overriding interest in preserving woodland as a hunting preserve, while sometimes the forest proved far more valuable to local industries and the community interest than conversion to arable. In the region of Feckenham Forest in Worcestershire, the royal park at Hanbury apparently had an "inhibiting" effect on assarting, as did private landowners there such as the bishop of Worcester, who supplied the nearby salt-works of Droitwich.[151] The very fact that deer were protected, it is claimed, would have discouraged assarting because there was nothing to prevent the animals from browsing on new-sown crops, since even driving off the deer with dogs was a punishable crime.[152] Aside from this, it is claimed that the English monarchy did not directly exploit its forests very intensively: Some hundreds of oaks and thousands of acres of underwood every year out of half a million acres of forestland. Mainly the crown valued forests for the exclusive products, such as venison or giant oaks, that it could grant away as special favors of the royal prerogative, although it was willing to suspend this generosity, as in 1257, when it perceived the conservation of timber and deer to be at risk.[153] Nonetheless, royal gifts and uses for its own works of timber, underwood, and other woodland products, such as can be glimpsed in great abundance from the Forest of Dean in the thirteenth century, made a vital contribution to the English economy and were put to diverse uses. These included: repair of castles and churches and construction of houses; shingling of roofs; repair of bridges; construction of park palings, fishponds, and fish weirs; operation of mines, charcoal industries, and iron forges; building of ships; and even making of catapults, ducking stools, and quarrels for crossbows.[154] Even if the forest law did no more than delay the inevitable, at least it gave the medieval woods a breathing space to allow them to be appreciated for their renewable resources.

In the end, it was left up to private landowners and the commons to manage the forest for what it could produce in terms of timber, firewood, palings, grazing, etc.; it seems likely that there was enough appreciation of the forest's productive capacity to place self-imposed limits on the exploitation of these resources in order to maintain the forest as a forest for future generations. Nevertheless, mature timber stands had an especially high value, until at least the mid-fourteenth century, that provided a tempting target for clearance in order to realize a quick sale, especially in cases such as monasteries or newly appointed bishops burdened with debts. There are numerous examples that can be cited from the fourteenth and fifteenth centuries of such private sales: Between 1333 and 1346, for example, Elizabeth de Burgh, Lady Clare, realized a profit of over £1000 from six separate sales from her woods in South Frith in Kent. However, the timber market suffered a collapse in the aftermath of the Black Death of 1348–49, perhaps due to a combination of much lower demand (many construction projects were halted owing to catastrophic drops in income and workers available) and a glut of more mature trees available from neglect of the woods. Coppiced or managed woodland and thickets, on the other hand, seem to have retained a robust value since they could be marketed for a variety of uses, including fence poles, laths for wattle-and-daub house construction, and as firewood and charcoal.[155] Unfortunately, the management of woodlots by private landowners—accomplished on a piecemeal basis parcel by parcel—is the one aspect of medieval forest history that is least documented in the surviving records, and consequently it is what we know the least about.[156]

## The management of woods elsewhere

Until recently, forests in the lands on the periphery of the British Isles—namely, Scotland, Wales, and Ireland—received far less attention than the better-documented forests of England. In Scotland, the myth of the great Caledonian Forest stretching unbroken from end to end of the Highlands seems to have been perpetuated by Roman authors who had little direct knowledge of the local terrain, which was but little penetrated by Roman armies.[157] Best estimates, largely derived from botanical archaeology, are that more than half of the Scottish forest may have disappeared, mostly due to human activity, even before the start of the Middle Ages.[158] What remained was perhaps as intensively managed as any forest in England, and likewise for a variety of uses, including building timber and associated wood products, grazing of domestic animals in wood pasture, and above all as hunting grounds and an environment sustaining wild animals. Scotland did have a legal framework and complex system of regulations that granted control over vert and venison to private landowners, but competing uses of the woods from other parties seems to have been allowed under a less stringent application of forest law than in England. By the fifteenth century, the Scottish parliament began passing laws restricting the felling of trees and encouraging plantations, evidently out of concern for scarcity of wood, at least in the Lowlands.[159] Ancient pollards and coppice stools that still survive in some Scottish woodland reserves, such as the Garscadden Wood just outside Glasgow, provide physical testimony to the forest management techniques that were used in the Middle Ages.[160]

In Wales, native laws in force from the tenth to the twelfth centuries forbade assarting or pannage without the owner's permission, although the odd timber could be cut at will for house construction. "Tree gilds," or compensation fines for cutting down different species of trees, valued oak the highest at 120 pence per tree (twice that for a cow), evidently reflecting its economic importance for producing timber. Beech—like oak, a mast-producing tree useful for feeding pigs—and apple trees that bore fruit were next in value at 60 pence, while trees not producing edible fruit or nuts, such as hazel, ash, willow, alder, and yew, were much further down the scale at 4–15 pence. With the Norman Conquest of Wales by about the mid-twelfth century, such local laws were supplanted by royal forest law and its administration—as in England. Thus we have the familiar pattern of preserving the forest mainly as a royal hunting ground, of conflicting claims to common woodland rights or estovers, and occasional flagrant violations of the forest law, such as the destruction of 500 acres of woods at Overton in 1309 or the felling of 3000 oaks at Coydrath and Rodewode in 1386. The forest hosted a kind of Robin Hood-like rivalry between the native Welsh and their Norman conquerors, with the former resorting to the dense woods for cover and ambush, while the latter sought to clear as much woods as possible wherever the military had to pass, and this pattern continued even down into the thirteenth century. Indeed, in those areas where the English campaigned, a free hand was given to almost anyone for the felling of trees, setting the whole run of the royal forest law at naught. Welshmen were also said to value forestland, according to one fourteenth-century petition, because "the greater part of their sustenance is derived from the woods," and the variety of uses that were extracted from them is impressive: timber for building; underwood and deadwood for fuel and charcoal; bark for tanning leather; foliage, fruits and nuts, and woodland pasture for grazing livestock; honey and wax from beehives; tamed hawks from eyries; soap and dye from wood ash; and so on. Forest management practices such as coppicing, operating on a rather long rotation of 10–14 years, were similar to those in England. As elsewhere, Cistercian monasteries in Wales allegedly were major agents of forest clearances, although later in their history the Cistercians were also leading practitioners of good forest management, perhaps as the woods became more scarce. By the late Middle Ages and on into the early modern period, forest management went into decline in Wales, accompanied it seems by the growth of secondary woods on abandoned farmland, due to the ravages of the Black Death and of the fifteenth-century revolt of Owain Glyndŵr.[161]

Evidence exists that Ireland's forest history was not much different than elsewhere—witnessed, for example, by the ancient oak coppice stools to be found in St. John's Wood in County Roscommon—but Ireland during the Middle Ages always had far less woods than other places in Europe, even compared to England, and consequently less of such evidence survives.[162] However, considerable evidence of intensive woodland management is available from the vicinity of Anglo-Norman Dublin (twelfth–fourteenth centuries), which was about 8 percent wooded in 1326, and that can be recovered mainly from archaeological excavations of sites that used wood products. Thus, oak was typically used as timber for building and harbor construction, although ash and alder were also used, and most structures reveal a combination of smaller, fast-grown trees felled in coppiced rotations and trucks of mature trees cut at 100–125 years of age (indicating the mixed "coppice with standards" approach). Hazel was the preferred underwood species for coppiced poles used for wattle-and-daub house walls and waterfronts, as well as for pathways, boundary fences, and ship components, although other coppiced species could include willow, yew, ash, alder, oak, holly, birch, elder, blackthorn, cherry, and fruitwood. Coppicing was usually done on a 5–8 year rotation, judging from tree-ring analysis. Some woods were also used for specialty items, such as coppiced ash-wood for lathe-turned bowls, and yew for finely carved tankards, spoons, pins, and gaming pieces. Remains of woodbanks such as at Glencree also indicate that woods were enclosed to protect against illegal grazing, harvesting, and poaching. Surviving account rolls and other documentary evidence show that woods were managed not only as part of the English royal forest system but also by private landlords, such as the Priory of the Holy Trinity in Dublin.[163]

In France and Germany, after the Carolingian era of the early Middle Ages, a long process ensued in which control over the usage rights to the forests—such as for hunting, wood for timber and fuel, and pasturage of animals—devolved upon monastic and aristocratic establishments and perhaps rural communities held in common, at the expense of the centralizing authority of the crown.[164] (Control over forest rights seems to have been less of an issue in Spain and Italy, since Mediterranean regions were less wooded.) This decentralization continued in the German Empire down through the late Middle Ages, but in France, the crown began reasserting its authority beginning in 1219, when King Philip II Augustus, perhaps influenced by royal forest policy in England, particularly after he took over the formerly English-held duchy of Normandy, halted the granting of new usage rights over the forests and ordered that all sales of wood be supervised by foresters. This was really no more than a resurrection of Carolingian policy dating back to the ninth century.[165] The thirteenth century saw the first regulations, in both France and the Low Countries, for coppicing of trees, particularly with regard to fell rotations and restricting the grazing of cattle that might eat the young shoots in spring; in addition, those conceded the right to cut down an oak or two usually had to plant replacements and protect these as well.[166] It was also in the thirteenth century that we see the creation of a new royal official, the *maître des forêts* (master of the forests), who first appears in an ordinance of 1291 under King Philip IV and who supervised the various officials in charge of individual forests within each district.[167] In 1319, 1346, and 1376, separate legislation known collectively as the *Ordonnances touchant les Eaux et Forêts* (Ordinances touching the Waters and Forests), which concerned the forest administration under the *maîtres* (who were to be seven in number by the fifteenth century), was issued and then re-issued in 1355, 1357, 1389, and 1402. The ordinance of 1346, sometimes known as the Ordinance of Brunoy, entailed a major reorganization of the French forest system as the result of an investigation, or *réformation*, of the forests by two royal officials in 1341. Its main provisions included no further granting of usage rights to the forests; the rendering of accounts of all forest officials including the *verdiers* (verderers), *châtelains* (castellans), and *maître sergens* (master sergeants) to the *maître des forêts* twice a year, with the *maître* himself reporting to the *Chambre des Comptes* (Chamber of Accounts) once a year; and the duty of the *maîtres* "to enquire into and visit all the forests and woods and supervise the sales there, in regard to which the said forests and woods can be perpetually maintained in good estate (condition)."[168] Evidently, there was some concern about corruption of forest officials: For example, the ordinance of 1376 set up a much closer supervision of the forests, specifying that the *maîtres* visit them at least twice a year, each time summoning all the forest officials to render account, and that the *verdiers, gruiers* (foresters), *and gardes* (wardens) or *maître sergens* visit their wards within the forest every *quinzaine* or approximately at two-week intervals. Each of these officials also had to submit a pledge or deposit of 500 *livres tournois* to the *Chambre des Comptes* to ensure their future good behavior. In the preamble to the ordinance, it was noted that all this was being done because a royal visitation of the waters and forests had uncovered that the crown had been "defrauded of our right by the culpability, fault, or negligence of the masters."[169] Plenty of examples of abuses of foresters, particularly in the records of the forests of the duchy of Orleans, can be cited, such as stealing of wood or extorting fines from people who were led to believe they could take wood legally; in some instances the forester spied an opportunity for extortion after being hosted in a home where an illegally cut log was burning on the hearth! Upheavals in France during the Hundred Years War from 1337 to 1453 often prevented the provisions of the forest ordinances from being carried into effect, especially when large parts of the country were in English hands, and this ineffectiveness shows in the necessity of re-issuing the same ordinance again and again. Also impeding enforcement was that too many *maîtres* were nobles who had no practical experience in forest administration; that forest positions were treated as sinecures instead of professional avocations since remuneration came from perquisites of the office, such as wood and fines, rather than a regular salary; another impediment was that the crown was inveterately opposed in its forest policy by the rest of the realm, particularly the nobles and clergy, as represented in the Estates-General and the *Parlement*, or high court, of France. Too often, in fact, the king had to make concessions in terms of not allowing the writ of his ordinances to run in forests outside the royal estates owing to the necessity of fighting a long and expensive war with England.[170] Thus, the late medieval history of the forests in France presents considerable parallels, but also some contrasts, with the history of the forests across the Channel. During the fifteenth century, France apparently followed the rest of Europe and England in placing less emphasis on direct forest administration: For example, no new legislation on the forests was issued for nearly a century after 1402, with the sole exception of the ordinance pushed through by the Cabochien Revolt in Paris in 1413, which was soon revoked.[171] This was due to a number of factors, including a decline, as in England, of the importance of forest revenues coming into the *Chambre des Comptes* after the institution of regular taxation, the easing of pressures on forest resources after the depopulations caused by the Black Death and the Hundred Years War, and the lack of interest by fifteenth-century kings in maintaining the forest as a hunting preserve.[172]

The Baltic States from the thirteenth century onwards apparently specialized in export of timber known as "Baltic oak," or boards cut from "large, slow-grown oak trees," which would have required careful management in order to maintain steady supply.[173] The ancient Białowieza Forest in Poland and Belorussia, where some surviving oaks are estimated to be 600–800 years old, perhaps give an idea of the type of trees that would have been involved in this trade.[174] Another part of Europe where the state of medieval forests has been illuminated by recent research is Hungary, which ended up with about 20 percent of its land still forested by the end of the Middle Ages and where the practice of coppicing was widespread, particularly in terms of pollards, since there was a strong tradition of woodland pasture for cattle grazing on the plains of the Carpathian Basin. Like elsewhere in Europe, Hungary had a flexible woodland management system designed to produce a range of products, including firewood and timber, while also following the pattern of having its common rights to the forest challenged by the late medieval trend towards privatization. After Hungary became a Christian kingdom around the year 1000, it also adopted a royal forest system modeled on those in western countries such as France and England, albeit adapted to local conditions, complete with "forest-guards" who were the equivalent of foresters. As in the rest of Europe, "forest" here was a legal and economic entity that did not always correspond with actual woodland. Yet it is credited with providing a stable management policy that preserved some ancient woods that survive up to the present day, such as the Pilis and Bakony royal forests in the center and west of the country. Even so, Hungary's royal forest system came to an end with the demise of the Árpád dynasty in the fourteenth century, throwing the fate of its various forests into uncertainty, with widely varying results.[175]

## Shaping the idea of wilderness

Attitudes towards the forest in the Middle Ages were shaped not only by the practical realities of administration and management, but also by medieval literature and other writings. For monastic and religious communities, the forest wilderness had a special place as both the means of spiritual renewal by providing an isolated shelter from worldly distractions, and as a test of spiritual resolve by being the haunt of demons and other temptations. Such associations go back to biblical stories of the desert and the experiences of the early Christian desert fathers: Even though in topographical terms a forest could be considered the complete opposite of a desert, the two did share connotations of solitude, otherworldliness, and an uncivilized purity that were essential to the ascetic exercise. Thus we read in Guibert of Nogent's *Autobiography* of the case of Evrard of Breteuil, viscount of Chartres, who in 1073 abandoned his noble way of life to become a hermit making his living as a charcoal burner in the forest, or in the next century of how St. Bernard of Clairvaux, one of the founders of the Cistercian order of monks, extolled the spiritual benefits of the woods in a famous letter to Henry Murdach, abbot of Vauclair: "You shall find something further in the forests than in books. The trees and rocks will teach you that which you will never be able to hear from the masters of science."[176] The Cistercians were to become renowned, of course, for their preference for desolate sites for their monastic foundations in an effort to reform their observance back to what was believed to be a pure, primitive Christian experience. They certainly had a reputation in the high Middle Ages as efficient clearers of the forest, as even their scathingly satiric critic, Gerald of Wales, grudgingly testifies when he writes: "Settle the Cistercians in some barren retreat which is hidden away in an overgrown forest; a year or two later you will find splendid churches there with fine monastic buildings."[177] Gerald's friend and fellow courtier, Walter Map, wrote of them that once the Cistercians acquired any piece of land, "the wood is cut down, stubbed up, and leveled into a plain." However, as already noted above, recently scholars have questioned how far the white monks actually settled in desolate or wilderness areas in the first place; it appears that, as in the *Life of St. Sturm* set in the forest of Saxony, many of their sites had a prior settlement history.[178] To the extent that they did make a Christian paradise in the midst of the forest, it is perhaps ironic that the Cistericans' very success in this endeavor helped contribute to the end of their cherished wilderness.

A similar dichotomy in attitudes towards the forest occurs in medieval romance literature. Taking their cue from classical antecedents such as by Ovid and Virgil, medieval romances from the twelfth through to the fifteenth centuries portrayed the forest in both positive and negative lights, presenting either thrilling opportunities for adventure and redemption or else posing sinister threats of violence and madness (and often both). Most of them concern the exploits of members of King Arthur's round table. In the twelfth-century romances of Chrétien de Troyes, such as *Erec et Enide*, *Yvain*, *Lancelot*, and *Perceval*, the forest enables the protagonists to redeem their chivalric identity, where it poses an enchanted and perilous contrast to the more civilized world of Arthur's court. The legends of *Tristan et Yseut* from the same century portray the forest at first as an idyllic setting for the lovers as they escape from the

*Figure 2.7* TRISTAN AND ISOLDE, misericord carving from Chester Cathedral, Chester, England, c. 1390. An illustration of the Tristan and Isolde romance legend, with King Mark spying out from a tree as Tristan and Isolde tryst in the forest. © Dominic Strange www.misericords.co.uk http://www.misericords.co.uk

suffocating court of King Mark to find freedom in their love bower or grotto hidden deep within the forest glade.

But later, when the love spell wears off, the forest comes to symbolize the harsh and cruel reality of their doom-laden adultery as they face deprivation and hardship in the wild. Such themes are continued in the thirteenth-century prose romances of the Arthurian legend, but with the forest now transformed from a pagan supernatural landscape into an allegorically symbolic Christian one, as in the *Queste del Saint Graal* (Quest for the Holy Grail) of the *Vulgate* cycle. By the fourteenth century, we have a variety of portrayals of the forest in the Middle English romances: In *Sir Orfeo*, a reworking of the classical tale of Orpheus and Eurydice, the forest serves as a marginal or transitional realm between the fairy and human worlds and as an entry point to the otherworld; in *Sir Launfal*, the forest is an idealistic land of wish-fulfillment that alone can permanently maintain the chivalric ethos, in contrast to the flawed court of King Arthur, which is symbolized by Launfel's permanent departure into the forest of fairyland at the end of the poem; in *Sir Gawain and the Green Knight*, the forest is once again a marginal and ambiguous realm midway between human and fairy land that also poses a challenge to the courtly romance ideal, and which is the setting for Bertilak's daily hunt in Hautdesert while his lady hunts Gawain back in the castle. The poems of Geoffrey Chaucer at the end of the century inject an element of social realism into the romance portrayal of the forest. Finally, the fifteenth-century prose work of Thomas Mallory, the *Morte Darthur*, serves as a grand summation of all these forest themes, just as it does in general for the entire Arthurian cycle.[179] For the most part, the medieval romance tradition does not provide particularly realistic depictions of the forest, with the notable exception of *Sir Gawain and the Green Knight*, where the poem supplies detailed references to the local topography of North Wales, which allegedly has the wildest forest in Europe.[180] One has the impression that such fictional forests would be largely unrecognizable to the denizens of the intensively managed landscape throughout much of the royal forest of England or France; rather, they seem to hark back to a more primitive time of the wildwood of a now mythical ancient past.

Another genre of medieval literature in which the forest features prominently is the Robin Hood ballads and other outlaw tales. Since these date to the late Middle Ages, primarily the fifteenth century, the royal forest, which is the setting for *A Gest of Robyn Hode*, would no longer have been the focus of popular discontent when, as we have seen, enforcement of forest law by the crown was in decline and disafforestment proceeding apace. Instead, what the audience of the ballads might have found most objectionable were enclosed chases and parks, whether these be owned by the king or in private hands, which were proliferating at this time and seem to have been the target of many of the trespasses haled before the courts. And yet, private parks go unmentioned in the Robin Hood ballads; any hint of discontent with the forest law that can be detected in them, such as with the severity of penalties for taking the king's deer that feature in most modern-day popular renderings of the outlaw, are likely to have been anachronistic relics of a distant past by the time the ballads were composed.[181]

Rather, the greenwood, in addition to providing a safe haven for Robin and his men, functions in the ballads mainly as a place where the corrupt justice of the outside world can be corrected and overturned into its rightful order. This is in spite of the fact that Robin has been declared by the king's courts as an "outlaw," that is, literally outside the king's law with a price on his head, or a "wolfshead," since it was the same bounty as that for a wolf. Thus, in the eyes of officialdom and other "law-abiding" citizens, as represented by the characters of the sheriff of Nottingham or Guy of Gisborne, the forest is a lawless landscape serving as a refuge for the most hardened criminals; meanwhile, from the perspective of the main protagonists, which was presumably that of the medieval audience as well, the forest was where true justice, honor, and virtue reigned. Therefore on the whole, the forest has a very positive connotation within the thought-world of these ballads. In addition, some of the ballads, which mostly date to the fifteenth century, evince an appreciation for the woodland greenery for its own sake. For example, here's how the ballad, *Robin Hood and the Monk*, preserved in a late fifteenth-century manuscript, sets the scene in its opening stanzas:

*In somer, when the shawes be sheyne,*
*And leves be large and long,*
*Hit is full mery in feyre foreste*
*To here the foulys song:*
*To se the dere draw to the dale,*
*And leve the hilles hee,*
*And shadow hem in the leves grene,*
*Under the grene-wode tre.*

In summer, when the woods be bright
And leaves be large and long
It is very merry in the fair forest
To hear the birds' song:
To see the deer draw to the dale,
And leave the hills high,
And shade themselves in the green leaves,
Under the greenwood tree.

Undoubtedly, these were judged to be popular sentiments that would resonate with the audience and hold its attention as a way to start off the ballad. As we will see in Part III, hunting was a popular pastime that only got more so as the fifteenth century came to a close; moreover, there were real-life outlaws such as the Folville and Coterel gangs who operated in the Midlands during the fourteenth century and who earned an admiring reputation as "fierce, daring, and impudent" men.[182]

Of course, there is another side to this story, one where the forest was the hideout of far less genteel robbers than Robin Hood who terrorized hapless travelers with their brutal violence. This is why periodically throughout the thirteenth century the crown ordered the paths through its forests be recut and recleared of new growth, so that they would present less of a danger to travelers being ambushed by "malefactors."[183] For example, in 1240 the king ordered the sheriffs of Kent and Sussex to see to it that all pathways through the forest be enlarged "due to the dense undergrowth of certain woods in your bailiwicks and the narrowness of the paths along which the common transit of men occurs, whereby no few damages to their goods as well as to their bodies are inflicted upon those traveling along the paths and through those woods, so that [henceforth] opportunities of doing harm may not present themselves to malefactors hiding out in the aforesaid woods."[184] In 1234 the crown ordered the sheriff of Hereford to go personally to the private wood of Roger de Chandos at Wulvivehop to clear the woods on either side of the highway, as it was reported that it was a particularly dangerous place, since "men are often found dead there and throughout the whole day there is imminent danger to travelers."[185] Likewise, in the 1230s the wood of Ogley in Staffordshire had to be cleared of oaks "on account of the number of outlaws" hiding out there, while in 1257 and 1262, the king commanded that the Hopwas Pass in the forest of Cannock and the highway between Maydenhith and Heneley within the forest of Windsor be cleared of undergrowth and widened on account of the homicides, robberies, and other crimes said to be "frequently perpetrated by many malefactors dwelling in the wood," all "to the grave damage of the countryside and to those passing through it."[186] In 1255, Henry III ordered a trench to be cut along twelve miles of road running through the Forest of Dean from Newnham to Monmouth, entailing the clearing of 288 acres of trees, and in 1236 and 1238 he issued a call to arms in the counties of Berkshire, Essex, and Wiltshire in order to root out malefactors said to be roaming the forests there or seeking refuge in them during the day.[187]

Yet another genre of literature on the forest during the Middle Ages are the works on natural history that treat of trees, in the tradition of Theophrastus or Pliny. Thus the early medieval encyclopedists, Isidore of Seville and Rabanus Maurus, in book 17 of the *Etymologies* and book 19 of *De Universo*, discuss the general nature of trees, listing their constituent parts such as the roots, trunk, bark, branches, leaves, etc., and they then proceed to explain the various types of trees and their uses for man. Isidore also discusses various names for groups of trees, such as shrub, thicket, grove, glade, and so on, and he explains some aspects of tree care or manipulation, such as allowing for second growth after clearing (*recidivum*) and grafting of young saplings.[188]

Later natural historians from the thirteenth century, such as Vincent of Beauvais, Bartholomaeus Anglicus (Bartholomew the Englishman), and Thomas of Cantimpré (along with his fourteenth-century redactor, Konrad of Megenberg), cite Isidore for much of their information on trees in, respectively, the *Speculum Maius* (Great Mirror), the *De Proprietatibus Rerum* (On the Properties of Things), and the *Liber de Natura Rerum* (Book on the Nature of Things), although they also cite plenty of other authorities, including Aristotle, Galen, Pliny, Dioscorides, St. Ambrose, Matthaeus Platearis, and Avicenna. There is also the late twelfth-century natural history by Alexander Neckam, the *De Naturis Rerum* (On the Natures of Things), which nonetheless has little new to say on plants and trees, dealing with the subject in 29 brief chapters. Neckam considers a couple of general questions, such as "Why do plants grow green?" and "Why do plants of contrary effects grow in the same earth?" The first question he answers by explaining that plants derive their nourishment from earth and water, in which earth, a cool and dry element, produces a black color, and water, a cool and moist one, is white, so that green is in between the two. The second he answers with the help of Aristotle, Virgil, and Ovid, explaining that the earth contains within it various elements, such as "fiery" and "airy parts," which produce both hot and cool plants. In the section on trees he discusses their various kinds, dwelling especially on fruit trees, such as apple and pear. He also explains grafting and how, if one is to cut into a tree and still wishes it to survive, one should do so during the half moon when there is less sap and consequently less chance for corruption, whereas during a full moon the sap is flowing.[189]

Beauvais follows Isidore in discussing the various parts of trees including the roots, bark, branches, leaves, flowers, and fruit, but he also has a much more extended section on the cultivation of trees (the fruiting kind in particular), including their germination from seed, transplantation, and grafting, as well as the various diseases to which they are susceptible and the harm that can be caused to them by men and beasts, such as debarking. Based on Ambrose and others he also explains how to distinguish trees by sex and age before discussing various specific trees and their "virtues" or uses, particularly in medicine.[190]

Anglicus, following Aristotle, tries to distinguish between trees and beasts, noting that "in trees is the soul of life ... but therein is no soul of feeling." Thus, a tree "feels no sore [hurt] when it is hewn or cut," even though some philosophers, such as Anaxogoras (c.500–428 B.C.E.), may maintain that they do, which "Aristotle reproves." Likewise, others maintain that "tree be more perfect than beasts" because they can generate, feed, nourish, and grow on their own, without any apparent waste products such as urine or shit. Here again Aristotle "reproves" with the argument that trees are bound to the soil and cannot move, nor can they see, hear, or use any other senses; thus beasts are more "knowing and complete" living beings because they have "more noble workings than a tree." In trees and other plants, on the other hand, the "virtue of life" is hidden and so they have "but a part of a part of a soul." Nevertheless, trees have humors, just like beasts and men, which can be seen when they lose their fruit or leaves, a sign of the humor feeding the tree getting "thin and scarce," the same as when humans lose their hair or nails. In great trees the humor is described as "milky," but in small ones it is "sticky" and unable to spread itself very far "in length and breadth," which better suits their size.[191]

Cantimpré, in his general introduction, follows Isidore in distinguishing between trees that grow from seed and those that are grafted, but he provides a much more detailed discussion of the latter. He also addresses certain "questions" about trees, such as why do some leaf out earlier than others, or why does the fruit of some apple trees have a hairy, rough, or wrinkled appearance, which he answers by referring to the influences of heat or moisture in their growth.[192] Beauvais, Anglicus, and Cantimpré all copy Isidore in dividing trees between those of the "common" variety, such as pine, oak, ash, birch, poplar, and maple, and those that are "fragrant," "aromatic," or "fruiting," such as aloe, cinnamon, cloves, myrrh, and nutmeg, although some of these we might classify more as plants or shrubs rather than trees proper. However, these later authors dwell far more on the medicinal properties to be extracted from trees than do their predecessors, who were more focused on the etymology of their proper names.[193]

We should not omit from this natural history on trees Albert the Great's commentary on the Pseudo-Aristotelean *De Vegetabilibus* (On Plants).[194] This could almost be considered a grand summation of all other entries on plant lore encountered in the encyclopedias. Albert discourses on trees and plants in no less than seven books, the first of which addresses some general philosophical questions on the existential status and relation of plants to other living things, along the lines also covered by Anglicus (Albert's almost exact contemporary). Thus we are treated to questions such as: Are plants alive; do they have souls or feelings; do they sleep during winter; are they superior to animals? Albert answers these questions much like Anglicus, although in considerably more detail. He concedes an animated principle and life, albeit hidden, to plants in that they nourish themselves, grow, and reproduce, in addition to having a life cycle much like humans and animals in that they eventually mature to an old age and die. However, he dismisses the claims of some ancient philosophers, such as Anaxagoras, Democritus, and Protagoras, that plants have intellect, intelligence, sensation, desire, or appetite as simply "absurd and unseemly" (*absurda et foeda*); so too he concedes to plants only "part of a part of a soul," not a complete one. Likewise, the opinions of some that plants are "more perfect" than animals or that they can sleep is declared to be "perverse" (*abusiva*). Although plants seem to have self-contained powers of nourishment, reproduction, regeneration, and evacuation, they lack, as Aristotle points out, animals' abilities of motion, individual form, and sensory input—the last of which is also a perquisite for differentiating between waking and sleeping, along with a complex digestion. Moreover, plant reproduction from seeds is entirely passive, unlike active animal copulation, which is capable of producing a distinctive type after generations of breeding, and there is the added fact that plants were created as food for animals, not the other way around. This question has obvious parallels with the proposition that brute animals are superior to men, since they can walk and eat from birth whereas human children are born helpless; but such a point is raised only by ignorant "common folk," who forget that what seems to be weakness in humans comes from a "more subtle [e.g., celestial] humor" that gives rise to our "noble workings of the senses and intellect," while animal potency is derived from an "ignoble, hard, and earthy nature."[195] The rest of book one in *De Vegetabilibus* is devoted to the more mundane details of plant anatomy. In like manner, the remaining books of this massive tome are more straightforward and pedestrian: Book two addresses the basis of plant diversity and the nature of flowers and leaves; book three concerns fruits and seeds and their tastes; book four discusses the four "virtues" of plants, how they reproduce, and their "accidents" or attributes of color and taste; book five describes various methods of grafting and transplantation, or converting wild plants to domestic ones and vice-versa, and plants' four complexions of hot, cold, moist, and dry; book six enumerates the various species of trees and also provides an herbal, which focuses in particular on plants' medicinal properties; book seven provides specific species of plants that are most useful for cultivation and domestication, such as grape vines.

Finally, from just the further side of the Middle Ages, or the early sixteenth century, comes the eminently practical advice on tree care and management from the pen of the Englishman, Sir Anthony Fitzherbert, in his *Book of Husbandry*. While natural histories and earlier estate management guides, such as the *Liber ruralia commodorum* (Book of Rural Benefits) by the early fourteenth-century author, Pietro de Crescenzi, had mainly discussed care and management of cultivated trees (such as how to transplant and graft fruit trees),[196] Fitzherbert gives us a rare glimpse into the everyday world of the woodsman in terms of exactly how trees and branches were harvested with a view to maintaining a healthy and productive forest, and how wood was marketed once felled. With regard to coppicing, Fitzherbert's advice was to fell underwood in winter, when it would do double duty as browse for cattle and as firewood once dressed in this fashion (although Fitzherbert also mentions the downside of woodland grazing, namely, that the animals can get ticks). For the same reason, winter was also the time when one should cut lops and tops off more mature trees: Timber trees should be cut three or four feet above the usable timber line, but other trees and hedge rows at 20–30 feet high (which is as tall as a 2–3 story building!). Fitzherbert explains the rationale behind this, in that "a tree hath a propertye to growe to a certayne heyght, and whan he [i.e., the tree] commeth to that heyghte, he standeth styll and groweth noo hyer but in brede [breadth], and in conclusion, the toppe wyll dye and decrease, and the body thryve." If such pruning were to be done every 12 or 16 years, Fitzherbert promised that the trees would bear more wood and bring "moche more profyte to the owner." As far as pruning technique, Fitzherbert advised cutting the lowest branches first, rather than the usual method of starting at the top, and cutting each branch first on its "nether syde" or underside, so that branches not break or fall in such a manner that the bark is stripped away from the trunk, which will cause tree rot. For this job a "lyghte axe" was best, which would allow such undercutting, and the bough should be cut a foot or two from the trunk so that it not take off bark as it falls. Fitzherbert also warned against cutting when the wind blew from the north or east or else when it was "sappe-tyme," i.e., the spring, when trees might go into shock. All this is advice still given in pruning seminars to this day. As for marketing and selling felled timber, Fitzherbert advised doing it onself or through a "dyscrete" agent, as there must have been much scope for petty embezzlement, and in the case of "small wode" or young trees, it was only worth it to cut them "by the hundredes, or by the thousandes." Felled trees should be sorted by type, such as ash versus oak, and also by size, small, medium, and large, and sold in lots of 10, 20, or 50. A large timber tree should be felled close to the ground, "for one fote next unto the erth is worthe two fote in the top," i.e., trees yield more wood per axe blow the lower they are cut, since they have a greater diameter at their base. In order to keep "sprynge-wodde" or young shoots regrowing from coppiced stumps, the woodsman should fence in the area the winter before he cuts, so that once felled the regrowth can be protected from browsing beasts, except for deer. A forest coppiced in this manner will "save it-selfe," i.e., regenerate, if left for seven years, but Fitzherbert advised waiting at least ten. The "under bowes" or lowest branches of a tree should be pruned, since they will die anyway from being shaded from above and moreover "take awaye the sappe that shoulde cause the sprynge [wood of the rest of the tree] to growe better." Again, this is all sage advice that would not be amiss today.[197]

In addition to the literary evidence discussed above and archaeological excavations of plant remains at human sites, another source of information about medieval greenery is the pictorial record, particularly religious paintings that depict an earthly paradise or the garden of Eden.[198] These give us a cultural context to plants that medieval people especially valued or deemed symbolically important, but they cannot pinpoint a particular provenance for the greenery as can the archaeobotanical record, nor do they tell us about the everyday, practical uses found for the greenwood as do the works of natural history. While it is harder to identify specific plants from paintings as opposed to the accuracy of archaeological analysis or nomenclature in natural history, they do give us a more species-specific guide to cultural attitudes to our botanical surroundings than does the canon of Arthurian literature, which treats of the whole forest as a mythological entity rather than of individual trees.

Clearly, the cultural record tells us that medieval people valued the greenwood for more than just timber, firewood, fencing, or plant extracts. There was a certain, magical ethos to the forest, a land of the other, a wild place where the beasts and the heathen gods roam, an alternative landscape where man could leave his civilized straitjacket behind. One imagines that there will always be room for the forest within our human value system so long as this mental thought-world about the woods from the Middle Ages persists.

*Figure 2.8* TREES, GROWING NATURALLY AND IN POTS, woodcut from Konrad of Megenberg, *Buch der Natur* (Augsburg, 1481). U.S. Library of Congress.

## A renaissance in regrowth of the forest?

It would be reasonable to assume that, with the advent of the Black Death in 1348–49, with an average death rate of 50 percent among Europe's population, this would represent a corresponding "reprieve" for the forest, as the trees advanced in the wake of the retreat and demise of human intrusion and demands upon the woods. A Lübeck doctor writing in 1411 made this very observation in his treatise on the plague, asking, "Where will the earth be cultivated, since the old, in whom is the light of knowledge and experience, are absent and the young are not to be found?" Quoting from the Old Testament, this observer concludes that his country must then necessarily "resort to a wasteland, where the ferocious beasts, wolves, and lions and dragons reign supreme in the region."[199] Such a resurrection of the wilderness is apparently just what happened in the forest of Hofgeismar along the Weser River in central Germany, where by c.1430 the forest had regrown to nearly what it had been in c.500, or in the Saintogne region of west-central France, where depopulation caused by the Hundred Years War gave rise to the saying, "The forests came back to France with the English."[200] In the immediate aftermath of the Black Death, properties lay vacant due to lack of tenants available to pay their rents, a fact starkly recorded by manorial court rolls.[201] Nor did human populations recover quickly from the plague, since the disease returned periodically throughout the latter half of the fourteenth and throughout the fifteenth centuries, keeping growth trends flat until the early modern period. Within England alone, it is estimated that 2000 villages were deserted in the course of the late Middle Ages (often with the only evidence for their former existence indicated by aerial photography), while thousands more were depopulated and reduced in size, as lay subsidy rolls amply testify.[202] On the Continent, best guesses are that between one-fifth to one-fourth of human settlements were abandoned and as much as a quarter of arable land lost to cultivation.[203] Surely the forest should have been the beneficiary of this human catastrophe?

The real picture is inevitably more complex. In England it has been found that so-called "champion" lands, i.e., those villages lying in fertile, open valleys or plains that were long denuded of any native vegetation, as opposed to woodland settlements that were recently carved out on the margins of the forest, bore the brunt of late medieval desertion. Moreover, we have seen that the process of assarting during the high Middle Ages was usually a very piecemeal, haphazard process depending on the decentralized initiative of local peasants rather than an organized enterprise directed by a lord or monastery, and the same was undoubtedly true of the reverse phenomenon of deliberate abandonment and reforestation. Although a woodland community might be designated as "marginal" by historians, its wood pasture economy could be quite as productive and prosperous as the more typical arable landscape, and such pastoral economies were better able to weather the ravages of the Black Death, since they required less human labor to maintain and could satisfy the growing demand for meat as peasant diets changed in response to more access to disposable income.[204]

Overall, every region in England experienced to varying degrees some reduction of arable land and decline of settlements, but this took place over a long period of time that began well before the Black Death and continued well into the fifteenth century. Other environmental factors besides plague played a role in this depopulation and desertion, depending on local conditions. In the North, Scottish raids across the border and widespread animal murrains during the Great Famine were important contributing factors; in East Anglia, South Wales, and the south-east of England, coastal flooding from sea storms and heavy rains played their part. Nor did the widespread conversion to pasture automatically mean of the woodland kind: Unless managed in such a way as to protect trees, such as by pollarding, pastures with sheep and cows on them could keep the woods at bay just as much as the plough.[205] Much of the evidence for forests growing back comes from areas where this was deliberate policy, through enclosure, because a woodland economy was now more lucrative in the scarce labor market after the Black Death than a cultivated one. By the fifteenth century, we even get cases of landlords delivering the *coup de grace* to long-dying villages, through eviction of the few remaining tenants, in order to create enclosures that might put the land to what was perceived as more productive use, a sign of trends to come.[206]

The case of the woodland village of Hanbury, on the edge of the Feckenham Forest in Worcestershire, illustrates some of the pitfalls in assuming a straightforward correlation between population decline and tree regrowth. Although Hanbury certainly did experience its share of human mortality during the Black Death as well as desertion of homes and fields, much of this was mitigated by a shift from an arable to a pastoral economy and by the concentration of larger landholdings in fewer hands, typically from the yeomen or gentry class. By the fifteenth century, new farms and cottages and even assarts were being established to make up for old ones lost.[207] Inevitably, some of the forest grew back in Europe during the late Middle Ages as the result of the decline of population pressures upon the woodland landscape, although at the same time, the advent of the Little Ice Age and cooler weather may have inhibited some of this regrowth. Exactly how much recovery there was of the forest that had fallen to the ax is almost impossible to tell.

By some reckonings, the Middle Ages, sandwiched between the ancient and modern periods, was the least destructive of the three to the forest, despite a relatively rapid increase in population, at least until the fourteenth century. Most devastating to the historical woodland landscape has been our own, contemporary time since 1945, especially when modern forestry practices advocated plantations of single species such as conifers that have resulted in a depressingly uniform tree topography. Close behind was the enclosure movement and Agricultural Revolution of the eighteenth and nineteenth centuries.[208] Whether we should attribute the medieval "achievement" in not destroying too much of the forest to wise management practices or sheer dumb luck of the Black Death (or to a little of both) we'll never know. What we can say is that probably the people of the Middle Ages were deeply appreciative of their woods, for they were heavily dependent on the many and varied products to come out of the trees. For them, the forest was not simply a romantic idea, but a vital landscape without which life would scarcely be possible.

# Part III

# Beast

Humans have been living side-by-side with beastly companions for thousands of years, going back to perhaps the very origins of the human species itself. The monumental depictions of animals such as bison, horses, aurochsen (wild cattle), and deer in prehistoric caves like those of Lascaux in France or Altamira in Spain certainly demonstrate early humans' awe of and perhaps reverence for animals, as well as a close observation of their behaviors. Some of these animals, especially those orphaned at an early age, may have been tamed for a few generations. However, only a few species, such as the dog, cat, goat, sheep, pig, cow, and horse, proved amenable enough to be domesticated on a large scale down through the millennia, beginning with the Neolithic period in around 8000 B.C.E. It is thought that what impelled this new relationship with animals probably had something to do with climate change, as the advent of a more inhospitable environment restricted the availability of wild game and forced humans to hoard their food supplies in the form of livestock. A sudden growth spurt in human population may also have necessitated a more steady and reliable source of sustenance. Yet domestication is a labor-intensive enterprise that was probably not the first choice of hunter/gatherer societies as a particularly easy or nutritious way to get their next meal. Moreover, once started down the path of domestication, man has ever since been caught up in an interdependent, cooperative cycle with his fellow creatures that perhaps will eventually change the very course of evolution itself, as even beasts that are now seen as living independently off in the wild will also come to be more intimately entwined with human beings if they are not to become extinct. This view sees domestication as more a case of certain animals "self-selecting" themselves to be tamed by man as an evolutionary survival tactic rather than as a purely human invention.[1]

The dawn of domestication had all sorts of implications for both animals and humans, including the exchange of microbes producing new diseases, changes in settlement patterns and eating habits, and, above all, new ways in how each related to the other. Among the more remarkable of ancient societies in terms of cultural attitudes towards animals was that of Egypt, where they were elevated and incorporated into the pantheon as animistic deities, either in animal form or as half-human, half-animal hybrids. This is aside from

Egyptians' more mundane uses for animals, including as beasts of burden or farmyard denizens (horses, cows, goats, pigs, sheep, geese); as pets (cats, dogs, monkeys, ferrets, falcons, pigeons, ducks); and as objects of the hunt or as a supplemental food resource (lions, leopards, gazelles, antelope, wild bulls and boar, crocodiles, hippos, fish, bees). Abundantly represented in their art, animals are claimed to have been accorded an unusual equality by the ancient Egyptians within their views of creation, especially when compared to the later Judeo-Christian tradition: The hymn to Amon, for example, has the god create in the same breath not only man but also cattle, fish, birds, mice, snakes, gnats, worms, and fleas. It is clear from their religious beliefs that the ancient Egyptians made a close study of animals, both wild and domestic. For instance, the protective and intimidating way in which a falcon might spread its wings or a cobra rear up and dilate its hood made them suitable symbols for royalty, as depictions of the god Horus or as the uraeus symbol on Egyptian headdresses, while the fearsome nature of the lion or of the crocodile that could not help but inspire respect likewise made them desirable objects of appropriation by the pharaohs, either by becoming the body of the sphinx or embodied in the god Sobek. Jackals roaming the desert robbing old meat from graves naturally associated them with Anubis who ruled the realm of the dead, while even the lowly scarab beetle was not beneath notice for the way it rolled along a ball of dung, which became a symbol of the god Khepri pushing the sun across the sky. Some of these animals, such as the cat, dog, jackal, falcon, ibis, horse, and gazelle, were even mummified and buried alongside their human masters, perhaps as intermediaries between men and the gods or just as companions in the afterlife.[2]

The first scientific study of animals came in ancient Greece, with the works of Aristotle (384–322 B.C.E.), who wrote no less than five books devoted to biology, including the *History of Animals*, the *Parts of Animals*, the *Progression of Animals*, the *Generation of Animals*, and the *Motion of Animals*. Some of Aristotle's information came second hand, from other works he consulted or from heresay, but some also came from his own direct observation, including the dissection of some fifty species of animals, including elephants and sharks.[3] Aristotle is considered the "father of animal ecology" or zoology, and he explained such forward-looking concepts as food competition among animals, population fluctuations, territorial behaviors, migration patterns, hibernation practices, and symbiotic relationships. He was also the first to conceive of a classification system by which to organize animals by species.[4] But even though Aristotle took an integrated approach in general towards the natural world, as evidenced in his *Metaphysics*, he still thought in hierarchical or "aristocratic" terms when assessing the intrinsic value of each animal species, with man naturally at the top or apex of the pyramid owing to his use of reason, which for Aristotle was the chief principle behind a state of happiness in humankind, as he makes clear in the *Politics* and *Ethics*. This is held to be in contrast to the modern scientific study of biology, which sees all organisms as occupying a complex and interconnected web of relationships that cannot be teased out into a simple linear order.[5]

There were, of course, other attitudes and approaches to animals in ancient times. Greek and Roman gods could be both protective of their creatures and demand their sacrifice: A good example is Artemis/Diana, the patron goddess of both wild animals and of hunters. She could talk to the beasts and was especially fond of their young and sometimes exacted terrible vengeance for taking their lives, so that some have called her an early kind of game warden presiding over the first game preserve, but at the same time she delighted in the hunt and gladly accepted the proffered slaughter of hunters. Undoubtedly this paradoxical aspect of the goddess merely reflected the conflicted feelings of her human worshippers; these were perhaps reconciled when ancient hunters were expected to propitiate the goddess with a share of their prey and treat it with respect, a custom that many will find reminiscent of Native American societies.[6]

Meanwhile, the Orphic school of philosophy, embodied in the teachings of Pythagoras (c. 570–495 B.C.E.) and Empedocles (c. 490–430 B.C.E.), held that all animals, not just humans, are possessed of souls and of some intelligence, and that therefore humans share an essential kinship with animals whose souls may, in fact, be the reincarnated embodiment of past human lives, a belief system known as metempsychosis. The Orphics consequently urged abstaining from hunting, animal sacrifice, and eating meat in an attempt to return to what was believed to be the true, original state of nature, where no hierarchies existed placing relative values on living beings, but all intermingled in a cyclical world with men and beasts being "tame and gentle to one another."[7] This obviously anticipates much of the ethos of the modern animal rights and vegetarian movements, and likewise has much in common with some Indian religions still adhered to today, such as Jainism.

The Romans are said to have evinced a paradoxical attitude towards animals, showing great admiration and affection for them on the one hand, as evidenced by their commemoration in a variety of artistic media, but on the other quite willing to indulge in sadistic pleasure in the wholesale slaughter of hundreds or even thousands of mostly wild animals in the amphitheater or arena.[8] However, by the late Roman Empire there emerged a fascinating defense of animals that was specifically employed to counter the rising religion of Christianity. This is the work of Celsus, a pagan Platonic philosopher who lived during the second century C.E. and who wrote an anti-Christian polemical work known as *The True Word*, which survives only in excerpts quoted by the Church father Origen in his riposte to Celsus, the *Contra Celsum* of c.248. In book four of *The True Word*, Celsus challenges the notion that humans were given special status and dominion by God over other animals by virtue of their closeness to God's image in terms of rationality, power, or any other quality. Celsus does this by, on the one hand, employing our derogatory view of some lowly animals in order to parody this concept of human exceptionalism: For example, he has worms holding an assembly "in some filthy corner" croak out sentiments very much akin to those of Genesis 1:26: "There is God first, and we are next after Him in rank since He has made us entirely like God, and all things have been put under us, earth, water, air, and stars; and all things exist for our benefit, and have been appointed to serve us."[9] On the other hand, Celsus comes to the defense of animal abilities and points to their more positive qualities when compared to humans. For example, in reply to the argument that God made men "rulers of the irrational animals because we hunt them and feast on them," Celsus responds that once upon a time, men were instead "captured and eaten by wild beasts and that it was very rare for beasts to be caught by men." Even now, men must hunt animals in the company of many other hunters using nets and weapons and dogs, whereas the wild beasts were granted by nature "weapons from the start in their natural powers, making it easy for them to subdue us." Other claims to human superiority, such as through social skills, powers of divination, and demonstrations of faithfulness and piety, are countered by pointing to animals who do the same, such as ants and bees, snakes and eagles, and elephants and storks. Celsus concludes with an Orphic appeal for equality of all God's creatures, each of whom has been uniquely "proportioned" for a particular destiny in the universe, as parts that make up a perfect whole.[10]

Even though Celsus may not have had the championing of animal rights uppermost in his mind when he made this critique—rather his main target seems to have been Christian assumptions of exclusive access to the divine compared to other, pagan religions—his challenge to humans' assumed rightful dominance over other animals was a powerful one. From a purely philosophical or rational standpoint, Origen's reply is less than convincing and hardly adequate in refuting Celsus. The essence of Origen's refutation is that humans are uniquely rational creatures, alone endowed with the capacity to reason, but his main justification for believing this seems to be biblical authority, which from a pagan's standpoint such as Celsus's would simply be a circular argument. Otherwise, Origen argues that man demonstrates his superiority by accomplishing with his intelligence and reason what animals do by pure instinct, and if Celsus wishes to elevate animals above humans, where would that leave his philosophical heroes like Plato, Socrates, or Pythagoras? Man also proves his dominance by being able to tame animals—even those much bigger and stronger than he such as elephants—to do his bidding, while those few cases where animals appear to do things humans can't, such as divination, are actually the work of demons to deceive man that take the form of notoriously "unclean" creatures such as the wolf, fox, serpent, eagle, and hawk. Those who would worship animals, such as the Egyptians, have merely lapsed into pagan idolatry and were rightly rebuked by Moses. But Origen still does not explain why God would have endowed humans with reason over other animals, or why it is better to do something—whether it be hunting or social organization—by rational abilities rather than by instinct? If man's capacity for reason gives him the ability to achieve virtue, justice, love, and goodness, then it has also given him the capacity to do superior evil, as Origen himself acknowledges by subscribing to the Christian doctrine that only a few are found worthy of communion with God. But if reason is what distinguishes man as having been made more in the likeness of God, Christian doctrine must also paradoxically deny that by reason alone man can approach the Logos, or divine mind, as pagan philosophers claimed. If humans believe they have "tamed" animals, could it not also be from the animals' perspective that they have successfully "trained" men to fetch their food, water and other necessities, as many modern pet owners will attest? To say that especially precocious animals are demons in disguise or that to pay them great honor and respect is idolatry is once again merely circular reasoning within a hermetically sealed Christian point of view. All in all, Origen appears flummoxed by Celsus's innovative challenge to human assumptions of exceptionalism, which can be broadened to a wider environmental critique of man's role in nature.[11]

We move on now to treat of animals in the Middle Ages, and within this broad purview I have opted to discuss each aspect of the subject topically, keeping to a chronological order within each section.[12] This will enable me to showcase the myriad ways in which medieval humans interacted with their animals, both wild and domestic, and seems the best fit for the available evidence. One notices that some of these topics seem familiar, such as "Animals on the Farm" or "Animals as Pets," since these have been perennial uses of animals through the ages up to our own times. But some will strike readers as quite strange, such as "Animals on Trial" or "Animals and Magic," that are perhaps unique to the culture of the Middle Ages.

Lately there has been some reassessment of human attitudes towards animals in the Middle Ages, particularly with regard to the assumption that medieval Christianity presumed human dominion over animals on the basis of Genesis 1:26–28.[13] One argument is that most medieval churchmen were not particularly interested in the concept of human dominion over animals or over any other part of living creation, since their main focus was upon the afterlife, or life eternal, rather than what went on in the mortal prison here below.[14] To borrow St. Augustine's idea, the city of God was of far more importance than the city of the world. Otherwise, some early attempts to establish human dominion over animals, such as Pope Gregory the Great's contribution to the 345th canon of the Church, can appear contradictory. Gregory argues that God's command to Noah to "be fruitful and multiply, and fill the earth" (Genesis 9:7) implied the codicil, "so that fear and dread of you may be upon all the animals of the earth" (Genesis 9:2). Elaborating further on this pronouncement, Gregory comments: "Man is indeed by nature lord over brute animals, but not over other men, and therefore it is said that men ought to be feared by animals. Wherefore it is necessary that rulers be feared by their subjects, when they apprehend that God is least to be feared by them."[15] If Gregory tries to prove man's rightful dominion over animals by comparison to human relations of power over each other, this then seems precluded by the very nature of man's lordship over brute beasts, which explicitly excludes other men.

Instead, most early Christian authorities drew distinctions between humans and animals initially as a repudiation of classical pagan attitudes that were seen as posing too close an affinity among the species: A good example of what was found objectionable is the Greek myth whereby Zeus instructs Prometheus to fashion more humans out of animals as a way of balancing out creation, except that the new humans still retain the souls of beasts. Throughout the Middle Ages, medieval philosophers, from St. Ambrose and St. Augustine in the fourth and fifth centuries to Albert the Great and St. Thomas Aquinas in the thirteenth century, differentiated humans from animals largely on the grounds that only the former possessed reason and had souls, or the divine spark, which allowed them to enjoy an afterlife and enter the heavenly realms, from which animals were excluded.[16] Any complex or precocious behavior observed in animals was put down to *estimativa*, a sixth sense that we might call instinct (and which animal lovers today argue is just as much evidence of intelligence as more ponderous human reasoning).

Medieval scholars derived their concept of *estimativa* from the eleventh-century Persian philosopher, Ibn Sina, or Avicenna, whose original term for animals' abilities to make value judgments or acquire learned behaviors was "intention"; in turn, Avicenna and other medieval philosophers owed much of their thinking about animals to Aristotle, who attributed powers of sensation, imagination, memory, and appetite to animals as analogues of human intelligence but without equating the two, or indeed suggesting that the species shared any of the same mental abilities. (Avicenna's ideas were expressed in a commentary on Aristotle's *De Animalibus*, translated from the Arabic into Latin by Michael Scotus in the early thirteenth century.) But despite being armed with Avicenna's theory of intention, the scholastics of the high Middle Ages did not quite know what to do with the conundrum of animal intelligence. The thirteenth-century English Franciscan, Roger Bacon, comes closest perhaps to allowing to animals the ability to reason, namely, to recognize universal truths and distinguish them from individual exemplars, although he still maintains an artificial distinction between animal and human intelligence on largely arbitrary grounds. Albert the Great devotes the most attention to the problem in book 21 of his own commentary on Aristotle's *De Animalibus* (Books on Animals), where he claims that humans alone have achieved perfection in terms of their soul, physical form, senses, and the use to which they put their bodily organs. However, he does have to admit other types of perfection to other animals, which in descending order include human pygmies, monkeys, quadrupeds, birds, fishes, reptiles, insects, and worms or vermin. Pygmies are the closest to humans in having the power to speak and learn by experience, even though they still lack the capacity of "drawing out" universal principles and thus possess only the "shadow of reason." Other animals, from monkeys and four-footed beasts to birds (especially those that imitate the human voice) and even fishes and reptiles, possess some degree of "teachability" or proclivity to being trained, largely on the basis of *estimativa*. Only insects or "ringed creatures" and vermin lack the ability to be instructed by human example, due to their incomplete sensory faculties, such as sight or hearing. Finally, Albert's fellow Dominican, Thomas Aquinas, differentiates between animals and humans on the basis of the latter's possession of free will, which allows men to resist their natural instincts, such as a proclivity to sin, whereas beasts are incapable of doing so and therefore behave in an almost automatic or predictable fashion, an idea that was taken up later by the sixteenth-century philosopher, René Descartes.[17]

It was therefore on philosophical grounds that medieval Christians elevated humans above animals rather than on the basis of biblical authority alone.[18] An interesting point of discussion in this regard is provided by the thirteenth-century encyclopedist, Thomas of Cantimpré, in the *Liber de Natura Rerum* (Book on the Nature of Things). In his general introduction with which he opens book three on four-legged animals, Cantimpré, perhaps in a nod to Celsus, asks the "playful" question of why man, since he is "more dignified" (presumably because of his reason) than all other animals, does not have some of their weapons, such as a boar's tusks, a lion's claws, or a bull's horns, or else their easy ability to flee, like the deer and the hare, by which "they avoid danger" and preserve themselves. One imagines that this is just the sort of question that could have arisen among a lecturer and his students in the schools. Cantimpé's answer, like Origen's, is to once again appeal to the presumed advantage that a human has as "a rational animal and therefore a social one, who is suited for two operations, which I call action and counsel. Thus in the business of war his daily conversation is of his desire to take up arms, but in time of peace he sets this aside." If man had been born with weapons like horns that he could not lay down at will, he would lose this rational flexibility when it came time "to negotiate a peace treaty." Likewise, on the other side of the coin, if he were able to flee at the slightest provocation and lost his constancy of purpose, such as "when it is necessary that he take up arms and forego entreaties for peace," then he would be reckoned among the foolish and the weak.[19] Such philosophical questions on man's relationship with animals continued to exercise learned men down into the late Middle Ages. For example, the fifteenth-century Lucerne physician, Johannes Hartmann, stated that "the soul of a rational [man] occupies a certain mid-way point between the souls of brute beats and [the souls] of heavenly [beings]." He made this statement by way of demonstrating that medicine and human health also should remain in a balanced state or aim for the mean, in accordance with Aristotle's philosophy, but with regard to men's souls vis-à-vis animals, he referenced the twelfth-century Moorish philosopher, Averroes, who apparently argued that "insofar as one is to reckon plants to be of a kind with animals," so too should "animals be of a kind with men."[20] Animals and humans do share a kindred spirit, according to Hartmann, in that both are created by the same God with the breath of life, but this can only go so far.

Nevertheless, it is argued that a paradigm shift in human attitudes towards animals emerged in the high Middle Ages. Beginning in the twelfth century, mental barriers separating humans from animals broke down, as people began to see more of the beast within themselves, and this, arguably, is the start of modern attitudes that grant almost equal rights to animals.[21] However, the evidence for this change in attitudes mainly comes from literature, such as medieval bestiaries and fabliaux, especially in legends of half-human monsters and shape-shifting creatures, such as werewolves.[22] It remains to be seen how much of the thought-world either of the philosophers who saw a rational and spiritual exceptionalism for humans or of the storytellers who elided the differences between species as they spun tales of fabulous hybrids was reflected in the real world of practical connections humans had with actual animals. For it is the latter with which we are primarily concerned here, both for the sake of brevity as well as consistency with the rest of the book.[23] Did the ordinary, day-to-day dealings of humans with their animals evince any seismic change in attitudes, and if so why did this change come about and how significant was it for the subsequent history of human relationships with animals? This is the theme that will be pursued through the various topics covered in Part III.

## Animals on the farm: the Early Middle Ages

Farmyard animals were among the first creatures to be domesticated in the Neolithic age: The most common of these (in the West) include sheep, goats, cattle, pigs, horses, donkeys, and chickens. They are usually descended from a single progenitor, in which Neolithic man managed to herd animals from the wild or else reared young orphans in captivity and then kept them in a perpetually juvenile, dependent state (known as neoteny). Out of all the animals available in the wild, however, only a very few meet the criteria deemed necessary for being "pre-selected" for domesticity, which includes being hardy, easy to tend, breeding freely, and having "an inborn liking for man."[24] This I know all too well from my own personal experience as a wildlife rehabilitator, where 70–80 percent of the wild animals I receive do not make it, as it seems nature intends. (Typically, the wiser but brutally honest animal parents have rejected their terminally ill offspring as lost causes.) Man's intervention here is an intrusion upon realms he clearly was not meant to trespass.

Sheep and goats were among the earliest animals to be domesticated, between 11,000–8000 B.C.E., based on remains uncovered in archaeological excavations: As in nearly all cases of formerly wild animals reared by man, a distinguishing feature of domesticated livestock is the smallness of their bones compared to their still wild cousins (an exception to this rule being the horse). In behavioral terms, sheep and goats are particularly well suited to domestication because they restrict themselves to a home range, do not aggressively defend their territory, are socially gregarious creatures who naturally congregate around a herd leader, are not as nervous or fleet-footed as other ungulates or hoofed mammals, and exhibit a submissive posture to man (at least in the Eurasian variety of sheep but not in the North American species). This explains why sheep and goats were domesticated and not other potential candidates such as deer, ibex, gazelle, and antelope, who lack these qualities (even though the Egyptians certainly tried). Sheep and goats have the further advantage of providing many uses to humans: clothing, meat, milk, tallow, etc., as well as helping to clear and fertilize scrubland through their ability to browse.[25]

Close behind was the domestication of the pig, descended from the wild boar, in around 9000–7000 B.C.E., which in Europe was to become a prime food animal during the Middle Ages. In their biology and behavior, pigs are very amenable to humans, much like dogs, and can easily be fed either on oak and beech mast on the forest floor or on scraps in the pigsty. They are also prolific breeders, giving birth to ten or more in a litter, and their ability to root may have been helpful in planting or finding edible forest products such as truffles. Cows and oxen, descended from the aurochsen, perhaps took a little longer to tame than sheep or pigs, especially given the intimidating size of the wild ox (extinct since the seventeenth century), which could reach six and a half feet at the shoulder and weigh in excess of 2000 pounds. Nonetheless, domestication of cattle was well worth the effort, since they are among the most versatile of animals in terms of their uses for man, perhaps even more so than sheep or goats: Nearly every single part of a cow can be used or consumed, providing meat (including marrow bones and organs), milk and other dairy products, leather hides, tallow and glue (from their hoofs), and so on; but in addition to this, their other main purpose in life (much more so for our ancestors than for modern Western societies) has been as draught animals to pull carts and ploughs. However, the milking of cows, which remains a chief function of the animal to this day, probably took many centuries to evolve, due to the challenges both of absorbing lactose by the human body and of milking a female cow by hand. Cattle also were important as sacrificial animals in religious rites, along with sheep and goats.[26]

Horses and donkeys were among the last animals to be domesticated in the Neolithic age, around 5000–4000 B.C.E. Initially, wild horses were hunted by Paleolithic man for their meat and hides, but because of predation and climate change, along with the steady advance of the forests, horses were pushed out of their usual ranges into the open steppes of central Asia (where they were to become the main ingredient for the rise of the Mongol Empire during the Middle Ages) and had to be re-introduced into Europe and North America by humans. Based on the archaeological record, horses continued to be used for food after domestication, but their main purpose soon came to be as a pack and draught animal, replacing or supplementing cattle in this capacity by 2000 B.C.E., and, eventually used as a mount for human riders after 1000 B.C.E. Since horses were then bred for speed rather than for meat and fat, domestic horses did not end up smaller than wild ones; indeed, it was often the other way around. A basic difference in body type, however, did emerge between North and South, with the shorter, stockier, and hairier Celtic ponies of Europe (the so-called "cold bloods") ending up quite different from the taller and sleeker Arabian-style horses ("hot-bloods") of North Africa and the Middle East, even though all types originated from a single wild progenitor. The donkey, which originated in Africa, mainly came to be used as a beast of burden and also for interbreeding with horses in order to produce the mule.[27] Chickens, prized apparently as "recreational entertainment" for their fighting abilities just as much as for their meat and eggs, were first domesticated in Asia around 6000 B.C.E. and later came westward through India, reaching Egypt by the second millennium.

Almost all these animals were known in most ancient societies, including those of Mesopotamia, Egypt, and ancient Greece and Rome. An inscription from the reign of the Assyrian king, Tiglath-Pileser I (c. 1115–1077 B.C.E.), speaks of "droves" of horses, cattle, donkeys, mules, and lambs carried off from various conquered peoples as tribute to be settled back in Mesopotamia, while wall paintings from ancient Egypt show tax collectors assessing herds of domesticated animals which included sheep, goats, cattle, pigs, ducks, and geese, as well as also depicting oxen hitched to plows and horses pulling chariots. It is clear from these paintings that the Egyptians had managed to develop distinctive breeds of some animals, including sheep, cattle, and dogs. But it was the Romans who were to go on to develop the fine art and science of breeding and who are commonly credited with producing sheep and cattle of a size not seen again until the Agricultural Revolution of modern times; it was also they who probably introduced sheep, goats, and mules into northern Europe from Mediterranean regions.[28] Agricultural works by such Roman authors as Columella (4–70 C.E.) and Varro (116–27 B.C.E.), both of whom wrote a definitive tome entitled *De Re Rustica* (On Rustic Matters), apparently still repay reading by modern students of farming and livestock. Columella was said to have been inspired by his uncle, Marcus, who conducted experiments on crossbreeding domestic sheep with wild rams imported from Africa. However, it was left to the Macedonians, who bordered on the Scythian kingdom in present-day southern Russia and whose famed horsemen perhaps gave rise to the myth of the centaurs, to fully exploit horsepower, especially in war. Farm animals were also an integral part of the pagan cults of Celtic peoples in northern Europe throughout the Roman era, either worshipped as animistic deities or employed in sacrificial offerings and burial deposits. Particularly important were the bull and the horse: The former symbolized in its wild state (the auroch) strength, ferocity, and virility, while its domesticated cousin meant agricultural wealth and draught power. The horse was a symbol of prestige among the aristocratic warrior elite and was worshipped in the form of Epona, goddess of fertility and horse breeding, who also had associations with healing and death, sometimes depicted holding a key to both the stable and the underworld. Epona was usually shown mounted on a mare with foal, while the figure of a male horseman was used to depict various war, sky, and solar gods. Evidence for the veneration of the bull and the horse comes in the form of bronze figurines, archaeological excavations of their bones (where their complete skeletons attest to their special status as whole offerings instead of being butchered like pigs and sheep), monumental cultic sites such as the famous White Horse of Uffington in Wessex, England, and mythological legends, such as that of the Brown Bull of Ulster told in the *Cattle Raid of Cooley*, where the kingdoms of Connacht and Ulster fight over an anthropomorphic bull capable of human reason. Lesser cults existed to other domesticated species or their ancestors, such as the boar, ram, goat, rooster, goose, and duck.[29]

In the early Middle Ages, Germanic and Anglo-Saxon societies favored cows, sheep, and pigs. Centuries earlier the Roman historian Tacitus noted that curdled milk was a basic foodstuff among the Germans and that gifts and fines were paid in the currency of heads of cattle. There are in fact several sources of information available on the farm animals kept by early medieval people, despite the reputation that this period has as being the "Dark Ages" in the aftermath of the fall of the Roman Empire. One piece of evidence that has been exploited by modern archaeologists and historians alike are the actual physical remains of early medieval animals recovered from sites in northwest Europe, such as northern Germany and France, Holland, and Britain, as well as in Italy.[30] As a general rule, excavations from early medieval settlements indicate that Europeans were deriving the vast majority of animal protein in their diet from domesticated livestock as opposed to wild game from the hunt, since the former comprise over 90 percent of the bones recovered while the latter make up 2 percent or less at all sites until the mid-ninth century. Over time, the average age of cattle bones gets progressively older, showing that cows and oxen were being kept more and more as draught animals and for milking, except for a small percentage of veal calves that were slaughtered as surplus animals that could not be overwintered. Pigs were exclusively meat animals, generally slaughtered at between one to two years of age, when they had reached their optimum growth, and they grew in importance in rural communities as there grew more mouths to feed, especially near forested areas where ample pannage was available. The Anglo-Saxon period of English history is generally thought to have represented the heyday of pig husbandry, coinciding with the widespread availability of forest pannage. Sheep, which vastly outnumbered goats, transitioned from primarily a meat animal to one allowed to mature for its wool, milk, and possibly fertilizer production by the eighth and ninth centuries. Horses survive in relatively few numbers at most early medieval sites, indicating that they had not yet attained the crucial role they were to play later in farming activities; however, in northern France their numbers double by the end of the early medieval period, indicating their increasing importance, and the circumstances of their burial both here and in East Anglia, such as the whole state and advanced age of their skeletons, indicates that they were being used as a draft/transport animal by this stage instead of being slaughtered for food as in the Iron Age. (Eating of horsemeat was proscribed by Pope Gregory III in 732.) Domestic fowls such as chickens (by far the most prevalent), geese, ducks, and pigeons also had a presence at this time, typically at less than 10 percent of bone assemblages, but they make up a greater percentage as the Middle Ages progresses; birds were valued especially for their meat, eggs, and feathers. At Apulia in southern Italy numbers of poultry, judging from their bones, exceeded that of cows, pigs, and horses until the end of the early Middle Ages, and chickens were mostly slaughtered as adult hens. It was said that the village of Villiers-le-Sec, owned by the abbey of Saint-Denis in northern France, had to provide more than a thousand eggs three times a year for the abbey's kitchens.

Cattle bones typically predominate at most early medieval sites, particularly market locations, indicating that they were the most valued farm and meat animal. Yet as we transition from the early to the high Middle Ages, sheep will become the more dominant animal over cows among domestic livestock, pointing to the growing importance of wool, and pig husbandry enters a long, slow decline, thought to correspond with the general retreat of woodland that provided pannage. Nonetheless, the size of some cattle longhouses, with enough stalls to house as many as 300 head during the winter months, speaks to the importance of cows at this stage as a marker of wealth and prestige and as a medium of trade, along with wool. However, most longhouses seem to have had the capacity to overwinter a dozen cattle or less. Evidence from northern France indicates that the Middle Ages experienced a drop by as much as a half in the average size and weight of livestock, including cattle, sheep, pigs, and chickens, as compared to what these had been in Roman times, perhaps pointing to a loss of breeding techniques, although by the late Middle Ages stock size begins to pick up again. In Anglo-Saxon England, cows were of comparable size to what they had been under Roman occupation, arguing for more continuity here in animal husbandry. Towards the end of the early medieval period, beginning in the eighth and ninth centuries, there are signs that crop cultivation was replacing livestock rearing as the main means of sourcing food, anticipating the great arable expansion of the high Middle Ages, even though pastoral farming increased then as well.[31]

The legal codes of the Germanic "successor kingdoms" that emerged in the aftermath of the fall of the Roman Empire by the late fifth century can provide a revealing window onto early medieval attitudes towards farm animals, particularly in terms of the relative values they had to society. That domestic livestock, particularly cattle, horses, and pigs, were "the most highly valued animals in the villages" is indicated by the fact that the codes imposed quite severe penalties on those who confiscated them as pledges for debts without royal approval, evidently because these animals were so crucial for performing the farm work that needed to get done if people were to grow the crops they needed as food or else because the animals themselves were their main food resource.[32] A number of Germanic tribes, including the Franks, Visigoths, Lombards, Alamanns, Bavarians, as well as the Anglo-Saxons of England, the Welsh, and the Norwegians, also set the equivalent of *wergild* (*nēatgild?*), or compensation prices, for killing, stealing, or harming another's livestock, which could include causing abortions in pregnant stock, castrating an animal, or knocking out the animal's eye or horn, and cutting off its ears, lips, mane, or tail.[33] The most detailed of these codes in this regard, the laws of the Salian Franks attributed to King Clovis between 507 and 511, devotes 20 clauses to theft of various kinds of pigs, followed by 14 each for cattle and horses, five for sheep, and a couple each for goats and roosters and hens. This would point to the relative importance of the pig in the Franks' diet, even though a cow or horse was generally valued at more than twice that of a pig of comparable age or gender, perhaps reflecting the dual service that a cow or horse could render as a draft as well as a meat animal. At the lower end of the scale, sheep, goats, and chickens were valued roughly the same, at three *solidi* or less. As we move into the tenth century, towards the close of the early medieval period, the laws of King Æthelstan of England (reigned c.924–39) clearly prize the horse as the most valuable livestock animal, at four or five times the value of an ox or cow, while the sheep has now outpaced the pig at 12 pence versus ten pence, perhaps reflecting a greater appreciation for the sheep's wool.[34]

It is creditably claimed that these law codes demonstrate that early medieval people mainly thought of their farm animals as property, especially in terms of their labor and products useful to man that could be extracted from them. Particularly revealing in this regard are the Visigothic and Frankish laws, which compensated owners if their animals were overworked or abused in another's care or stipulated an additional payment for time lost in using animals when assessing their value should they be stolen. Even in death, animals were contested in the codes, for their hides could be turned into leather for shoes and harnesses, or if young enough, as fine vellum or writing parchment.[35] However, one must not lose sight of the fact that quantification of every individual's worth, even that of humans, was a general principle operating throughout these codes, and so from this perspective virtually everyone "had their price" as a form of property with a certain value to society. Perhaps equally revealing are those aspects of the codes that assign punishments for when animals cause harm to persons or property. The attitude seems to have been that the animal itself was not to be held to blame for the "crime," but rather the human owner. Thus in the Burgundian code it is explicitly stated that any animal that "causes death to man" does so by "chance" or accident and hence does not fall under "the ancient rule of blame" whereby the death is compensated or avenged. Animals who killed other animals, however, were liable to penalties, but these fell upon the owners, who had to hand over their animals to the aggrieved party as compensation. A similar principle applied to animals that were found to have entered upon and trampled crops or vineyards.[36] In the time of the Lombard king, Liutprand (712–44), a case arose whereby a man borrowed a mare with foal for hauling and the colt kicked out and killed a child as they were passing by; when the relatives of the child sought restitution, it was decreed that the owner pay two-thirds of the wergild and the man who borrowed the animals pay one-third, since he "was a rational being and able to warn the child, if he had not neglected it, to watch out for itself, [for] then this accident would not have occurred." It is quite evident from all these laws that animals were adjudged to be without reason, a view remarkably in sync with that of later medieval thinkers such as St. Thomas Aquinas. Yet by the late Middle Ages judges were putting animals on trial and holding them to account for their offenses, which seems directly counter to what their early medieval predecessors in the law had deemed reasonable (see "Animals on trial").

Another source of information about early medieval attitudes towards animals are the works of the first encylopedists, such as the seventh-century Spaniard,

*Figure 3.1* ADAM NAMING THE ANIMALS, from the Aberdeen Bestiary, composed in the north Midlands of England during the twelfth-century. Aberdeen University Library, Scotland / The Bridgeman Art Library

Isidore of Seville. Book 12 of the *Etymologies*, which Isidore devotes to animals, was to serve as the model for later medieval bestiaries and sets the tone by quoting in its first sentence Adam's naming of the animals during creation from Genesis 2:20.

This not only establishes Isidore's general principle of deriving each animal's significance from its Latin and Greek names but also implies man's rightful dominance over other animals by virtue of his reason, since Isidore's very definition of "livestock" is "any animal that lacks human language and form." And yet, Isidore is also driven by etymology to admit that "animal" is derived from *animans* and *animare*, Latin terms that signify "animate beings" that "are animated by life and moved by spirit." This would seem to imply that even non-human animals have souls, albeit on a lower scale than humans. Following the fourth-century precedent provided by St. Ambrose of Milan, Isidore places animals into two basic categories, i.e., those that are tame and those that are wild, or what he calls "livestock and beasts of burden" and what he terms simply "beasts." The former he defines as those animals that are "either suitable for food ... or are suitable for use by humans," namely, that they "assist our labor and burdens by their help in carrying or plowing"; wild beasts, on the other hand, "enjoy a natural freedom and are driven by their own desires—for their wills are free and they wander here and there, and wherever their spirit leads, there they go." In the former category Isidore includes the typical farm animals you might expect: sheep, goats, pigs, cows, oxen, donkeys, and horses; but he also includes animals that today we would more usually classify as wild, such as deer, boar, hare, and buffalo, as well as the ancestors of domesticated cows and donkeys, the aurochs and onager. For Isidore, these latter animals occupy a sort of third, amorphous category, the "quadrupeds" that walk on four feet and are similar to livestock, but "are nevertheless not under human control," being neither entirely wild "beasts" nor "beasts of burden" that can "assist the useful activities of humans."[37]

## Animals on the farm: the High and Late Middle Ages

With the start of the high Middle Ages, an important piece of documentary evidence about medieval livestock comes down to us in the form of the Domesday Book, compiled in 1086 as an inventory of taxable wealth in England. Tallying farm animals in the east and southwest of the country, the survey found that oxen predominated as the work animal or plough beast on most manors, as they had done in Anglo-Saxon times based on charter evidence, with oxen outnumbering horses by an average of 19 to one, with work horses making up about 5 percent of the livestock. Typically oxen were employed in plough teams and for hauling, with an average of eight oxen to a team, while horses were used as pack animals and for harrowing. Anglo-Saxon kings did have their horse studs, and Domesday testifies to some wild herds of probably Celtic-style ponies that roamed the woodland pastures and were used by some estates as their breeding animals, but the Normans greatly "improved" on their stock by importing breeds from Spain, for they greatly relied on a mounted force for their successful invasion in 1066. Domesday Book also records sheep in great numbers: In Norfolk, Suffolk, and Essex, they outnumbered pigs, the second largest pool of livestock, by 4 to 1, a trend that was already indicated in the ninth century on the Carolingian estates, where sheep were beginning to edge out pigs. However, this trend may be distorted by the fact that peasant holdings in swine go unrecorded in the survey, except indirectly in terms of pannage dues.[38] At the height of the medieval period during the thirteenth century, tax assessment records indicate that the average peasant may have owned a dozen sheep, 3–4 cows, and a couple horses.[39] Chickens seem to have been kept by the peasantry in large numbers, despite the fact that they are usually excluded from tax records since they were not counted as moveable goods. Nonetheless, hens and eggs played a large role in the payment of rents and fines in kind: The bishop of Durham, for example, was entitled to receive a total of 2,798 birds and 17,095 eggs from his tenants each year during the late twelfth century.[40]

One of the most important changes in livestock farming that occurred in the high Middle Ages and on into the late medieval and early modern periods was the rise of the horse as an alternative plough and hauling beast to the ox. This trend really began to emerge in England by the end of the twelfth century, when according to manorial records the percentage of horses on large estates rose to 10–15 percent overall, or one horse to every seven oxen, with some areas such as East Anglia, which had light, flat, sandy soils ideal for working with horses, rising to 30 percent; here there was perhaps some cross-fertilization of new techniques from the Continent, where horses seem to have been introduced into plough teams even earlier. On the smaller, peasant farms this trend was even stronger, with horses making up on average 50 percent of stock by at least the fourteenth century, with East Anglia averaging 75 percent; by the late fifteenth century country-wide averages rose to 55 percent of stock dominated by horses and over 60 percent during the sixteenth century, when fully two-thirds of farms were run exclusively with horses. Horses were particularly attractive to the peasantry because they were a versatile farm animal: They could be used for ploughing, harrowing, hauling, pack-work, milling, riding or personal transport, etc.—the only thing an ox apparently could do that a horse could not was offer itself up towards the end of its useful life for food, whereas horsemeat was taboo, at least in England. Horses were also a more efficient work animal, allegedly pulling 50 percent faster than an ox and having more stamina, able to work two hours longer.[41]

Undoubtedly horses were an integral part of the so-called "agricultural revolution" of the high and late Middle Ages, especially as they went hand in hand with technological developments such as the wheeled plough, the collar harness, the swingletree and pivoted axle on carts, and so on. Yet even though horses were associated with the intensive farming systems of the late Middle Ages, particularly in the eastern regions of England and in northwest Europe, it is not at all clear that they were valued primarily for their speed and efficiency, especially as the majority of farms seem to have used them in mixed plough teams with oxen. Rather, their value was in their cost effectiveness, allowing for fewer animals per team and fewer teams overall, translating into less animals to keep and less ploughmen to hire. For example, a two-horse team or a mixed team of two horses and two oxen could do the work of an eight oxen hitch. This also meant that farms could divert more of their resources into non farm-work animals, such as dairy cattle, sheep, or pigs, which in turn could increase crop production by increasing the amount of manure available for soil fertility. An equal, if not more, important consideration was that horses increased accessibility to urban markets from rural areas, since the distance covered by a horse and cart could be four times that of an ox. The typical work horses not used primarily for riding, such as "stots" and "affers," were also a cheap investment for most peasant families, although by the end of the fourteenth century carthorses were costing nearly double that of oxen and plough horses nearly as much. While horses did require more expensive feed, such as oats, when worked intensively, peasant families may have been able to get away with cheaper fodder as their horses stood idle much of the time. Horses also seem to have been less susceptible to disease: The great cattle plague of 1319–21 during the Great Famine seems to have determined more than one farm to switch over entirely to horses. Nevertheless, there were good reasons why some allegedly conservative farms, particularly in the southwest or the north of England, stubbornly held on to the slower all-ox teams, mainly because their heavier, clay-like soils were more suited to a more powerful animal, one that could also pull the higher-capacity two-wheeled *plaustrum* cart and could return value even at the end of its life as meat.[42]

The other big development in high medieval farming, of course, was the rise of the wool industry and trade, particularly in England, which became a leading exporter of wool, especially to the woolen factories in Flanders and Italy. At the height of this trade in the early fourteenth century, England was exporting annually about 35,000 sacks of wool, which probably represented the clip from 9–10 million sheep.[43] Perhaps the biggest threat to this trade were the devastating sheep murrains, especially among large flocks where diseases had greater opportunities to spread (see also the section on "Animals and disease"). Aside from the "liverfluke" that so afflicted flocks during the Great Famine, there was also scab, the sheep pox, and a rather mysterious disease known as the "red death." Liverfluke is caused by sheep ingesting snail parasites as they graze on grass, which migrate to the animal's liver and cause it to degenerate. Scab and the pox, although not potentially fatal, affected the skin and therefore destroyed the sheep's ability to produce wool, resulting in the animal being culled from the herd for slaughter—medieval shepherds usually tried to treat these diseases with "greases" made up of a mixture of compounds including sulphur, verdigris, mercury, tar, and tallow or animal fat.[44] Even in non-plague years mortality in sheep-flocks could be substantial: In the East Midlands over the course of the fourteenth century, deaths of adult sheep, i.e., ewes and wethers, averaged between about 5–17 percent, while those of lambs averaged between 10 and 27 percent. This could mask wild fluctuations from year to year, with some years witnessing a mortality rate of 60 percent or more that practically wiped out the entire flock. With lambing rates averaging between 64 and 90 percent, i.e., at less than one lamb per ewe, this made it nearly impossible for flocks to be self-reproducing and necessitated the added expense of periodic re-stocking: Even on the well-run estates of the bishop of Winchester in southern England, for example, the entire flock of ewes and wethers needed to be replaced every five years. All this made sheep-rearing a risky business, especially in the late Middle Ages when sheep murrains seem to have become more common. One other trend in livestock management that should be mentioned that emerged by the late medieval period is the increasing custom of fattening pigs in pigsties or in town gardens as opposed to forest grazing on oak and beech mast. This undoubtedly was a more efficient, if not necessarily more healthy, way of raising pigs, but for landlords it did lead to precipitous declines in incomes from pannage dues, and illuminations of woodland pannage scenes in books of hours perhaps speak to a wistful nostalgia for a bygone era.[45]

Turning to records of a more literary nature, we come now to what natural histories and husbandry manuals have to say about domesticated farm animals. In general, twelfth-century bestiaries as well as the great thirteenth-century encyclopedias of scholastic authors such as Alexander Neckham, Bartholomaeus Anglicus, Vincent of Beauvais, Thomas of Cantimpré, and Albert the Great rely heavily on the earlier work of Isidore of Seville and other ancient authorities, especially Aristotle, Pliny, Solinus, and Avicenna.[46]

Beauvais stands out for his emphasis on the medicinal properties to be derived from the various parts of domestic animals, which includes their flesh and internal organs, such as the brain, lungs, gall bladder, liver, and kidneys, as well as their horns, teeth, fat, blood, milk, and, in the case of sheep, their wool, which has the added advantage of absorbing and retaining other medicinal compounds. What must strike modern readers as strange and quite revolting is that Beauvais even recommends use of the animals' urine and feces, often on the authority of Avicenna: Thus pig shit drunk with water, wine, or vinegar is beneficial for blood-spitting, pains in the sides (pleurisy), and upper arm weakness, while cow dung is administered for phthisis (tuberculosis) and other lung disorders and when used as a plaster heals sciatica and apostemes behind the ears.[47] We can't assume that these bizarre recommendations were all theoretical. A surgeon's manual by Jean Pitard, physician to King Philip IV of France, lists a number of remedies containing domestic animal products, such as pig lung, horse manure, and split-open chicken applied to the head for lethargy and frenzy; goat and dog milk to induce vomiting for poison; goat dung for wasp stings; ground-up pork for rabid dog-bite; boiled cow meat and broth for intestinal worms; goat and cow milk for diarrhea; boar and pig fat for gout; and goat blood to dissolve kidney stones.[48] Plague treatises advised drinking sheep or horse milk for those afflicted with boils, or applying plasters containing pig, goose, or chicken fat and cow, dog, duck,

Part III. Beast   159

*Figure 3.2* FOUR-FOOTED BEASTS, woodcut from Konrad of Megenberg, *Buch der Natur* (Augsburg, 1481). U.S. Library of Congress

and pigeon dung, that last of which could easily be harvested since many a medieval farm had a dovecote.[49]

Albert the Great, in Book 22 of *De Animalibus*, has an extended entry on horses, most of which is devoted to diagnosing and curing the various diseases to which horses are prone: lesions in the mouth and on the hide and legs (some caused by overwork), strangles, "farcy" or glanders, *infundatura* or what is now called colic (i.e., stomach problems), and lameness or injuries to the hoof, tendons, joints, etc. As in human medicine, Albert prescribes for most of these ills either various poultices or else bleeding, which in horses was done by cutting across the vein "like a saw cuts wood."[50] Clearly, Albert is indebted to veterinary works, which in the Middle Ages were almost exclusively devoted to horse care. Among those Albert could have consulted include the *Mulomedicina* ("Mule-medicine") by the fourth-century Roman author, Vegetius; the *Hippiatrica* (Horse-medicine) compiled in the tenth century under the patronage of the Byzantine emperor, Constantine Porphyrogenitus; and possibly the *De Medicina Equorum* (On the Medicine of Horses) by Jordanus Ruffus, knight-farrier to Emperor Frederick II during the first half of the thirteenth century. In the later thirteenth and during the fourteenth centuries, Jordanus' work was to serve as a model for various "Marshalcy" books, mostly by Italian and Spanish authors; England was not to produce one until the fifteenth century.[51]

Of more practical value to farmers, perhaps, were animal husbandry manuals, usually written by those with experience in the field and specifically geared for those charged with managing agricultural estates. A prime example from the Middle Ages is *Le Dite de Hosbondrie* (Sayings of Husbandry), written in colloquial French (the language of the gentry and upper classes of England) in c.1280 by Walter of Henley, who mentions his own experience as farm bailiff. A famous and oft-cited quote of Henley's is his comparison of horses to oxen as plough beasts: For Henley, cost of upkeep of the animal was just as important a consideration as work efficiency; for this reason, oxen had the advantage over horses since they did not require as expensive feed (namely oats) and shoeing. Henley backs up this statement with detailed price estimates for the upkeep of both animals. In addition, Henley repeats the truism that "when the horse is old and worn out then there is nothing but the skin," whereas an ox can be fattened for the larder or sold for as much as it originally cost. While this is usually taken as reflecting the more conservative farming practices of England, Henley does seem to be in tune with the majority of the country in preferring a mixed team of horses and oxen, combining the surefootedness and speed of horses with the pulling power of the ox. Key to the success of maintaining a healthy herd, no matter what the species—cattle, horses, sheep, or pigs—is making sure to cull the herd once a year, for "bad beasts cost more than good," either because other animals have to take up their slack if they are plough beasts or because they can spread disease, and one should get what one can for their meat before it is too late. Presumably for the same reasons Henley recommends laying out money to treat sick animals. Much of Henley's advice actually concerns the humans who are around the animals rather than the animals themselves: Thus one has to watch for malicious ploughmen who "will not allow the plough [of horses] to go beyond their pace"; for oxherds who steal barley meant for the cattle; for dairymaids who skim off the milk, butter, and cheese; for shepherds who overdrive their flocks when they are angry. It is evident that Henley has closely observed animals, for how else can he tell us of how sheep that are driven out at night in bad pasture come into the manger in the morning starving and choking back their hay; of how pigs delight in digging and lying long in the morning in the mud; and of how oxen benefit from being bathed and curried, since "they will lick themselves more"? While it may be nice to think that Henley always has the animals' welfare uppermost in his mind, there is always in his advice a strong undercurrent of the bottom line, as when he recites the proverb when arguing for investment in their medical care: "Blessed is the penny that saves two."[52]

Of slightly later date (1304–9) is the estate management guide of the Bolognese jurist, Pietro de Crescenzi, who discusses the care and management of domestic animals in book nine of his *Liber ruralia commodorum* (Book of Rural Benefits). Mostly, Crescenzi seems to rely on the agricultural works of classical authors, particularly the *Opus agriculturae* of Palladius, but some of his knowledge may have been gained from personal experience running his estate, the Villa Olmo, outside Bologna. Here Crescenzi not only discusses the usual domesticated species such as horses, mules, donkeys, cows, sheep, goats, pigs, dogs, and chickens, but he also addresses the enclosure or housing and feeding of semi-domesticated wild animals, such as hares, deer, fishes, peacocks, pheasants, partridges, geese, ducks, turtle-doves, thrushes, as well as pigeons and bees in dovecotes and beehives. In addition, Crescenzi discusses the identification and cure of various illnesses in livestock, which Henley left unspecified. Thus, he mentions over 40 different kinds of diseases and injuries in horses; at least five in cattle, not including "many other illnesses" that come from overwork and that are diagnosed when they go off their feed; and goiter, inflammation of the spleen, and fevers in sheep. Other useful topics that Crescenzi covers include the breaking in and training of horses and how to recognize which have good conformation, the stabling of cows during winter, and the pasturing of sheep.[53]

Although technically postdating the Middle Ages, *The Book of Husbandry* attributed to the judge and legal scholar, Sir Anthony Fitzherbert, printed in 1523 and written in the author's native English, must reflect much of late medieval practice. Fitzherbert, like Crescenzi, has as one of his main concerns the treating of diseases in livestock: He identifies six separate illnesses in sheep, eight in cattle, and no less than thirty-eight in horses. His remedy for the "turne" in cows involved a trepanning operation in which a piece of the brain-pan or skull was removed in order to evacuate "a bladder full of water two inches longe and more" that was weighing on the brain, perhaps caused by head trauma such as head-butting or a kick to the forehead. This was as daring and complex a surgery as any performed on humans,[54] let alone animals, but it was apparently done successfully, for Fitzherbert tells us that he has "seen many mended" by the operation, although never in sheep that were similarly afflicted. Sometimes, however, there simply was no cure: In the case of cattle murrain, Fitzherbert's advice was to cull the afflicted animal before it infected the rest of the herd. After flaying the hide and selling it to the tanner rather than taking it home "for peryll that may fal," one should dig a deep pit and cast the beast in and cover it over lest dogs and other scavengers seek out the carrion and also be infected. For charity's sake, one should also cut off the cow's head and stick it "upon a longe pole and set it in a hedge, faste bounden to a stake by the hyghe-waye syde, that everye man that rydethe or goeth that waye, maye se and knowe by that signe that there is sycknes of cattell in the towneshyp." His advice to invest in "severall closes and pastures" for beasts constructed with palings ("quicksets"), ditches, and hedges instead of relying on herdsmen, swineherds, and shepherds was well in tune with the enclosure movement that was taking place in England at this time.[55]

It is clear from manorial records, mostly surviving from England, that herds and flocks of domestic animals, namely cattle and sheep, increased markedly during the late Middle Ages, with much of the initiative and innovations coming from enterprising gentry and yeomen farmers who took advantage of opportunities for land acquisition opened up by the Black Death. We have already seen in Part II how more and more arable was converted to less labor-intensive pasture in the later medieval period, but even so, the increase in herds and flocks was such that many village commons became over-grazed. Private flocks of two or three hundred sheep are now listed in court rolls (where owners were charged with overburdening pastures or trespass), whereas they averaged around a hundred before, and cow and pig herds now numbered between 20–40 as opposed to perhaps a dozen on average prior to the Black Death.[56] Much of this rise in livestock farming was driven by a greater demand for meat, not just among the upper classes but even among the peasantry, who could now afford and expect more meat in their diets. For example, amongst the harvest workers on the manor of Sedgeford in Norfolk, England, between 1256 and 1424 the average meat intake per person went from three and a half ounces per day, representing 2 percent of their total caloric intake, to a pound of meat per day, or 23 percent of caloric intake.[57] Account books for aristocratic households show that the average member during the fifteenth century consumed 2–3 pounds of meat, making up 4,000–5,000 calories per day, which is twice the present-day recommended daily caloric intake for men and women in the West at 2,000–2,500 calories per day. However, meat consumption could vary considerably from household to household, and within a single household it could also vary depending on the season of the year (such as Lent, when only fish was eaten), local availability of foodstuffs, and personal taste of the individual diner, with some privileged few enjoying access to as much as 8,000 calories of meat per day, much of it in the form of more exclusive young and tender veal, lamb, piglet, kid, and poultry (also recommended as medicinal meats by health regimens). In addition household records document a clear trend away from eating pork in favor of beef and mutton, which was perhaps due as much to changes in pig farming, such as from mast to sty feeding, as to changing perceptions of the pig as a food animal.[58]

Part III. Beast    163

With the rise in meat-eating by the average person during the late Middle Ages, more animals were sent to slaughter after being culled from herds, which we have seen was the focus of much advice from husbandry manuals. Rural supply and marketing of meat to urban areas became more prevalent and more commercialized, as village inhabitants seemed willing to sacrifice prime, young stock for sale and make do with older carcasses for themselves. By the end of the thirteenth century, more than 60 percent of peasant farmers' taxable wealth was now in livestock. Much of this shift from "corn to horn" by the late Middle Ages was made possible by the contraction of arable land by as much as 30 percent by the end of the fourteenth century and the increasing use of more efficient work animals such as the plough horse. This freed up more land for pasture and more animals for non-working uses: The roughly 60–40 proportion of working to non-working farm animals in c. 1250 reversed itself two centuries later by c. 1450, representing an astounding 160 percent increase in non-working (i.e., meat, milk, or wool producing) livestock, anticipating the gains to be made in the Agricultural Revolution of the modern period. The rise of the market and of national and international trade also made possible greater regionalization or specialization in livestock farming, as farms no longer had to be as self-sufficient as in the earlier medieval period. Many farms now got their replacement animals from markets, fairs, and dedicated breeding manors, with the animals often traveling distances greater than ten miles to reach their destinations. Cattle droves of 400 head from vaccaries in Wales across the border into England took place on a yearly basis from the mid-fourteenth century. The expense of this transport could often outweigh the animals' worth, but this may have been justified by the larger size of cattle, sheep, and even pigs now being achieved by livestock breeders, who were clearly breeding the animals for meat rather than farm work or wool production, since animals were now being slaughtered at a younger age.[59] On the Continent, great cattle drives by the end of the Middle Ages of 10,000 head or more supplied northern Europe from as far away as Hungary, Poland, and Denmark.[60]

In addition to using more farm animals as food, the one other trend that may have changed how late medieval people related to their livestock were the new ways in which the average farmer housed his animals. After the Black Death, it seems to have become standard practice for nearly all peasant dwellings, even in the case of humble cottages, to house animals separately from their human owners, whereas previously the longhouse had been the rule, with humans and animals housed under one roof albeit at opposite ends from each other. Evidence for this trend comes not only from archaeological sites but also from documentary records, such as building and maintenance agreements negotiated between tenants and their lords in the more fluid rental environment created by the Black Death. Such outbuildings could consist of byres for cattle, sheepcotes, stables for horses, and dovecotes for pigeons. Indeed, sheepcotes had already experienced a heyday of construction during the thirteenth and early fourteenth centuries, coinciding with the

*Figure 3.3* BUTCHER SLAUGHTERING OX, misericord carving from Worcester Cathedral, fifteenth-century. © Dominic Strange www.misericords.co.uk http://www.misericords.co.uk

height of the wool trade. Among the wealthier class of peasants, the house and farm buildings would be arranged around a central courtyard. Even in the case of longhouses that continued to be constructed in the late medieval period, which experienced a building boom in peasant housing, animals could be compartmentalized in walled-off spaces, even to the point of having separate entrances for animals and humans, in which case they were known as "blockhouses".[61] This change in human and animal housing implies a lower degree of intimacy, or comraderie if you will, when animals were banished from sharing the same space as their human masters. Much of what may have impelled this separation were fears of disease contamination from the dung or manure of animals: A number of plague treatises, for example, warn of the stench from "stables of draught animals" or dung-pits that could be a near cause of the corruption of the air during a pestilence.[62]

Farm animals represent a kind of grand summation of medieval attitudes towards beasts in general, since these run the gamut from a callous disregard for animal welfare for the sake of being at the disposal of human use and consumption, to according them almost equal status with very human-like qualities. Sometimes medieval slaughterhouses, like some modern abattoirs of today, could needlessly prolong their victims' suffering: In London, Winchester, and Exeter, city regulations actually required butchers to bait bulls with dogs before slaughtering them, in the belief that this rendered the meat more

Part III. Beast   165

tender; it was a jury of London butchers who in 1349 objected to this practice by declaring that it had no effect on the quality of their cuts.[63]

Also unspeakably cruel was the plague remedy of medieval doctors, said to have become "indispensable" by the fifteenth century and which apparently originated with the papal physician, John Jacobi, chancellor of the medical school at Montpellier during the 1370s, of taking a live hen or rooster, plucking the feathers from around its anus, and then holding or tying the animal down over the bubo or glandular swelling of the plague victim either on his groin, armpit, or neck, so that it might suck out the "poisonous matter" contained therein. Eventually the chicken was believed to die from the ingested poison going straight to its heart, but how long this could take apparently varied: One doctor reported the animal dying before the *Pater Noster* or Our Father prayer could be recited; another that the chicken only needed to be held down for "the seventh part of one hour" (i.e., about ten minutes); others recommended leaving the chicken for an hour or three to four hours (tied down). It is safe to say that the real cause of death for these poor animals was the stress and agony induced by this process, which was sometimes increased by rubbing the anus raw with ground-up salt and mustard, whose "corrosiveness," it was hoped, would increase the anus' drawing power, or else the chicken's anus was "split open" (*fisso*), presumably with a knife or scalpel, in order to allow an easy application over the aposteme. All this was to be repeated with new victims, "exchanging one for another," until a chicken no longer expired, indicating that there was no more poison left in the boil to draw out. The fifteenth-century Burgundian physician from Besançon, Theobaldus Loneti, testified that he once went through thirteen birds in this manner and still the patient died, because it was the middle of the night and no more chickens could be found to sacrifice to his cure. Nonetheless, he recommended it as a "very mild" cure for his patients, even infants and children, except in the case of "excessively fat" people, whose "overhanging fleshiness" made it difficult to palpate the plague swelling and attach the chicken. It is safe to say that all these animals were wasted in terms of their meat, for they were believed to be contaminated by the plague poison such that anyone who ate of them was also poisoned. The fifteenth-century physician, Hermann Schedel of Nürnberg, believed the efficacy of this practice could be increased by forcibly holding shut the bird's beak, so that it would be forced to "inhale through its anus"; evidently this doctor knew little of bird anatomy, for he seems to have forgotten that birds breathe through their nares at the base of the beak! Other creatures that could be applied in place of chickens included live leeches, snails, green frogs, toads, and even crabs tied down over the plague tumor, but in the last case this may have been equally uncomfortable or even painful for the human patient as it was for the animal![64] Clearly, chickens, a common-enough animal on the farm that perhaps occupied the lowest rung on the chain of beings, were viewed as entirely at the disposal of medieval man and his medicine. But before we judge the medieval physician too harshly, we should remember that animal testing and experimentation in by declaring that it had no effect on the quality of their ores.

On the other hand, it is clear that medieval people could also greatly empathize with their domestic animals. From the *Etymologies* of Isidore of Seville in the seventh century to the *Buch der Natur* of Konrad of Megenberg in the fourteenth, bestiary authors, encylopedists, and natural historians all commented on how animals could have feelings and express emotions for each other and even towards their human owners. The horse, for example, was said to shed tears of grief when its master died, the only animal to do so, while the ox possessed an "extraordinary affection" for its plow mates at the yoke, lowing with apparent forlorn should its comrades be missing.[65] Albert the Great tells us that a sheep "has compassion for its own species, so that when a healthy sheep sees a sick one laboring in the heat, the healthy one places itself in front of the sun and makes a shadow for the sick one."[66] Less attractive human attributes, however, are said to be exhibited by the mule, which Alexander Neckam describes as crafty, spiteful, and disobedient, in that it will often pretend to be lame in order to halt its journey prematurely until the owner sees through its stratagem. To illustrate the mule's vengeful nature further, Neckam tells the story of how a youth on a journey for his lord incessantly goaded his mule to the point that the vexed animal took its first opportunity to vent its "indignation": When they came to a wooden bridge they had to cross in order to get over a raging river, the mule with stealthy steps took an "oblique line" away from the safety of the middle of the bridge to the very edge, bearing its rider along the "narrowest plank," until the youth broke down in tears, pissed his pants, and was "nearly driven out of his mind" with fear. At the urging of a watching crowd, the youth gave free reign to his mule, which, its honor satisfied, stole back to the more traveled path. The youth, however, was less forgiving than the mule and vented his own anger at the animal by stabbing it to death with his sword once they arrived at their destination.[67]

Undoubtedly the most entertaining and amusing (if chauvinistic, from a modern perspective) comparison between beasts and men is that by Anthony Fitzherbert, who took an especial liking to horses, since he describes himself as a "horse-mayster," by which he means he specializes in the breaking, training, and re-selling of wild and young horses, as opposed to a "corser," who simply acts as a middleman for horses already trained, and a "horse-leche," who attends to the veterinary care of horses (even though we've already seen Fitzherbert dabbling somewhat in that art). In a section of his husbandry manual devoted to the "properties of horses," which is an expansion upon the 15 listed in the hawking and hunting treatise, *The Book of St. Albans*, printed in 1486, Fitzherbert lists these as no less than 54, of which 42 are compared to those of other animals, including the badger, lion, ox, hare, fox, and ass or donkey, mostly in terms of physical attributes. Of the dozen remaining, two qualities of a horse remind Fitzherbert of a man, namely, the animal's "proude harte" and that it is "bolde and hardy" (the same as in *The Book of St. Albans*). Last are listed the ten properties that a horse allegedly shares with a woman:

to be "mery of chere" (i.e., warm and welcoming when at the door); "well paced," by which we assume he means having a lively gait or stride; a "brode foreheed"; "brode buttockes"; "harde of warde," or in other words, not easy to maintain in terms of the expense of their upkeep; "easye to lepe upon"; "good at a longe journeye"; to be "well sturrynge under a man"; "to be alwaye besye with the mouthe"; and to be ever "chowynge on the brydell," or the bit. (In *The Book of St. Albans* the properties are listed as three: fair breasted, fair of hair, and easy to leap upon, with an added double-entendre, "wel travelid women ner welle travelid hors wer never goode.") It is hard to know if all this is more flattering for the horse or more insulting to the woman! Fitzherbert concludes by observing that he could list as many "defautes" or defects of horses as their good properties, but then he would be breaking a promise he made at "Grombalde Brydge," when he went one time to buy colts at Ripon fair.[68] Clearly, this man made a great study of, and had much affection for, the animals under his care; one might even hazard a guess that he got along better with horses than he did with people, particularly of the opposite sex! (Fitzherbert was married twice and had a large family by his second wife, Matilda Cotton.)

Medieval authors and artists could even imagine and recreate a view of the world from the perspective of their domestic animals. One of the better-known literary attempts at this is the *Speculum Stultorum* (Mirror of Fools), or *The Book of Burnel the Ass*, by the twelfth-century satirical poet, Nigel Wirecker, otherwise known as Nigel de Longchamps. In the poem, a donkey named Burnel ignores the advice of his doctor friend, Galen, to be content with Nature's lot and goes off in search of a way to lengthen his tail to match his long ears. His adventures take him to the medical city of Salerno, the university schools of Paris, and various religious orders, all of which provide ripe targets for satire. Along the way, Burnel hears a cautionary tale from Galen about two sister cows whose tails are trapped in the ice as they lie in their winter pasture, one of whom, Bicornis ("two-horned"), decides to free herself by cutting her tail off after hearing her new-born calf lowing in the barn, while the wiser of the two, Brunetta ("the dark one"), waits for the morning sun to melt enough of the frost to allow her to free herself, thus preserving her tail with which she can swat the maddening gnats and gadflies in the coming heat of summer, as well as hide her private parts from the "public gaze." Burnel's story ends with him being recaptured by his woodcutter owner, Bernard (after twenty years' worth of freedom), and taken back to Cremona, where he falls back into his usual routine of pulling logs and carrying loads for his master; even though his ears are docked as punishment, he is reconciled to his fate by the fact that at last his tail and ears "agree so well."[69] The poem is arguably a unique contribution to the animal fable tradition because it has Burnel interacting with humans at their level while at the same time retaining his essential animal character as a donkey (e.g., at the end of his studies in Paris, all he can recite is "hee haw"). The intended audience of some of Wirecker's stories is also not just humans but other animals: The two cows, Brunetta and Bicornis, provide a higher moral lesson along the lines of Galen's dictum, "What Nature denies, no one can supply," but at a more mundane, basic level they in addition speak to their fellow domestic animals on the utility and necessity of tails, to which humans obviously cannot relate (except perhaps the English, whom Burnel joins at Paris on the basis of the medieval rumor that they had tails).[70]

Another, lesser-known medieval literary work featuring domestic animals is *The Debate of the Horse, Goose, and Sheep*, by the fifteenth-century English poet, John Lydgate. Here, the three animals debate their usefulness to man before the court of the Lion and the Eagle, a variation on the animal debate genre whose early exemplar is the twelfth-century English poem, *The Owl and the Nightingale*. The poem is revealing for showing us how medieval society perceived the uses to which each domestic animal could be put: The horse serves as a steed of war and of the joust as well as a beast of burden, pulling ploughs and carts and carrying grain, wood, wine, hay, water, lead, and stone to building and storage sites and to the market; the goose portends the onset of winter by its migratory flight, while its fat and dung are medicinal ingredients in remedies for gout and burns, and its feathers make for writing quills, down stuffing for pillows and mattresses, and fletches for arrows; the meek sheep (using the ram as spokesman) points to the value of its wool, the basis of English wealth, that is turned into garments and its skin into writing parchment, while every other part of the animal finds a use for man, such as the meat for nourishment, the horns for hooks, the bones for handles, the tallow for plasters, the dung for fertilizer, the guts for harp strings, and the head boiled down into a restorative ointment. In the ensuing back and forth argumentation of the debate, the animals also point out each other's drawbacks: The goose, for example, argues that horses are of no use when the arrows it has feathered pierce them, as was demonstrated at the English victory at Poitiers in 1356 (specifically fletches from the white goose, since the arrows were said to fall "like snow"); the horse claims that the sheep's wool is often a cause for war, not peace, as in the duke of Burgundy's siege of Calais in 1436, and in any case, war is the basis for many men's livelihood; the sheep chastises the horse and the goose for their contributions to war, which is less desirable than peace. Both the goose and the sheep also marshal in their defence a self-sacrificial quality to their natures, in that even in death they are useful to man as food for the table, which the horse cannot match because its meat is taboo. (The owl makes a similar argument, that in death it can be used as a scarecrow, to the nightingale in *The Owl and the Nightingale*.) This debate ends the only way it can, with none of the animals winning: All are urged by the Lion and the Eagle to be content with what Nature has ordained for them—the horse to work, the goose to swim, and the sheep to graze—and to live in harmony and not "despise your neighbor," a constant refrain of the moralizing section at the end of the poem.[71] Lydgate thus makes domestic animals serve as the mouthpiece for his nostalgic longing, apparently, for a return to the "right order" of the world in which rulers did not oppress their subjects and everyone knew his place, an especially poignant message in the wake of all the social turmoil and upheaval after the Black Death and other catastrophes that plagued the late Middle Ages.

The last word on this subject should be left to a marginal illustration in a Flemish book of hours from c. 1300, preserved in the library of Trinity College, Cambridge. It is a remarkably daring and playful jest on the "world turned upside down" theme, where a horned cow is shown milking the breasts of a naked women while holding a wooden milk pail to catch her breast milk.[72] What closer identification of man with beast can one hope to find in the artistic commentary of the Middle Ages? This clearly crossed the line separating the species, where animals became humans and humans became animals, even if only in the transgressive margins of the medieval imagination.

## Animals as pets and companions

Around the same time as animals began to be domesticated for farm work and as food, some were tamed mainly for the sake of companionship, or what we today know as "pets." The earliest species to be adopted for this purpose seems to be the dog, descended from the wolf or perhaps a wild version of *Canis familiaris*, around 15,000–10,000 B.C.E. Wolves exhibit submissive behaviors and can socialize on an individual level that would have made them ideal companions for prehistoric man. They are also likely to have intersected with early humans on hunting expeditions, since both species hunt similar prey and do so in a similar fashion, working in groups or "packs."[73] In addition to providing companionship, dogs aided man on the hunt and as guardians of the household and of other domestic animals, such as sheep.

The other main pet animal we will focus on here is the cat, which was domesticated from the wild around 7500 B.C.E. Alone among other domestic species, cats are naturally solitary, aloof, and shy animals, a trait that readily distinguishes cats from dogs, as any modern pet owner will attest. But because cats are also territorial by nature, staying close to a home range, and because they can establish a mutually beneficial relationship with humans— receiving food, warmth, and shelter from man and in return condescending to be a receptacle for a universal human need to show affection (indeed, somehow filling their owners with a sense of gratitude just for letting themselves be petted!) and ridding a household or barn of its vermin, particularly rats and mice— cats have tolerated domestication as a kind of "exploited captive." Nevertheless, cats live far more on the edge of wildness than any other domestic animal, and the domestic species of cat show relatively few genetic and physical differences from its wild cousin, *Felis silvestris*, which flourishes even today in Europe, Africa, and Asia.[74]

Both cats and dogs were well established in ancient societies, even though humans might have also experimented with other animals as pets, such as ferrets, monkeys, birds, and even lions and cheetahs among the royalty. Dogs seem to have been the favored animal among the ancient Assyrians, Greeks, and Romans, while the Egyptians of course had a religious cult devoted to the cat, evidenced by the 100,000 cat mummies excavated in the nineteenth century from Beni Hasan, even though it is also clear from tomb paintings and other evidence that they kept dogs as pets as well, who were given names and sometimes buried in a place of honor beside their masters. The ancient Greek historians, Herodotus and Diodorus Siculus, describe the thousands of live cats kept in Egypt at the cult center of Bubastis, sacred to the cat goddess, Bastet, and of how owners would mourn their pet cats by shaving their eyebrows or execute those who dared to harm their cats, even if by accident.[75] Eventually the Egyptians exported their cat fetishes northwards to ancient Greece and Rome. The presence of cats and dogs in Anglo-Saxon, Carolingian, and Viking villages during the early Middle Ages is attested by archaeological excavations of sites in Britain, Holland, and Denmark, where the bones of these animals have been found.[76] At the village of Villiers-le-Sec in northern France, the presence of dogs steadily increases from the seventh to the eleventh centuries, but cats barely register at all, which seems generally true of archaeological sites during this period, although there is plenty of evidence for their existence during the earlier, Gallo-Roman period.[77] Excavations of later medieval sites reveal a range of size in dog skeletons, from those approximating the modern Alsatian at the large end to the toy poodle at the small end, with the latter clearly bred exclusively as pets rather than for herding or hunting.[78]

More revealing evidence for early medieval pets is the testimony of various law codes enacted by Germanic societies throughout Europe between the sixth and eighth centuries. In the Alamannic, Bavarian, and Frankish laws, whole chapters are devoted to dogs of various kinds that illuminate the uses to which these animals were put in Germanic society, which mainly consisted of hunting and guard duty.[79] In the Burgundian code, a man who stole a hound, hunting dog, or running dog was given the choice of either kissing "the posterior of that dog publicly in the presence of all the people," or else paying a fine of seven *solidi*, five of which went to the dog's owner.[80] It seems that this punishment was designed to humiliate the thief by forcing him to play-act like a dog and thereby degrade himself to its level, for sniffing out the anus is a commonly observed way by which dogs are known to greet and identify each other. The Alamannic code likewise gives the owner of a dog that kills a man the choice of descending to the level of an animal in order to be spared full payment of the fine. If the dog is hung 9 feet above his entrance and remains there until "it becomes completely putrefied and drops decayed matter and its bones lie there," then the owner only has to pay half the wergild for the deceased man. But "if he removes the dog from the entrance or enters the house through another door," he must pay the wergild in full.[81] This law has been interpreted as forcing the dog to share in the man's punishment and thus holding it somewhat culpable for its crime, in contrast to how other domestic animals were treated as merely property extensions of their owners.[82] But while it is certain that the dog suffered cruelly as it died a slow death from strangulation, I think it more likely that the punishment was designed, as in the Burgundian code, to degrade the man to the level of a dog by forcing him to face up to rotting flesh and bones, since it was well known that dogs gnawed on corpses, a trait shared by pigs and wild beasts, for the code specified that this was the reason why it rewarded those who buried abandoned dead bodies out of "human kindness."[83] Rather than eliding the distinction between animals and man, therefore, this particular law actually reinforced that difference by forcing the owner to act like a dog as a form of humiliation. The only code I am aware of that mentions cats are the Welsh laws of Hwel Dda of 948, which place a relatively high value on the cat, apparently for its fur, at three pennies and record it as one of three animals that "reach their worth at a year [old]," along with the sheep and the dog. Theft of the prince's cat was punishable by a fine of a sheep plus enough grain to cover the cat from its outstretched tail to its head lying on the ground; the cat was also included, together with a rooster and a bull, as among the animals that make up the lawful compliment of a hamlet.[84]

There are indications in the Germanic laws that other animals besides cats and dogs may have been kept as companions of some sort by early medieval people. For striking or stealing a domesticated stag that had reached maturity (which it was said one could tell by its ability "to roar" during rutting season), the Lombard code prescribed a penalty that matched that of the highest valued livestock.[85] The Alamannic and Frankish laws imply that a host of wild animals, including bear, bison, buffaloes, and deer, were domesticated since they were kept in "shackles" or bore a "brand." In Bavaria, people kept a variety of hunting hawks as well as other wild birds that were domesticated and tamed "by human effort ... so that they can fly and sing in the courtyards of nobles."[86]

As in the law codes, dogs are singled out for almost exclusive attention among pets in the early medieval compendiums of knowledge, such as the *Etymologies* of Isidore of Seville. Of all animals, Isidore seems to have the highest regard for dogs, largely because of their close association with humans:

> No animal is smarter than the dog, for they have more sense than the others. They alone recognize their own names; they love their masters; they defend their master's home; they lay down their life for their master; they willingly run after game with their master; they do not leave the body of their master even when he has died. Finally, it is part of their nature not to be able to live apart from humans.

Cats receive an all too brief entry in the *Etymologies*, being noted simply as "mousers" for their ability to catch mice, as well as being able to see in the dark and being "clever." Thus cats are apparently not valued for their companionship like dogs, but only for their usefulness in vermin control.[87] For Rabanus Maurus in his *De Universo* (ninth century), dogs have a negative connotation as symbolizing heretics and Jews because of their backbiting and bullying qualities. Cats are associated by Rabanus with witches, which presciently anticipates accusations of having cats as familiars during the witch trials and hunts of the late Middle Ages (see "Animals and magic").[88]

Twelfth- and thirteenth-century bestiaries and encyclopedias repeat this information from Isidore but add much more in the way of anecdotes and legends, particularly about dogs, on the authority of a host of mainly classical authors, which now include Aristotle, Pliny's *Natural History*, the *Hexameron* by either Basil of Caesarea or St. Ambrose, Solinus' *De Mirabilibus Mundi* (On the Wonders of the World), Avicenna's *Canon*, the prose romance, *Alexander*, and *the Physiologus*, a Greek work from the second century C.E. that served as the basis for the medieval bestiary tradition.[89] The thirteenth-century encyclopedist Vincent of Beauvais has by far the longest entry on dogs, providing separate sections on the differences between the sexes of dogs, how they give birth to and raise pups, and their average life-spans and diseases to which they are susceptible. From Avicenna and other authorities, Beauvais also prescribes the medicinal properties of various dog products, such as its blood and milk to prevent hair growth, its urine for warts, its shit for "old ulcers" (provided the dog has been fed on bones), its bile for gout, and dog testicles for treating various apostemes, ulcers, and abscesses. If a man eats the larger testicle of a dog, he becomes more manly; if a woman eats the smaller one, she grows more feminine. When eaten, moist dog testicles are also said to increase the desire for coitus, but when eaten dry, take it away. Cantimpré asserts that shoes made from dog-hide can cure gout, but warns that live dogs will want to pee on them.[90]

Albert the Great, probably basing himself on an earlier veterinary work, greatly elaborates on the diseases to which dogs are prone and how to cure them. Dog ills are listed as nine in number: They include "scabies" or "leprosy," which is probably mange; worms; "swellings"; thorns stuck in the limb (undoubtedly common in hunting dogs); rabies; emaciation; sluggishness; fleas; and constipation (said to be common in ladies' lap dogs).[91] Only late medieval hunting manuals devote more attention to illnesses in dogs.[92] Rabies was clearly a feared disease that receives its own separate entry in encyclopedias, especially the one by Beauvais. Dogs were believed to get rabies from eating dead bodies or an overabundance of black bile (especially common in autumn and spring), and on the basis of Avicenna and the ninth-century Jewish physician, Isaac Israeli ben Solomon, otherwise known as Isaac "the Jew" (Isaac Judeus), the signs to look for in a rabid dog were disorientation as if drunk, a foaming mouth, hanging tongue, refusal to eat or drink, fear of water, contorted body, inability to recognize its master, and its avoidance by other dogs. A recommended remedy for a rabid dog, attributed to "Armeria, king of Valentia," was to immerse the animal for nine days in warm water, then shave its head and anoint it with beet juice. If that didn't work within the space of a week, then the dog should be killed "because it will not get better." For humans, rabid dogbite remedies (largely derived from Pliny) include cutting around the wound down to the "living flesh" and leaving it open for a year, or else applying goat's liver, the ashes from a dog's head, ground-up cock's comb, duck fat, chicken dung, ground-up crabs, fish brine, and wild rose roots. One can also ingest goat, cuckoo, or swallow dung mixed with wine and the liver from the rabid dog; however, administering the saliva slime taken from under the dog's tongue, which was believed to be efficacious for the symptoms of lunacy, can only have made things worse. According to Avicenna, victims sometimes died within a week, sometimes in six months, but the average morbidity period was forty days.[93]

Cats still get short shrift in the bestiaries and encyclopedias when compared to dogs: One suggested reason for this is that the domesticated weasel took the place of the cat in many medieval households. In the *De Naturis Rerum* of Alexander of Neckham, an entry on the weasel (*mustela*) seems to replace the cat, where the animal is mentioned as a "skilled huntress of mice, to a quantity far exceeding the weasel's little body," and which "is accustomed to lay its kill at the feet of the lord and lady whose house it inhabits."[94] But in other, thirteenth-century encyclopedias, such as those by Bartholomaeus Anglicus, Vincent of Beauvais, Thomas of Cantimpré, and Albert the Great, entries on cats give considerably more information than that they are just mousers, much of which will be instantly recognizable to modern cat lovers. Even though we are told that cats are unclean and "odious" or "poisonous" animals, we are also informed that they love to be petted by the hand of their human owner, "whereby they express their delight by their mode of singing [i.e., purring]." They also can be "most easily lured into play by men" and seek out the "blandishments" or attention from humans as well as warm places, such as in front of a fire, "whence they burn their hides due to their extreme laziness." As a kitten the cat is especially "swift, pliant, and merry," leaping "on all things that are before him, and is led by a straw and plays therewith," but in old age he becomes a "well heavy beast … and full sleepy." They also play with the mice they capture before eating them, but with poisonous toads and snakes, they are more cautious and must quench their thirst afterwards. Among the cat's special abilities is to be able to see and hunt in the dark at nighttime with "glowing eyes," to be able to "fall on his own feet" and land unhurt "when he is thrown down from a high place," and to clean its face by licking its paws first. However, if cats' whiskers are cut "they lose their audacity" [perhaps because then they can no longer judge tight spaces], and their feces "stink full foul," so that they "hide it under the earth and gather thereupon a covering with feet and claws." Yowling male cats fight ferociously among themselves with teeth and claws over mates and food, and it is noted that in the wild, the cat is "a cruel beast" that hunts woodland prey such as rabbits and hares. Indeed, medieval authors acknowledged that the domestic cat lives on the edge of wildness, for they noted that in mating season it will seek solitude and turn "into a wild creature, as if it were exhibiting shame." To prevent the animal from turning completely feral, they recommended cutting its ears, since then they will seek out shelter against the rain. Cantimpré recounts the legend that cats are so enamored of their own kind that they will stare at their reflection in a well and even leap in to join their imagined comrade, which is obviously a moralizing lesson against vanity, although this is said to happen especially among female cats burning with lust for a mate or among young cats that lack experience with life's dangers. Cat coat colors are noted to be solid white, red, or black, but also striped and speckled "like to the leopard [or perhaps the tiger?]," which is how they usually appear in the wild. The "fair skin" of cats could also be their downfall, as skinners were said to lie in wait to ambush them, and for this reason it was advised that pet cats be clipped of their fur.[95] The only medieval medicinal remedy I have seen derived from cats is a fourteenth-century French recipe that calls for the "fat of a male cat" in a prescription for gout.[96]

That cats were indeed sometimes skinned for their fur is known from the excavation of cat bones at some sites, such as at York, Norwich Castle, Colchester, and Cambridge in England, at Canne in the Apulia region of southern Italy, and at Odense in Denmark dating to the early and high Middle Ages: At all of these sites the bones bear traces of cut or chop marks with a knife, in particular scoring along skulls, which is not terribly compatible with the relationship most of us imagine having with our animal companions. At Odense, about 68 cats were killed by having their necks broken with a wrench delivered to the head, while at Cambridge, a total of 79 cats (one of the largest finds ever recorded) had their throats slit. The Odense cats seem to have been bred in captivity for this purpose, since most were female around a year old, while the Cambridge cats were probably feral and caught as such, since there was an equal distribution of the sexes and the cats, mostly young adults and juveniles, were smaller than both wild and domestic species elsewhere. Judging from other butchery marks besides those from skinning, the Cambridge cats were also used for food: Normally, medieval cultural norms dictated that cats and dogs only be eaten in dire necessity, such as during a famine or siege, but cat meat is also said to taste similar to hare, so it is possible that the cats, after being skinned, beheaded, and having their tails and paws chopped off, were sold and disguised as such. (The excess body parts were disposed of by being thrown down a well, from which they were recovered during the modern excavation: After the skull and mandible, the next most common bones were the phalangeal and metapodial bones of the feet and the caudal veterbrae of the tail.) At Norwich Castle, it appears that five of the cats, aged about a year and a half, were skinned while still alive and then thrown down a well, where they then crawled into putlog holes in the side of the well in order to die.[97] This obviously gives a whole new meaning to Cantimpré's story of the cat falling down a well: In real life, this was entirely due to human agency and denotes sins such as wanton cruelty that are far more disturbing to the modern mind than mere vanity. All this may speak to a strictly utilitarian medieval view of cats, but there must have been many more domestic cats that were kept as mousers and companions, among which we can perhaps count the cats found at other sites bearing no butchery marks.[98]

Stories of dogs' fidelity to humans were not just retailed in academic encyclopedias, but also penetrated medieval popular culture as well. From the twelfth-century *chanson de geste* tradition comes the story of the dog of Montargis, who avenges his master's murder in a judicial duel at Charlemagne's court, later transferred in sixteenth-century versions of the tale to the court of King Charles V of France at Montargis in 1371.[99] The thirteenth-century French inquisitor, Etienne de Bourbon, encountered the "superstition" of St. Guinefort after visiting a village in the diocese of Lyons, where he heard the confessions of "numerous women" who admitted to invoking the saint to cure their sick children at a special shrine in the woods. At first thinking that the saint was human, Bourbon eventually learned that it was in fact a "holy greyhound," which was "martyred" by its master in the belief that the dog had devoured his child in its cradle, when in fact the blood on the beast's maw was from a "huge serpent" that the dog had fought and killed in defending the child, upending the cradle in the process. After realizing his mistake, the dog's owner buried Guinefort in a water-well planted round with trees, which later became a kind of miraculous healing shrine visited by the local peasantry.[100] (The legend is the basis for the French film, *Le Moine et la Sorcière*, or *The Sorceress*, from 1987.) Finally, there is the late medieval legend of St. Roch, a patron saint of plague victims, who was fed bread by a dog as he lay recovering from the disease in the woods: A German woodcut illustration from c.1480 depicts the dog jumping up onto the saint's leg as an "angelic doctor" heals his plague bubo.[101]

There is also plenty of historical evidence of real-life pets owned by medieval people. Despite their vow of poverty, medieval nuns in particular seem to have delighted in keeping pets, which could include dogs (the favorite animal), cats, monkeys, birds, rabbits, and squirrels. The nunnery of Romsey in Hampshire had to be warned in two separate visitations during the thirteenth and fourteenth centuries against keeping dogs, monkeys, birds, and rabbits in private chambers or even bringing them to church during the recitation of the offices of prayer, to the distraction of the service and of fellow nuns. Similar injunctions had to be issued to several nunneries in the diocese of Rouen by Archbishop Eudes Rigaud during the thirteenth century.[102]

The Luttrell Psalter depicts pet lapdogs and squirrels fondled by elegant ladies, while in several late medieval tomb effigies and monumental brasses, the deceased, usually women, are depicted with their pet dogs at their feet or hidden within the folds of their dresses. On at least two brasses the dogs are given names, Terri and Jakke, who sit obediently at the feet of respectively Alice Cassy and Sir Brian de Stapilton in Deerhurst, Gloucestershire and Ingham, Norfolk. A lapdog wearing a belled collar is depicted with its owner, Lady Margaret Roos, in a fifteenth-century stained-glass window at York Minster. Household account books record pet popinjays brought to Queens Joan of Navarre and Elizabeth of York in 1419–21 and 1502, while cats were recorded as purchased by Eleanor de Montfort, countess of Leicester, in 1265.[103] In the early fourteenth century, a page of the chamber for Lady Elizabeth Despenser was paid for a six-days' journey in order to bring her white greyhound that was "heavy in whelp," i.e., pregnant, from Whorlton to Cowick in Yorkshire.[104] The *Ancrene Riwle*, a thirteenth-century guide for female recluses or anchoresses, advises against keeping any pet animals "except a cat," perhaps for practical reasons to keep down mice in the anchorhold, but it may also be because cats do not require as much attention as dogs that would distract a holy woman from her devotions.[105] Another reference to Church-sponsored cats comes from Exeter Cathedral, whose obit accounts from the fourteenth and fifteenth centuries record cats on the payroll at a penny a week, while the physical evidence for perhaps the only salaried pets in medieval history can still be seen in the form of a cat-hole carved in the north transept door of the cathedral.[106] Cats actually played a vital role in preserving the integrity of Church ritual, as they were expected to keep down the mice population that might sacrilegiously gnaw on communion wafers: A Chi-Rho page from the Book of Kells of c.800 shows two cats catching by their tails two mice nibbling on a communion host.[107]

Dogs are still very much valued in the high and late Middle Ages for their hunting and guarding abilities, as evidenced by hunting and estate management manuals.[108] Yet there does seem to be a transition taking place at this time in which dogs are beginning to be more appreciated for their companionship qualities, especially in terms of their loyalty or devotion to their masters. It is also evident that dogs are given more human attributes, and rabid men dog-like ones, thus eliding the difference between the species, in the bestiaries and encyclopedias of the thirteenth century. Likewise, cats are for the first time recognized as receiving and returning affection from their human owners. This is an important shift from attitudes insofar as these can be judged from law codes and natural histories from the earlier medieval period.

## Animals of the hunt: origins of medieval hunting

The hunting of wild game is an activity that goes back to the very dawn of human civilization. Prehistoric cave paintings from Los Caballos and Remigia in Spain show bowmen hunting deer, an ibex, and a boar, while other potential game animals such as bison, aurochens, rhinoceros, and wooly mammoths are also depicted. Stone Age artifacts also include weapons and tools clearly designed for the hunting, trapping, and butchering of animals.[109] The tradition of pictorial depictions of the hunt continued in wall paintings and other mediums such as relief carvings and mosaics down through the ancient civilizations of Mesopotamia, Egypt, Greece, and Rome. Already from the earliest societies, hunting acquired its aristocratic associations, rituals, and symbolisms—particularly as a source of power, prestige, and patronage—that were to characterize it during the Middle Ages. In the *Epic of Gilgamesh* from the Sumerian civilization of Mesopotamia (c. 3000 B.C.E.), Gilgamesh, king of Uruk, and his friend, Enkidu, slay the Bull of Heaven (perhaps a wild aurochs) in an epic hunt in which Enkidu grasps the bull by the horns and by the base of the tail, but defers the honor of delivering the final death thrust to Gilgamesh, who plunges his sword in between the nape and the horns. For this deed the hunters are welcomed back as heroes and feasted as master craftsmen marvel over the massive horn trophies which they fashion into libation vessels, but meanwhile the temple priestesses to the goddess Ishtar bemoan the death of the animal that had been created at her command and for which she exacts the terrible punishment of Enkidu's illness and death. Evidently the hunting of wild animals already evoked ambivalent feelings in early human societies.

During the succeeding Assyrian Empire, hunting became a fully fledged form of royal propaganda. Despite the impressive friezes depicting what appear to be spontaneously energetic showdowns between man and beast, such as that of Ashurbanipal (reigned 668–631 B.C.E.) stabbing a male lion at close quarters, these royal lion hunts were apparently stage-managed affairs: A caged lion was released and shooed by beaters towards his royal executioner while an audience stood by and watched the whole spectacle. Clearly the hunts carried religious and ceremonial significance as demonstrations of the gods' favor upon the monarchy and of the king's ability to protect his people. An inscription from the reign of Tiglath-Pileser I (c.1115–1077 B.C.E.) boasts of a slaughter of 900 lions killed from a chariot and 20 on foot, 10 wild buffaloes in the land of Kharran, and 4 wild bulls, "strong and fierce," slain with iron-tipped arrows in the country of Mitan, all under the auspices of the god Abnil-Hercules. The skins and horns of the wild buffaloes and bulls were taken back to the capital city of Ashur to be displayed as trophies, while four wild buffaloes were also taken back alive, perhaps as an experiment in domestication. The inscription hints at much other slaughter besides this, for Tiglath-Pileser claims to have "extirpated all wild animals" that existed "throughout my territories," whom he tracked by chariot over easy terrain and by foot on more uneven ground. For example, "droves" of wild goats, ibexes, sheep and cattle are described as being hunted in the "depths of the forest," while their orphaned offspring, "the delight of their parents' hearts," were led away alive like tame livestock to be offered up as sacrifice to the god Ashur.[110] Wall paintings and carvings from ancient Egypt likewise depict royal chariot hunts and the use of trained hunting dogs, as well as fishing expeditions along the Nile on reed boats using harpoon spears, where the most dangerous game suitable for royalty were crocodiles and hippopotami. On a more pedestrian level, the papyrus fragment, *The Pleasures of Fishing and Fowling*, describes the trapping of "thousands" of marsh birds such as geese and ducks using rope nets, while innumerable perch are speared and thrown into bags, even though the fisherman does not find the prospect of gutting and cleaning them all so pleasurable.[111]

Aristocratic associations with hunting continued in ancient Greece and Rome, as well as the political and social implications of the hunt. Hunting was pursued for recreation, physical fitness, and as a supplemental source of meat among the classes that had the leisure and wherewithal, such as dogs, slaves, and equipment, to do so. Boars, hares, deer, and foxes seem to have been among the Greek hunters' favorite quarry, and vase paintings depict older male lovers bringing home their catch to share with their boy lovers as inducements to amorous encounters, a form of "saying it with hares and foxes" instead of flowers. Also transgressive, at least from our modern point of view, was the adoption of a female deity, Artemis/Diana, as the patron of a very masculine enterprise in which women were generally not allowed to participate. Although Artemis/Diana demanded sacrifices from the hunt, there were also islands and groves, such as Delos, that were considered sacred to the gods and where hunting was forbidden, and consequently where wild animals, such as hares, proliferated to the point of becoming a nuisance.[112]

During the Hellenistic era, Alexander the Great (reigned 336–323 B.C.E.) encountered in the course of his world conquests the magnificent game parks, or "paradises," of the Persian empire, which were managed sanctuaries where royalty could pit themselves against fearsome game such as lions and bears. However, the next century witnessed a remarkable exception to royal patronage of hunting on the eastern border of Alexander's former empire. Under the Buddhist convert Ashoka, who ruled India from 269 to 231 B.C.E., edicts inscribed on stone pillars erected throughout the land officially protected a host of wildlife including ants, fish, birds, mammals, and indeed of "all four-footed creatures that are neither useful nor edible." Castration and slaughtering of some domestic livestock were also proscribed, particularly nursing mothers and young less than six months old, nor were animals to be gratuitously killed for sacrifices or to be fed to one another. This protection extended to habitat, for fires were forbidden to be lit in the forests in order to smoke out animals. Ashoka was perhaps the only ruler in history to voluntarily give up his hunting privileges, and he has been called the first animal conservationist. An edict that proudly proclaims, "Our king killed very few animals," indicates some emulation of his example, but fines for poaching deer in royal preserves show that hunting and fishing still went on regardless among the general populace, albeit undoubtedly in diminished volume.[113]

Finally, at perhaps the opposite end of the scale, the Romans seem to have given rise to the professional hunter, who trapped and ensnared wild animals live for transport to the Colosseum in Rome for staged hunting spectacles, known as *venationes*. These were of course designed by their Roman donors for political gain as well as entertainment of the populace, in which the more exotic the animal, the better the response from the crowd (e.g., giraffes, elephants, hippopotami, rhinoceros, etc.). For example, in 50 B.C.E. Cicero's friend, Coelius, an aedile of Rome, asked for a shipment of panthers from Cilicia, where Cicero was proconsul, to help boost his reelection campaign, but his friend replied that the beasts had been hunted out of his province. Thus, the Romans' appropriation of wild animals for the arena apparently led to the extinction of some exotic species in their native habitats, which is to be expected when sometimes hundreds or even thousands of them were slaughtered at a time, such as the 9000 killed over the course of 100 days at the inauguration games at the Colosseum in 81 C.E. On the other hand, sometimes the Romans attempted to tame wild animals for entertainment: The poet Martial claimed that hares were trained to lie down in lions' mouths as their warren, or that eagles caught up young boys in their talons and flew a short distance before releasing them as a reenactment of Zeus' rape of Ganymede on Mount Ida.[114]

In Celtic Europe roughly contemporaneous with the Roman era, wild animals were venerated, sacrificed, and depicted in manifold art forms, sometimes in association with hunting cults and deities, among which the two most important were the boar and the stag, which were also the most common game animals. Sometimes these animals are depicted on hunting weapons and armor, alongside the remains of the animals themselves, at high-status burial sites and funerary deposits.[115] Boars were natural symbols of war and hunting owing to their fierce reputation when cornered, but they also denoted feasting and hospitality since they were a popular food animal, as evidenced by their butchered bones that show up in archaeological excavations.[116] Roman authors such as Strabo and Diodorus Siculus also comment on the Celts' fondness for boar flesh, with revelers competing for the "champion's portion" at the feast. In some figurines Celtic deities are depicted with boars, such as the bronze image of the goddess Arduinna shown astride a boar with a hunting knife in her hand that was found in the Ardennes Forest in eastern France. Stags were naturally associated with Cernunnos, the horned god who was "lord of the forest," as well as with various hunter gods who adopted a complex and ambivalent protective posture towards their prey. This seems a veiled reference to the Divine Hunt, in which enchanted beasts like stags and boars lured hunters who sought to slay them to their own journeys to the underworld, linking the hunt not only with death but also rebirth and immortality. Based on such legends, it is asserted that "the Celts revered the beasts they hunted" and believed that "the success of the hunt depended on harmony and the correct relationship between hunters and hunted." The seasonal attributes of stag horns, shed each winter and regrown in the spring and sometimes depicted complete with their velvet, would have only strengthened such associations. A variety of other wild animals were revered by the Celts, including bears, snakes, fishes such as dolphins and salmon, and a number of birds including eagles, cranes, crows, ravens, doves, peacocks, and swans. Bears were associated with the goddess Artio, native to Switzerland, invoked for both the hunting and protection of the animal, while snakes were associated with rebirth, based on their ability to shed their skin, and with healing, perhaps as a holdover from the Greek god Asclepios. Dolphins were appealed to for protection in maritime travel and salmon for their wisdom, while birds played an important role in divination: The majestic eagle was an auspicious symbol, but crows and ravens who feasted on carrion were portents of death. A fascinating aspect of these animal cults is the metamorphosis or shape-shifting that blurred the boundaries between human and beast: Among the animals that could partake of human form, particularly in female guise, were cranes, crows, eagles, ravens, swans, and deer.[117] Such animal cults persisted down into the early Middle Ages, as attested to by disapproving Christian authors, although sometimes they reappear in a Christianized format.[118]

Early Germanic societies had a tremendous hunting culture, as the Roman historian Tacitus testifies as early as the first century C.E. He writes in the *Germania* that whenever the Germans "are not fighting, they pass much of their time in the chase," and that fresh game constituted one of their main foods. The importance of hunting and access to wild animals is also reflected in various Germanic laws: The codes of the Franks, Alamans, and Lombards all prescribed penalties for killing, stealing, or hiding another's game animals including deer, boar, bear, bison, buffalo, fish, and birds. It's clear that at least some of these were domesticated, especially in the case of stags, who were used for the purpose of luring other wild animals of its own kind during the hunt, the tame decoys being distinguished by brands or shackles. Fines were also assessed for pilfering someone else's game, birds, and fish from traps and nets, but there was a two-edged sword to such shenanigans, for the Lombard, Burgundian, and Visigothic codes absolved the trap-setter of any responsibility for someone's death from running into the trap or being killed by the wounded beast within it. In the Burgundian and Visigothic laws, locals had to be forewarned of any traps containing tripwires that caused them to release their arrows at the target, such as was used for wolves. The Burgundian code also gave the thief who stole a hunting hawk or falcon the choice of allowing the bird to "eat six ounces of meat from his breast" if he did not want to pay the full fine of eight *solidi*. As in the case of dogs, the code once again shows itself well in tune with the particular attributes of the animal concerned, for the falcon would have used its razor-sharp talons to grip the man's chest (lying presumably in a supine position) as it ate the meat with its razor-sharp beak, all of which would have caused excruciating pain and thus administered the requisite punishment for the crime.[119]

Already in Anglo-Saxon and Carolingian times, royal hunting preserves were established in England and on the Continent, although this does not seem to have impaired free men's ability to hunt in the forest outside the preserve. However, even at this early stage, it seems that hunting and fishing were largely aristocratic activities, for archaeological excavations of rural sites in the north of France during the Carolingian era reveal few wild animal bones among their remains.[120] In Italy, game are largely absent from all sites until the eleventh century, when it coincides with the first great era of stone castle building, where it seems to become part of the new status attained by the aristocracy.[121] Meanwhile, the custom in Anglo-Saxon England seems to have been to allow free access to hunting outside the royal preserve, in line with the old Roman principle of *res nullius*, that a wild animal, until caught, was "nobody's property." Nonetheless, excavations of Anglo-Saxon sites, such as West Stow in Suffolk, also show that hunted wild game contributed very little to the diet, making up less than 1 percent of animal bone assemblages.[122] Norwegian laws from the twelfth century are notable for their elaborate regulations with regard to fishing, both along rivers and streams and at shoreline. Fishing nets, for example, had to allow free passage for salmon swimming upstream to spawn, which was also true of the Visigothic code in Spain.[123] It is clear that by this stage Norway had a thriving cod and herring industry: The tithe on cod, for example, was to be paid at the "drying frames," where the fish were processed for shipment, while the Norwegians sought an indulgence from the pope to let them fish for herring whenever they "seek the shore" on all except the holiest days. The Norwegian laws also regulated whale hunting, which was done both out on the open sea using harpoons and along the coast with bows and arrows shot from the shore. In the former case the whale belonged exclusively to the hunter but in the latter the whale had to be shared by both shores of a fjord. If a hunter drove a whale into someone else's cove or up onto his shoreline, he had to share it with the owner, while very elaborate regulations governed the partition of whales that were found beached.[124] In the rest of Europe, whales were evidently a rare enough sight that the chronicler Rodulfus Glaber took the sighting of "an enormous whale" at Berneval-le-Grand off the coast of Normandy as one of the apocalyptic signs of the millennium in the year 1000.[125]

For the high and late Middle Ages, one of the most important sources we have on medieval hunting are the records of the enforcement of forest law in England, which was set up shortly after the Conquest in 1066 by Duke William of Normandy. William's obituary in the contemporary *Anglo-Saxon Chronicle* claimed that the king "loved the stags as much as if he were their father," which was not a compliment, for William's "great protection for the game" meant a forest law that was deemed oppressively harsh by his native subjects, decreeing "that whoso slew hart or hind should be made blind," although almost always the crown preferred the more financially rewarding expedient of levying a monetary fine.[126] The administration and enforcement of English forest law has already been discussed in Part II. By law, no one could hunt in the royal forest except the king and those receiving his special warrant, or license. In the case of forests in private hands, the landowner had the duty to enforce the law at his own expense by hiring a woodward, although in a few instances (apparently no more than two dozen at any one time), "free chases" were granted in which the landowner could legally hunt game on his own land. Any deer in a free chase, even if it wandered in from the forest, was considered to be the private property of the owner, for it was ruled that any "deer of the forest can only be considered to belong to him in whose forest they are found, since they cannot be ear-marked and know no boundaries."[127] (Today the same reasoning makes all wild animals public property within the United States.) Free chases were therefore highly coveted privileges, as the long, three-year dispute in 1275–78 between Thomas of Cantilupe, bishop of Hereford, and Gilbert de Clare, earl of Gloucester, over hunting rights to the chase of Malvern in Worcestershire, testifies.[128] From the very beginning of the Conquest, however, there were also private "parks" in which landowners were authorized to hunt all game but which had to be enclosed, typically with a ditch and a barrier such as a wooden paling or stone wall or hedge. (Landowners who failed to keep up their enclosures could have their parks confiscated by the crown.) Penalties for poaching in private parks were even higher than in the royal forest, being prosecuted in the common-law courts where fines were assessed on behalf of both the crown and the landowner, or else entailing three years in prison. Often the grant of the right to empark land came with a gift from the crown of so many live deer to populate the park, which were caught using nets and transported on carts: One of the largest of such animal transfers was the 100 live fallow bucks and does that King John ordered be sent in 1202 from Windsor Park in Berkshire to Langley Marish in the neighboring county of Buckinghamshire for stocking Richard de Muntfechet's park there.[129] Inside the park grounds were usually a hunting or park-keeper's lodge and an observation platform.[130]

The keeping of wild animals within a park—mainly consisting of deer but sometimes also wild boar and wild cattle—turned them into semi-domestic beasts, which had to be managed on an ongoing basis in order to prevent widespread die-off from starvation and disease. Records of such "deer farming" tell of reserving grass pastures and supplying cut browsewood and hay and oat fodder in winter, building shed shelters, maintaining a steady water source such as an artificially constructed pond (which could also double for keeping fish and waterfowl), disposing of dead and diseased carcasses (sometimes by burning or hanging them from trees), and even suckling orphaned deer calves on domestic cows.[131] In addition, a staff of foresters or woodwards and scouts or watchers during fence month and rutting season had to be employed in order to protect the deer from poachers and trespassers in the park. All this, of course, entailed great expense, which is why parks were usually owned only by the upper aristocracy, such as the great barons and churchmen, including the king himself. The main justification for establishing a park was to have a ready and constant source of game meat, a "live larder" if you will, in an era when having game at table was a rare and exclusive privilege used to mark special feasts and occasions. (Game was not usually available on the open market, although a "black market" for this meat did exist in some towns.) But there was already at this time an element of natural recreation and landscape appreciation behind the construction of parks, which was to become its main *raison d'être* in modern times. A rare glimpse of a great medieval landowner taking leisure in his park is provided by the account roll for the manor of Downham owned by the bishop of Ely: In 1345–46, payments are recorded as made to certain men who helped "drive deer from one part of the park to the other so that the bishop may see them."[132] Parks were to proliferate during the high Middle Ages as disafforestation of royal forest land gained momentum and seigneurial incomes from arable farming were at their height: By 1300 there were over 3000 parks throughout England, mainly concentrated in the Midlands and the South and Home Counties, with an average size of about 200 acres. After the Black Death, parks went into decline or were converted to other uses that had always coincided with game management, but by the end of the Middle Ages they seem to have made something of a comeback with the return of agricultural propserity.[133]

Other types of enclosures included "hays," which either denoted park-like spaces or small, fenced-in and netted areas (also known as "buckstalls") designed to trap deer for slaughter, and "warrens," which were smaller enclosures than parks designed for hunting of smaller game, especially hares and rabbits (coneys). The latter were especially valued for both their meat and fur and initially required construction of special stone-lined burrows known as "pillow mounds," at least before their escape and propagation in the wild.[134] Licenses to establish warrens were only granted on land outside the royal forest, and their boundaries were enforced by a fine of ten pounds for trespass.[135] An idea of the scale of such warrens and the animal populations they could sustain is provided by the plea that Bishop John Trillek of Hereford entered against a poacher, Walter Moton, in the king's bench in 1354, in which the bishop accused Moton of taking no less than 500 hares, 1000 rabbits, 1000 partridges, and 200 pheasants from his four warrens at Ross, Upton, Ledbury, and Eastnor, worth £1000 in total: This would represent an average of 675 animals per warren.[136] Obviously, the case also illustrates the paradoxical dilemma that concentrating animal populations within a park or warren posed, since it provided tempting targets for poachers, as is often illustrated by the court rolls.[137]

## Animals of the hunt: deer and other game

The main game animal hunted during the Middle Ages and which shows up almost exclusively in the forest eyre rolls that record poaching is the deer, of which there were three species in England: Fallow deer, which were introduced by the Normans after the Conquest as a more sedentary animal particularly suited to stocking parks, red deer (similar to the elk of the United States), and roe deer, the smallest of the three deer species.

Hunting of deer was supposed to be off limits during "fence month," consisting of the two weeks on either side of midsummer when does were said to be having their fawns (and when cattle were supposed to be "fenced in" to keep them from disturbing the deer), and during rutting season in the autumn time when deer were mating; at both times foresters were often ordered to be on extra alert or special "watchers" were hired on private estates. In 1339 the roe deer was declassified by the king's courts as a protected beast of the forest on the grounds that it was a nuisance and driving out the other deer; from then on it officially became a beast of the warren.[138] The other animal of the chase covered by royal forest law was the wild pig or boar, although hares also seem to have received the crown's protection as "honorary deer" in the royal warrens.[139] Boar hunting, while common among the elite in France and Germany throughout the Middle Ages, as evidenced by the recovery of their bones from archaeological sites there, does not seem to have been indulged in much by the nobility of England at any time during the medieval period, judging from the almost complete lack of their bones at all sites. Nonetheless, the archival record does indicate that wild boar were present during the

*Figure 3.4* STAG HUNT, misericord carving from Gloucester Cathedral, fourteenth-century. © Dominic Strange www.misericords.co.uk http://www.misericords.co.uk

mid-thirteenth century in the forest of Pickering in the north of England and in the forest of Dean in the west: Writs issued out of chancery from the crown to royal hunters instructed them to take a total of 300 swine from both places for the king's table in 1251 and 140 boars from the Forest of Dean in the 1260s. The only instance of boar poaching I have found in the forest eyre rolls is when two men took a wild boar on October 28, 1266 that had followed some domestic sows into the village of Rous Lench in Worcestershire.[140] Even if present in England at this time, the wild boar declined rapidly thereafter since it subsequently disappears altogether from the historical record, probably due to habitat loss rather than overhunting.[141]

Bear and beaver seem to have become extinct in Britain by the Roman and Anglo-Saxon periods, respectively, although bone remains of both beaver and brown bear were found in the early Anglo-Saxon settlement at West Stow in Suffolk.[142] The wolf was systematically hunted to extinction in England throughout much of the Middle Ages: Bounties were still being paid for wolf pelts in the thirteenth century, and an "engine" or trap for catching wolves is mentioned as being used in the forest of Macclesfield in Cheshire in 1303, but thereafter reliable records of the animal seem to disappear, although wolves continued to survive down into the eighteenth century in Scotland and Ireland.[143] Hunting of wolves was done with pits, nets, snares, poisoned and booby-trapped bait (containing tension-tied needles designed to pierce the wolf's intestines once swallowed whole), and with an assortment of traps, including hinged leg jaws akin to the modern spring-loaded variety, except made with wooden boards fitted with iron teeth.[144] Part of the reason for the ruthless hunting of wolves was that they had a fearsome reputation in the Middle Ages for not just attacking whole flocks or herds of domestic livestock, but also for hunting and consuming humans, even though actual documented cases of medieval wolf attacks are quite rare.[145] Cranes, herons, grey partridges, and spoonbills also seem to have been on their way to extinction in England by the end of the Middle Ages, judging from attempts by royal statute in the early sixteenth century to protect them and the absence of their bones in archaeological sites.[146]

On the Continent, boar, bear, and wolves seem to have survived the Middle Ages, judging from their continued inclusion as game animals in hunting manuals, albeit their numbers undoubtedly dwindled due to both hunting and habitat loss. In Scandinavia, wolves were not hunted to nearly the same extent as in Britain and they survived in good numbers down into the nineteenth century, largely due to the fact that wolves were not perceived as great a threat to game deer and livestock (particularly sheep) since human inroads had not penetrated as deep into the wolf's wilderness habitat.[147] Beaver were practically eradicated from the West due to drainage of wetlands and targeting as "pests," but otter seem to have thrived due to proliferation of fishponds. Bear were even a protected animal in Portugal during the fifteenth century under King Alfonso V, whose hunting ordinances assessed the enormous fine of 1000 *libras* for anyone killing a bear. This was hardly for the animal's sake, however, but rather to reserve the beast for the king's hunting, which apparently was done even in the spring when bear were just emerging from hibernation.[148]

Hunting in warrens typically targeted the hare and rabbit, but other small mammals hunted in a warren included the fox, badger, otter, wildcat (possibly lynx), marten, and squirrel, as well as game birds such as partridge, pheasant, plover, and lark. Some of these animals were classed as "vermin" and hunted as nuisance pests to other game and fish: The fourteenth-century hunting manual, *Le Livre des Deduis* (Book of Delights), attributed to Henri de Ferrières, has a fox and otter divide up the natural world between them as the "masters of waters and forests" (*maistres des yaues et des forés*), a pun on the title awarded the French king's master huntsman.[149] Even though the animals of the warren were typically exempted from the king's special protection accorded to the four beasts of the chase, hunting them in the royal forest was nonetheless a privilege that required a special warrant, which was eagerly sought after by a wide array of subjects of the English crown, ranging from dukes, barons, abbots and abbesses to town burgesses and men who seemed to specialize in such game.[150] The animals were mostly hunted for their pelts, since their meat (except for hare, rabbit, or birds) was considered inedible, although wolf's fur, which could be rather coarse, does not seem to have been traded much, since its hide was held to stink even after being treated or "tawed."[151]

The records of deer poaching in the forest eyre rolls provide our best window onto how hunting was actually carried out in medieval England. Since only a privileged few ever received licenses from the king to hunt in the royal forest, the range of those caught poaching was a very broad one, numbering in the hundreds for each roll and extending from lowly peasants and hardened criminals and outlaws to great nobles and high churchmen, with practically everyone else in between.[152] Most of the poachers seem to have been local men who came from the gentry class or above, for whom hunting was part of their social ethos and who perhaps had the financial wherewithal to absorb fines if caught and the leisure to indulge in a sporting pastime. This class of poachers usually hunted with bows and arrows and dogs (usually greyhounds and mastiffs but occasionally bercelets and spaniels) and did so quite openly, in the company of others, sometimes bagging large numbers of game and often resisting arrest when encountering foresters. Yet noble hunters by no means disdained more pedestrian forms of hunting, such as with nets, which still required dispatching the animal such as by slitting its throat with a hunting knife. In one instance, a noble hunter, Almaric le Despenser, lord of Oldberrow, stooped to the ruse of using a "stalking horse," i.e., a decoy "made up in the form of a horse" under which the hunters crouched and shot at their target, in order to poach two fallow bucks in Feckenham Forest in Worcestershire on October 18, 1279.[153] Nor did aristocrats refrain from taking fawns when the opportunity offered instead of more mature game, which can hardly have provided much sport of the chase or tested the hunter's mettle and skills.[154]

A substantial number of poachers, however, came from the lower classes such as the peasantry, who were often pardoned their fines due to poverty.[155] They usually operated singly or in pairs, although sometimes they occur as part of larger hunting parties, especially hardened or "notorious" poachers who served as guides perhaps for a fee and share in the bag or in exchange for protection and "maintenance" from their upper-class colleagues.[156] Hunting methods of choice for peasant poachers were snares or traps (*laqueus, trappa*, and *ingenium* in Latin), which used rope or horsehair cords and sometimes wooden frames hung up on high or along the ground to trip up deer by the antlers or legs.[157] One poacher, Robert le Noble of Sudborough, a chaplain, was caught by two walking foresters in the very process of making just such a snare, coming out of the woods at dusk carrying in his hand "a branch of green oak and an axe." A search of Noble's house the following morning also turned up the "the woodwork of a certain trap, with the string of the trap broken into two parts, and upon the string was hair from deer."[158] At the 1280 eyre for the New Forest in Hampshire, it was presented that all the men of the village of Langeley, numbering twenty in all, were accustomed to placing rope snares in order to take deer coming to browse in their grain and hay; perhaps this was a way to both protect their crops and gain some extra meat.[159] Other hunting weapons included slingshots, axes, sharpened wooden stakes, and even "large stones," which John Burmer used to bring down a stag on February 22, 1305 in Pickering Forest.[160] Snares and slingshots had the advantage of being easily transportable and hidden, while axes were allowed to be carried openly in the forest, although they usually could only be used against deer calves or deer that were weak, wounded, or semi-tame and thus easily approachable.[161] Bows and arrows entailed a fine simply for having them on one's person within the forest, since the assumption was that one was using them for hunting, and in any case these too often merely wounded the animal and required packs of dogs to track it.[162] Peasant poachers also differed from aristocratic ones in the often "opportunistic" way in which they came by their meat, scavenging the carcasses of deer wounded by other hunters or even ones diseased and half-eaten by wolves and dogs.[163] They were also more than happy to sell their game to willing buyers, such as to abbeys (which are frequently mentioned as harboring poachers) and to butchers in market towns, like Bristol and Gloucester near the Forest of Dean, Southampton and Winchester near the New Forest, and Droitwich near the forest of Feckenham.[164]

Unlike the image of medieval hunting provided by hunting manuals and tapestries, most poachers seem to have hunted on foot, since dogs are practically the only animals recorded as confiscated from poachers (by law, all dogs owned within the forest had to be expeditated, meaning three toes cut off of one forefoot, leaving only the ball, or else a fine of three shillings paid every three years); poachers' horses that are mentioned in the eyre rolls were usually used as pack animals to carry game out of the forest.[165] Nor did poachers have any regard for the official seasons of hunting, since they seem to have taken game at all times of the year, although sometimes they were obviously looking to bag a deer in preparation for a festive occasion like Christmas or Easter. A big difference between then and now is the medieval habit of hunting fawns, which would hardly be countenanced today (see Table 1).[166] But one has to remember that medieval people, at least those who could afford it, were used to eating young domestic livestock, such as veal calves, kid goats, lambs, and piglets, and a fawn would probably be regarded as a delicacy and a most tender cut of meat, if rather a small one. Some encylopedists like Vincent of Beauvais and several late medieval plague treatises recommended the flesh of deer-calves, along with that of immature farm animals, as particularly good meats to eat in order to maintain health.[167] Nonetheless, hunting fawns was obviously a very short-sighted hunting policy, since it did not allow these animals to grow up and help sustain the population by breeding.

In a different category altogether were the king's professional huntsmen, who numbered six or seven in the thirteenth century at the height of venison consumption by the royal household. These men were expected to cull hundreds of deer a year from various forests and parks across England, sometimes with only a few weeks notice, in order to provide venison for the king's festive meals, for which King Henry III was famous, although a considerable portion of the catch Henry typically gave away to select recipients of the royal favor. The huntsmen worked in teams with a staff of "valets" (berners and fewterers) in charge of packs of more than fifty running dogs and greyhounds, and the cost of such weeks-long expeditions could run in excess of £6. Most noblemen

Table 1 Deer Poached in the Royal Forests of England, 1245–1494

| Dates | Forests | Fallow | | | Red | | | Roe | Venison Offenses or Offenders unspecified[i] | Average per year |
|---|---|---|---|---|---|---|---|---|---|---|
| | | Bucks | Does | Fawns | Harts/Stags | Hinds | Fawns/Calves | | | |
| 1245–1255 | Rockingham (Northants.) | 87 | 53 | 8 | 15 | 3 | | 10 | 29 | 21 |
| 1248–1279 | Feckenham (Worcs.) | 52 | 57 | 5 | 5 | 6 | | | 160 | 9 |
| 1249–1285 | Cannock (Staffs.) | 58 | 53 | 11 | 13 | 13 | 3 | 8 | 66 | 6 |
| 1250–1285 | Kinver (Staffs.) | 7 | 4 | 2 | 38 | 40 | 9 | 2 | 34 | 4 |
| 1262–1287 | Sherwood (Notts.) | 67 | 120 | 8 | 43 | 53 | 4 | 2 | 35 | 13 |
| 1247–1326 | New Forest (Hants.) | 65 | 55 | 10 | 16 | 13 | 7 | | 885 | 13 |
| 1484–1494 | New Forest (Hants.) | 101 | 180 | 9 | 16 | 75 | 5 | | 94 | 48 |
| Total/Average | | 437 | 522 | 53 | 146 | 203 | 28 | 22 | 1303 | 16 |

Sources: *Select Pleas of the Forest*, ed. G.J. Turner (London: Selden Society, 1901), pp. 27–38, 79–116; *Records of Feckenham Forest, Worcestershire, c. 1236–1377*, ed. Jean Birrell (Worcestershire Historical Society, new series, 21, 2006), pp. 43–44, 60–76, 90–91; *Collections for a History of Staffordshire. The Forests of Cannock and Kinver: Select Documents, 1235–1372*, ed. Jean Birrell (Staffordshire Record Society, 4th series, 18, 1999), pp. 42–58, 64–71, 77–85, 112–36; *The Sherwood Forest Book*, ed. Helen E. Boulton (Thoroton Society Record Series, 23, 1965), pp. 110–40; *A Calendar of New Forest Documents, 1244–1334*, ed. D.J. Stagg (Hampshire Record Series, 3, 1979), pp. 69–74, 77–92, 96–114, 167–77; *A Calendar of New Forest Documents: The Fifteenth to the Seventeenth Centuries*, ed. D.J. Stagg (Hampshire Record Series, 5, 1983), pp. 1–24, 30–36.

i In these instances, either: the species of deer is not specified, listed simply as "beast"; offenders are described as entering the forest with the "intent" to offend but no firm result is recorded; or the accused is described as a "habitual" or "common" offender but no exact figure is given of how many deer he actually poached. Deer wounded by arrows and trapped by snares are counted as poaching kills even if they are recorded as escaping for the time being.

in fact seem to have employed huntsmen on a part-time basis as needed, or had a household servant do double duty when the occasion required.[168] The favored hunting technique for such large culls of the deer herd by the royal huntsmen seems to have been the "bow and stable" method or the "deer-drive," in which the deer were driven by dogs and beaters past set ambushes of archers who would shoot as many of the animals as they could as they ran past, with the dogs finishing off those merely wounded.[169] This might yield 20–30 deer at a time and seems to have been the most efficient way of harvesting deer as quickly as possible, but the huntsmen would also use nets, particularly when they had to trap deer live for transport to a recipient's park.[170]

How effective can we say the royal forest system was in conserving the deer? This is a question that is asked not only by modern scholars but was also asked by the medieval crown as well: One of the "chapters of enquiry" by the justices of the forest eyre was how "the king's venison is kept," and a typical "view of the forest" ordered by the crown, such as took place in the forests of Cannock and Kinver in Staffordshire in 1235, noted not only trees and underwood felled that provided cover and browse for deer but also the general number of animals themselves, such as that few or no deer were seen in the hay of Alrewas and the wood of Bernesmor, while a "reasonable number of beasts" were viewed in the hay of Teddesley.[171] When Sir Richard Cholmeley was arraigned on a bill of complaint for his stewardship of Pickering Forest between 1499 and 1503, he claimed that he left the deer "in fare better condicion" than they were at the time of his entry into the office, a defence that the jury found credible, since eight men making a circuit of the forest viewed 140–60 red deer within two hours, representing, they said, an increase in the red deer population by 200 and in that of fallow deer by 160. Sir Richard also made it his policy to give away a certain number of deer to "lords and gentylmen" living on the borders of the forest so that they "shuld be loving and favorable to the Kynges game there," basically coopting them from poaching.[172] This same policy was used by Sir Richard's fourteenth-century predecessor, John Dalton, to justify the taking of a total of 292 deer of which he was accused in 1334, and it probably also explains some of the extraordinary harvest of 2000 deer in six years (over 300 deer a year), between 1216 and 1222, that was charged to William de Ferrers, earl of Derby and bailiff of the Forest of High Peak in Derbyshire.[173]

Clearly, people in the Middle Ages at least had the concept of conservation of sustainable deer populations. In 1271 King Henry III suspended deer hunting in Feckenham Forest in Worcestershire for two years, overriding any permits he had issued in the meantime, due to its diminished deer population, and in 1489 King Henry VII did the same in Pickering Forest in Yorkshire for three years, owing to the fact that it had come to the king's attention that "our game of dere and warenne within our seid Honnor is gretly diminished by excessive hunting." During the mid-fourteenth century, the Black Prince was concerned enough about the deer in Delamere Forest in Cheshire to require the abbey of St. Werbergh in Chester to reduce its annual deer quota by half, and for similar reasons he enjoined that more lodges and chambers be built for foresters and that timber-cutting privileges not be abused. In 1328 the Prince's father, King Edward III, was advised to exchange the tithe of venison enjoyed by St. Mary's Abbey in York from the forest of Galtres with free hunting of game in Spaunton Forest for a period of five years, the reason being that Galtres was "burdened beyond measure by too frequent hunting therein for the use of the king and by other persons with [his] leave."[174]

Nonetheless, it does not seem that deer populations in the royal forests were excessively burdened by poaching or other forms of hunting. Table 1 gives some figures based on the eyre rolls for deer poached from six royal forests distributed across England and over an extended period of time, from the mid-thirteenth to the first quarter of the fourteenth centuries and from the very end of the Middle Ages.[175] One notices immediately that yearly averages of deer taken could vary quite a bit from forest to forest; much of this can be attributed simply to the size of the forest and thus the number of deer available to be poached. Rockingham Forest in Northamptonshire, at about 350 square miles, was nearly twice the size or more of other forests such as Feckenham Forest in Worcestershire, Sherwood Forest in Nottinghamshire, and the New Forest in Hampshire, at 185, 156, and 150 square miles respectively, and the much higher yearly average of deer taken at Rockingham seems to reflect this fact.[176] It is also obvious that the category of unspecified venison offenses constitute a big unknown: For example, the extraordinarily large and unusual number of 885 offenses for the New Forest in the thirteenth century can be mostly accounted for by the 500 beasts that were charged to the steward of the forest, Walter de Kent, for which he was fined an astounding £5000, or £10 per beast, in addition to 151 beasts, or "beasts without number," laid to his account at the 1276 and 1280 eyres. It is not clear whether this was a politically motivated punishment for a courtier who fell from grace or an exposure of a truly corrupt forest official, but the importance of the case is revealed by the fact that the king, Edward I, and his queen, Eleanor of Castile (who officially owned the New Forest), made a personal appearance in the court sitting at Lyndhurst in order to empanel a new tribunal of justices on August 24, 1276.[177] In any event, the total yearly average of 16 deer from each forest should have been easily sustained by deer populations that probably numbered in the thousands. If we extrapolate this for the 70 forests within the royal forest system, this would give us some 1120 deer poached each year throughout the whole of England. Moreover, the fact that the vast majority of deer poached were of the fallow species indicates that most of them were taken in royal parks or hays (as often noted in the record), where hunters and trappers took advantage of the fact that deer were confined in concentrated populations and had customary feeding grounds, so that even poaching was a heavily dependent by-product of the semi-domestic management of deer herds, i.e., deer farming.[178] (An exception seems to be the forest of Kinver in Staffordshire, where poached red deer greatly outnumbered that of fallow.)

By comparison, the royal household of Henry III utilized at the height of its consumption an average of 811 deer per year, of which the vast majority were fallow deer (607), followed by red (149) and roe (45). Nearly half this total went to the king's table (at which the number of game could reach into the hundreds, and in the case of hares, even over a thousand); a third was given away in the form of carcasses and another sixth in live form for stocking parks (all gifts being of fallow deer).[179] If we also include in the royal column licenses granted to hunt—which sometimes could be for extraordinarily large quantities of venison, such as the 70 deer that the earl of Cornwall's son, Edmund, was allowed to take from four forests in October 1283, then this could easily have equaled or exceeded what poachers took according to the eyre rolls. Like those that were poached, the deer taken for the royal household were fairly evenly distributed across the various royal forests, since the crown set specified quotas for its huntsmen from each forest in order not to "cause destruction" or overburden the deer population of any one of them.[180]

Naturally, official poaching figures from the eyre rolls are said to grossly underestimate the real scope of illegal hunting, since the judicial machinery of the time could hardly be expected to catch everyone in its net.[181] This assumes that behind every "common" or "habitual" offender of the venison who is counted but once in Table 1 lies a whole slew of crimes of which he can be accused, but realistically, how many deer does this represent?[182] The most successful poachers were those who worked in large gangs, but these were exactly the kind of hunting occasions that were most likely to draw the attention of forest officials and of the forest communities obligated to report them, and they are duly recorded in the judicial records.[183] Otherwise, poachers could most easily evade detection by working alone or in small groups, but their chances of success were then drastically reduced. Even modern hunters and trappers, with their technologically more advanced weaponry and locating devices, mostly fail to get their bag.[184] In the case of one medieval hunting party in Sherwood Forest led by two rectors, the poachers spent a total of twelve days in the woods, from July 4 to July 15, 1277, without evidently taking anything, since a servant of one of the rectors had to bring them food.[185] (The rectors must also have found proxies to cover for them in church on Sunday!) One also has to keep in mind that any medieval poacher entering the forest had to keep one eye out for forest officials *hunting him* at the same time as he was hunting his quarry. Fines for the hardened offender could run into the hundreds of pounds, and merely to be caught in the vicinity of a dead deer could result in years of imprisonment, while some notorious poachers were arbitrarily beheaded or shot to death by their pursuers.[186] Some poachers had to go to great lengths not to be caught or recognized, such as by adopting aliases and donning masks and other disguises: Thus, Roger, reaper for the abbot of Peterborough, gave his surname (surely in jest) as "Grim" to authorities after being caught stalking four hinds with dogs, and Hugh le Scot adopted the guise of a woman in order to escape the church where he had been holed up for a month seeking sanctuary.[187] Certainly there are cases of communities obstructing the pursuit of poachers and of forest officials who colluded with or themselves participated in poaching, but there are also numerous instances of foresters who bravely performed their duty even when badly outnumbered, risking death from arrow wounds or being subjected to beatings and trussings or having their dogs strung up and their horses shot out from underneath them.[188] The effectiveness of forest law in stopping at least some poaching is also evident from the fact that spikes in lawlessness occurred whenever parts of the forest were disafforested or at times of political unrest and uncertainty.[189] Like with the medieval Inquisition, the forest courts were able to wield their written records, the eyre rolls, as a potent weapon against offenders, noting instances of when suspects were previously convicted or bailed in order to clinch convictions or increase fines against them.[190]

Since medieval hunters were haled before the courts and adjudged as common offenders of the venison just for entering the forest with the *intent* to take game, without there being any necessity of proof that they actually bagged their quarry, unspecified offenses may not always underestimate the number of animals actually killed. Even if we take one historian's advice to multiply our poaching figures by a factor of three in order to account for undetected kills (thereby giving us a total yearly average of 3360 for all 70 forests), this would still yield only one deer taken for every 150 acres, assuming that about half a million acres made up the entire royal forest system; a further threefold increase to account for all hunting, both illicit and licit (i.e., conducted by the king and his huntsmen and those receiving his warrant), would result in one deer from every 50 acres, a highly sustainable cull rate.[191] A roughly equal amount of acreage was probably encompassed by private deer parks (where most illegal hunting may actually have taken place), and which seem to have been able to sustain about one deer for every two acres. Even though deer could not be managed and cared for to the same degree in the royal forest, it is more than likely that medieval England's deer population was well in excess of half a million beasts. Poaching therefore culled a bare fraction, less than 1 percent, of the year-to-year deer herd, far less than the 9 percent yearly rate at which park owners culled their private herds for their own consumption.[192] The records of hundreds or even thousands of deer dying from murrain in royal forests during the late thirteenth, early fourteenth, and late fifteenth centuries point in fact to an overpopulation of deer that was unsustainable by the natural ecology.[193] If deer populations were in decline by the dawn of the fourteenth century, as reflected in declining consumption and gifts of deer by the crown, then this seems to have been largely a function of reduced habitat from arable expansion at the expense of the forest, although on occasion hundreds of acres of assarts were literally stopped in their tracks because of the deer.[194] With the advent of the Black Death in the mid-fourteenth century, some of this habitat should have been restored as well as much of the hunting pressure relieved due to the drastic collapse in the human population.

Hunting seems to have undergone something of a proletarianization in the later Middle Ages. This was in spite of game laws like the one enacted by the

English Parliament in 1390, which tried to restrict hunting among townsmen and the lower classes, such as by forbidding "artificers," "labourers," and any other layman owning property worth less than £2 a year from keeping greyhounds or any other hunting dogs, as well as ferrets, "heys" or buckstalls (enclosures into which deer were driven), nets, "harepipes" or traps, "cords" or snares, and other "engines" for hunting deer, hares, or rabbits.[195] Peasant poaching was still very much a concern by the end of the medieval period: In 1485, yet another statute complained of the "rioutouse" manner in which poaching in forests, parks, and warrens was taking place, as carried out by "divers persons" decked out with "paynted faces" and visors and arrayed "in maner of werre," who nonetheless could not be apprehended because their identity, owing to their disguises, could not be known.[196] The statute's complaint that this sort of hunting was leading to "heynous rebellions, insurreccions, rioutres, robberies, murdres, and other inconveniences," was not far off the mark when we read the rather understated entry for the New Forest roll of 1488, that 500 deer were slain during the turbulent reign of King Richard III (1485–88) by "Northern men and servants of the king" who had been brought down to assist in Richard's usurpation of the crown.[197] (These are not counted in Table 1, since it seems clear that they were at least tacitly allowed by the government.) Even without this over-large cull, the yearly average of deer poached in the New Forest during the decade 1484–94 was the highest recorded at any time in the country.

Similarly restrictive game laws were promulgated on the Continent, such as an ordinance issued by Charles VI of France in 1396, which forbade laborers and other "non-noble persons" from hunting without royal permission game including red and black deer, rabbits, hares, partridges, pheasants and "other beasts and birds" using dogs, ferrets, snares, nets, and other engines of the chase.[198] Such laws testify to both a perceived greater practice of hunting by the peasantry which necessitated them, and a cultural expectation that this was not the norm in medieval society, although some claimed that the laws were a novel attempt to turn hunting into an aristocratic privilege, overriding ancient rights of men of whatever class to take game in common fields and woods. Exactly the same dynamics were at work in labor and sumptuary laws enacted in the wake of the Black Death, and the tensions created between expectations and reality, among both the ruling and subject classes, naturally exploded in the form of peasant uprisings and other expressions of discontent and defiance of such restrictions, like poaching and breaking into private parks and warrens.

The eyre rolls for the New Forest in Hampshire during one of the closing decades of the Middle Ages, between 1484 and 1494, provide a remarkable picture of how hunting came to be dominated by tradesmen and the lower classes of late medieval society, even in the face of restrictive game laws. Table 2 gives the breakdown for 143 separate offenders listed in this roll by profession or social class. By far the most common poachers came from the yeoman class, who account for almost a third of all offenders. Even though the members of

# 194   An environmental history of the Middle Ages

*Table 2* Frequency of Professions or Social Classes Among Deer Poachers in the New Forest, 1484–1494

|  | Number of Offenses | Percentage of Total |
|---|---|---|
| Gentleman includes one "gentlewoman," Elizabeth Bedyng, of Newton Bury. Includes one "gentlewoman," Elizabeth Bedyng, of Newton Bury. | 18 | 12.5 |
| Squire | 5 | 3.5 |
| Knight | 1 | .7 |
| Yeoman | 43 | 30.3 |
| Groom | 1 | .7 |
| Servant | 6 | 4.2 |
| Bailiff | 2 | 1.4 |
| Priest included in this category are men described as vicars, parsons, rectors, and chaplains. | 10 | 7 |
| Monk | 3 | 2.1 |
| Husbandman | 16 | 11.3 |
| Laborer | 10 | 7 |
| Butcher | 5 | 3.5 |
| Tailor | 4 | 2.8 |
| Baker | 2 | 1.4 |
| Mercer | 3 | 2.1 |
| Saddler | 2 | 1.4 |
| Fuller | 2 | 1.4 |
| Merchant | 1 | .7 |
| Fletcher | 1 | .7 |
| Faucet-Maker | 1 | .7 |
| Bucket-Maker | 1 | .7 |
| Shoemaker | 1 | .7 |
| Vagabond | 2 | 1.4 |
| "Singleman" | 1 | .7 |
| Total | 143 | |

Source: *A Calendar of New Forest Documents: The Fifteenth to the Seventeenth Centuries*, ed. D.J. Stagg (Hampshire Record Series, 5, 1983), pp. 1–24, 30–36.

this class could represent a wide range of wealth and social status, at the lower end a yeoman, typically defined as a freeman holding a farm or estate worth at least £2, or roughly 30–40 acres, barely qualified to hunt with dogs according to the 1390 statute. Certainly below this cutoff would have been the husbandman farmers, whose numbers nearly equal those of the gentleman poachers that Parliament would have deemed more socially acceptable as hunting participants. Most remarkable in this list is the number of trades represented whose members would have practiced in the town or city: Altogether 13 such professions are listed, some, such as "turner," i.e., one who manufactures wooden implements with a hand-turned lathe, or "faucet-maker," who made the spigots used to tap barrels of beer or wine, seem to be new arrivals on the medieval artisanal scene. No doubt these were exactly the kind of "artificers" that Parliament had in mind when seeking to limit access to hunting; although the numbers in each category are quite small, when taken in aggregate the poachers from the thirteen trades together make up nearly 18 percent of all those hunting in the New Forest. Motivations for the more widespread popularity of hunting may have included not only a practical desire to put meat on the table, but also a recreational incentive to engage in "moderate exercise" which hunting offered, as the fifteenth-century plague treatise of Johann Widmans recommends, as does Gaston Phoebus in his *Livre de Chasse*, who mentions the sweating induced by hunting that purges sickness.[199] As seems to have been equally true in earlier centuries, poachers here also crossed class boundaries in order to hunt: Thus Elizabeth Bedyng, described as a gentlewoman, hunted with her fellow residents of Newton Bury, John Newman, a husbandman, and Stephen Houker, a laborer, in order to bring down a beast at a place called "La Lee" on July 3, 1489, while two Londoners, Edmund Bremous, tailor, and John Fredsam, esquire, stood surety for Walter Tradde, a yeoman of Beaulieu, who was convicted of a venison offense at the eyre in 1487.[200]

## Animals of the hunt: romance vs. reality

How do the hunting manuals that proliferated during the late Middle Ages compare with the reality of medieval hunting, as this is revealed to us in the eyre rolls? Such manuals, along with literary romances and decorative artworks that feature hunting like *Sir Gawain and the Green Knight* and the Devonshire Tapestries, form the basis for most modern studies of medieval hunting.[201] Yet the picture we get of hunting from the manuals, romances, and tapestries is a highly ritualized and ceremonial one (consisting of no less than eight stages to the hunt), whose main value seems to be that it reflects the cultural attitudes and image that the aristocracy preferred to project of itself, rather than showing us how hunting was actually practiced in real life. In keeping with this, the manuals and other stylized depictions of medieval hunting emphasize the hierarchical dominance of the aristocracy over both their social inferiors among humans and over the natural world of animals.[202] This image was a fantasy, a literary convention that had practically no relation to the hard realities of medieval hunting in everyday life as it was practiced by most people, including the aristocracy.[203] One scholar who has made a study of poaching in medieval England based on the eyre rolls has concluded that there is nothing in them "to suggest that the hunt was conducted according to the complex procedures described in late medieval hunting literature," such as that the hunt should be carried out in "distinct stages" with assigned roles to each participant, including the final ceremony of the "unmaking" of the deer's carcass (i.e., gutting it).[204] Indeed, one would hardly expect that poachers of whatever class would indulge in such rituals as horn calls and voice cries when seeking to avoid detection by the royal foresters. Those who foolishly did so got caught: Roger le Scot, the serving man of Lady Isabel de

Brus, was apprehended in the forest of Essex on July 18, 1242 as he stalked a doe wounded in the thigh because the forester, Roger of Wollaston, recognized his horn call, while James of Panton was convicted at the Rutland eyre in 1269 of poaching four does beyond the two he was allowed by warrant because his beating of drums flushed too many deer out of the forest.[205] Any gutting of a deer carcass was a nerve-wracking job that had to be done in a hurry, since it was then that poachers were so often apprehended by foresters, and there must have been the additional concern of leaving bloody traces for officials to find later. Thus, a poacher's bloody surcoat from butchering a roe deer, even though turned inside out, gave him away to the very steward of Rockingham Forest, Sir Hugh de Goldingham, as he came riding towards the woods after dinner on April 10, 1251.[206] When carcasses were cut up and distributed by poachers at the kill site, it was often driven by other considerations that were far from the niceties of ceremonial, such as the necessity of concealing the game or bribing onlookers with a piece of meat.[207] Despite the portrayal of noble unmakers such as we get in the prose romance, *Sir Tristram*, in real life noblemen seem to have left the messy job of gutting deer to their servants: When the mastiffs of William de la More conveniently chased a doe right to his moated manor near Cannock Forest, he had his groom, ploughman, and shepherd carry the beast out to the moor to be skinned there, since apparently he didn't want the beast to be seen "inside his courtyard."[208]

Perhaps hunting ritual could be indulged in by the king when he hunted in person, which is sometimes glimpsed in the records. Henry III, one of the more enthusiastic huntsmen among England's kings, is recorded as coming at least four times between 1248 and 1255 to the forest of Rockingham in Northamptonshire in order to hunt deer "at his pleasure," on two occasions accompanied by his queen, Eleanor of Provence.[209] In 1323, Edward II employed both nets and the chase when hunting in the forest of Pickering in Yorkshire.[210] Yet on the whole royal hunts were extremely rare affairs, said to be about as common as a coronation, that were not at all typical of most medieval hunting. Indeed, the stage-managed nature of the royal hunt is illustrated when preparations were made for the coming of King Henry VII to the New Forest around the time of his coronation in 1488, when 17 deer were driven into New Park armed with a "stable" or buckstall, and, as to be expected, all but two of the deer were subsequently slain. Similar preparations were made for Henry's son, Henry VIII, for a hunt in Hatfield Chase in Yorkshire in 1541: 20 bucks were to be captured live "to be ready for the king's sport" by the chief forester of Hatfield, Francis Talbot, earl of Shrewsbury, but in the end the hunt was aborted.[211] Instead, the crown usually relied on its professional huntsmen to provide its venison, which we have seen was done in the most efficient manner possible, and one would hardly expect the larderer that followed in the huntsmen's wake to observe ceremony when dressing dozens of carcasses in preparation for shipment back to the royal household.[212] The statement made in *The Master of Game* that deer in England "are not slain except with hounds or with [bow] shot or with strength of running hounds" is flatly contradicted by the eyre rolls, which reveal that prosaic but highly effective hunting methods such as netting deer were common practices even among "knightly and baronial hunters."[213] When Lord James Audley took a doe's fawn in the forest of Cannock on June 21, 1360 (during fence month), it is hard to imagine him then going back to his castle or manor-house to listen to the fierce and daring hunting exploits of Bertilak from *Sir Gawain and the Green Knight*: Such a thing is more worthy of a satiric scene from Cervantes' *Don Quixote* than anything portrayed in a noble hunting manual.[214]

It is claimed that excavations of medieval castle sites in England support the notion that *par force* hunting and unmaking rituals such as are described in hunting manuals were actually performed, yet this evidence is actually quite inconsistent and inconclusive. The basis for the claim is that there is a prevalence of hind limb bones, which would have been kept and carried back to the castle by the noble hunter in accordance with hunting ritual, and a corresponding absence of the parts of the skeleton, such as the pelvis and meaty forelimbs, which would have been ceremoniously disposed of or granted away at the kill sites.[215] Nevertheless, one must be wary of extrapolating too much from the archaeological record without the support of documentary (as opposed to literary) evidence. If the pelvis or *"corbyn* bone" was indeed cast away at the kill site as a ceremonial offering to ravens and crows, then why weren't these found in great numbers by foresters and cited as evidence of poaching in the eyre rolls, since by law any part of a deer that had been hunted illegally in the royal forest had to be kept intact and presented as evidence before the court? The most common part of the deer that was actually found in the house or on the person of the poacher was the skin or hide; other confiscated body parts included the shoulder, side, and haunch, which were of course highly prized, but in addition poachers were caught with the head, neck, ears, and entrails or offal, which the manuals considered fit only for feeding to the dogs.[216] Even noble hunters must have realized that leaving any part of a deer carcass for royal officials to find and trace back to them was indeed a very foolish thing to do; instead, deer were usually carried away whole back to the redoubt of the hunters, where perhaps there may have been time, leisure, and inclination to dispose of them ritually and gift them out to various parties as ceremony demanded. But of course in no way does this constitute evidence of ritualized hunting out in the field. Most of the incomplete faunal assemblages in England seem to come from red deer, which is indeed the deer species that serves as the focus of *par force* hunting in the manuals, but by the late Middle Ages, which is when the vast majority of the classic hunting manuals or treatises were produced, it is the fallow deer that overwhelmingly predominates at castle, urban, and religious sites.[217] Fallow deer were also the most common species hunted by poachers in the royal forest, according to the eyre rolls (see Table 1).[218] Yet fallow deer were frowned upon as an object of *par force* hunting by the manuals due to their poor stamina in the chase, which is precisely why they were the favored beasts to be kept in the confined space of deer parks, generally too small to allow for the re-enactment of the classic *par force* hunt.[219] But since red deer tend to be the species most prevalent at early medieval sites, I would suggest that by the time of the late medieval heyday of the hunting manual, the ritualized, *par force* hunting of them was a holdover from a "mythical" time that was enshrined as a nostalgic memory, rather than as a description of the contemporary reality of the business of hunting (usually done in most aristocratic households by servants).

Ironically, perhaps the greatest danger and thrill of the chase for aristocratic poachers in the royal forests was provided by the king's foresters, whose attention was naturally attracted by large hunting parties.[220] In nearly every forest examples can be found of bloody duels to the death between well-armed poachers and foresters. One such incident occurred on May 2, 1246, when four foresters ambushed a party of poachers following five greyhounds and who were armed with a crossbow and bows and arrows in the "lawn of Beanfield" at the edge of Rockingham Forest; in the ensuing fight, Matthew of Thurlbear, forester of Brigstock Park, was wounded by two "Welsh arrows," one under his left breast about a hand deep "slantwise," the other in his left arm "to the depth of two fingers," wounds that proved fatal.[221] More lucky were two walking foresters who met up around midnight at Weybridge in Huntingdonshire with a party of a dozen poachers armed with an axe, a "long stick," and bows and arrows: After exchanging arrows, which included on the poachers' side both barbed and "genderated" varieties (the latter of which was armed with a lead ball designed not to penetrate deeply but to break bone), the poachers disappeared into "the thickness of the wood and the darkness of the night."[222] Perhaps even more disturbing to the crown than such assaults upon their forest officers was a case of outright, brazen defiance of the forest law by a gang of 19 poachers, which included an earl's bailiff, a chaplain, and a woodward, in the forest of Rockingham on August 24, 1272: After shooting with bows and arrows in the forest all day, killing a dozen deer, they cut off the head of a buck and a doe and stuck them on stakes with a spindle and a "billet" (i.e., a writ or document) stuck in their mouths, an act that was held to be "in great contempt of the lord king and of his foresters."[223] The 31 poachers who took 43 red deer on Blakey Moor in Pickering Forest in 1334 likewise cut off the heads of nine of the animals and "fixed them upon stakes in the moor." One of these men, Hugh de Neville, also a former bailiff of the forest, had set up the head of one of six harts he had poached upon a pillory at Pickering in the sight of Henry of Grosmont, earl of Lancaster, and was then arrested and imprisoned in the castle for six weeks in iron fetters.[224] Unmakings of this kind sent almost the exact opposite message from that in hunting manuals: Rather than maintaining a rigid division or hierarchy to society, medieval poaching tended to elide such social differences, bringing together "respectable" or mainstream as well as unsavory and fringe elements of the hunting community together in a mutual effort at connivance in pursuit of their shared love of hunting, in spite of the king's monopoly. Since other countries with a strong hunting culture, such as Scotland, France, and Spain, also had in place and enforced royal hunting monopolies, one would expect that similar dynamics among poachers and hunters operated there as well.[225]

It should be almost superfluous to say that peasant deer poachers were completely untouched by the ritualized hunt: They evidently acquired hunting skills in their own fashion, passed down from father to son, which were especially geared to evade detection and valued craftiness and stealth, wholly foreign to the ethos of the noble hunting manual.[226] (Deer bones at town or village sites generally do not bear evidence of the unmaking ritual.)[227] Despite women being portrayed in manuscript illuminations and in some hunting literature as participating in stag hunts or the unmaking of deer, the records of the forest eyre rolls generally show them as receivers of poachers and their game: Most often these women were the poachers' mothers or noble patronesses, such as Lady Agnes of Bednall in Staffordshire, who allowed four poachers to set out from her house with her greyhounds, which they used to take a buck in Cannock Forest and returned with it to the house on July 26, 1258, for which Agnes was fined two marks, or a little over £1, paid in court by her husband, Nicholas of Orton.[228] In this case it seems clear that the lady was the chief instigator and beneficiary of the hunt, even if she did not participate directly herself. But sometimes women are cited as the ones doing the actual poaching. On September 28, 1332, Lady Blanche, Baroness Wake of Liddell, together with her men took in the forest of Pickering at Emmeldburg a soar (a four-year-old fallow buck) and two hinds; undoubtedly this lady felt herself entitled to hunt here, since her father, Henry, was the present earl of Lancaster who owned the forest and indeed, the records state that he "directed his justices to stay all further proceedings" against his daughter.[229] No doubt Amice de Clare, countess of Wight, also felt herself entitled by baronial privilege to take two beasts in the New Forest on January 20, 1253, on her return from the court of the queen, Eleanor of Provence.[230] More typically, ladies are portrayed as participating in falconry or hunting the beasts of the warren: The abbess of Barking in Essex was granted license in 1221 to hunt foxes and hares in Havering Park, while Matilda de Bruys was said to be accustomed to hunting hares (and catching them) in Pickering Forest, for which she was fined five shillings.[231] Another problematic aspect of the late medieval hunting manuals is their derivative nature and heavy moral and allegorical tone, which they share with bestiaries. For example, Henri de Ferrières' *Livre des Deduis* of 1354–76, also known under the title, *Livres du Roi Modus et de la Reine Ratio* (Book of King Method and Queen Reason), since it is paired with an allegorical work by the same author, served as the model for Gaston Phoebus' *Livre de Chasse* of 1387–89, which in turn was copied and translated by Edward, duke of York, for his *Master of Game* of 1406–13.[232] Observing the "fewmets," or scat, preparatory to hunting the hart, as Gaston Phoebus recommends in a chapter from his *Livre de Chasse*, is a practical piece of advice that any hunter would recommend today, but his accounts of the bear licking its cub into physical form at

200  *An environmental history of the Middle Ages*

birth or of the hare as a hermaphrodite are clearly legends going back to Pliny and the *Physiologus*.[233] This requires a close and careful reading of the manuals to separate out whatever fact can be gleaned from fiction, which is not always possible to do as the two are so inextricably intertwined. Similar observations can be made about the bestiary genre as a source for how medieval people related to wild animals in general.[234]

## Animals of the hunt: falconry and fishing

Two variations on the theme of hunting that also deserve a mention here are falconry or hawking, under which we can also include fowling, and fishing. Both falconry and fishing were driven throughout the high and late Middle Ages by the demand for birds and fish as food, although considerations of sport ranked equally high among aristocratic practitioners as reasons for indulging in them. Bone assemblages from medieval archaeological sites, particularly those at castles, record an impressive number and variety of wild birds and fishes that were eaten. Consumption of certain species, such as swans and herons or salmon and sturgeon (reserved by French and English law from the thirteenth century for exclusive use of the king) became a way for the medieval aristocracy to maintain social distinctions in eating habits, particularly as

*Figure 3.5* A FALCON WITH A PREY BIRD IN ITS CLAWS, misericord carving from Chester Cathedral, c. 1390 © Dominic Strange www.misericords.co.uk http://www.misericords.co.uk

more meat became available to all classes during the late Middle Ages. Such was the desirability of wild birds at feasts that special farms dedicated to raising them, known as swanneries and heronries, arose to fulfill demand.[235]

Medieval falconry inspired in some authors a more "scientific" study of nature based on direct, personal observation and experience.[236] This is particularly true of the Emperor Frederick II's *De Arte Venandi cum Avibus* (On the Art of Hunting with Birds), as well as Albert the Great's section on falconry from the *De Animalibus*, both dating to the thirteenth century.[237] For example, Frederick's advice to leave nestling falcons as long as possible with their parents, for "no one can feed them like the parent bird," is borne out by my own and others' experiences in wildlife rehabilitation: Naked hatchlings almost never survive, while it is unheard of to successfully hatch wild birds from found eggs, since there is a very narrow window of time before the embryo grows cold and dies.[238] Equally astute is the emperor's observation that falcons never really become "tame" in terms of developing a close attachment to men in the same way as dogs and other mammals do, which is why modern falconers carefully monitor the weight of their birds, in order to keep them always a little perpetually hungry so that they will return to the lure.[239] Albert the Great provides detailed descriptions of various species of falcons that seem derived from his own observations, independent of earlier authorities, while he cites "a very experienced falconer" from the Alps as his source for how the eyries of peregrine falcons built into the hollow of sheer cliff-faces were accessed, which involved the hair-raising method of lowering a man secured to a rope as much as 300 feet long down from the top of the cliff.[240]

Albert's extensive list of cures for diseases in falcons, which included herbal concoctions (typically infused into the falcon's meat), special diets, theriacs, plasters, bleeding, cauterization, and so on, testify to the high level of veterinary care that falcons received, on a par with that for horses and dogs and even humans.[241] On a more mundane level, fowlers, who provided most of the birds set on noble tables, made use of wide variety of nets, traps, snares, projectiles, decoys, baits (which included "bird-lime," a sticky substance that held the bird fast wherever it perched), bird-calls, and even camouflage.[242]

Medieval fishing was done in a variety of ways, either by hand using the "tickling" method in which fish were enticed close enough to be grabbed (especially when the hand was smeared with a fragrant "salve") or by using various nets, traps, baited hooks and lines, spears and harpoons, even piscicides made from herbal ingredients and quicklime explosives that stunned large schools at a time.

Much of the commercial trade in fish during the Middle Ages, which focused on preserved salt-water varieties, especially herring and cod, was driven by rising demand as towns began to dramatically increase their populations in the high Middle Ages and the Church effectively imposed its religious fasting requirements, while overfishing and man-made alterations to the environment of inland freshwater resources necessitated turning to marine supplies of fish. By the thirteenth and fourteenth centuries, Denmark was exporting in

*Figure 3.6* BIRDS, woodcut from Konrad of Megenberg, *Buch der Natur* (Augsburg, 1481). U.S. Library of Congress

excess of a hundred million herring a year and Norway several thousand tons of cod. The durability and bulk of processed herring and cod made them accessible and affordable to all classes of medieval society.²⁴³ Artificial fishing environments, such as fishponds and fish weirs set into river beds and estuaries, tried to ensure a steady supply of especially fresh-water fish, but these

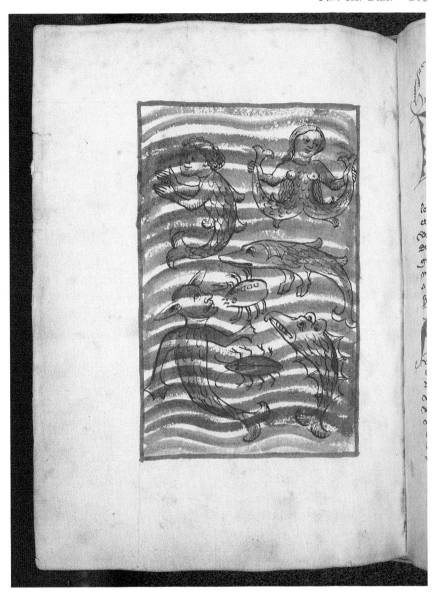

*Figure* 3.7 FISH AND OTHER AQUATIC CREATURES, woodcut from Konrad von Megenberg, *Buch der Natur*. Johannes Hartlieb, Kräuterbuch (Cod. Pal. germ. 311), fol.160v, Universitätsbibliothek Heidelberg

were also subject to much regulation and contention as they impeded river traffic and fish migrations and caused flooding of neighboring woods, crops, and pasture land. Fishponds at their most elaborate entailed a whole complex of holding areas for fish at various stages of their life-cycle, from young fry to

*Figure 3.8* MISERICORD CARVING OF FISH, probably pike, at Exeter Cathedral, thirteenth century © Dominic Strange www.misericords.co.uk http://www.misericords.co.uk

mature stock (equivalent to state fish hatcheries, such as the one in my hometown of Roxbury, Vermont), and they were used by wealthy patrons such as the English crown to reward favored subjects with coveted samples of bream, tench, roach, perch, dace, eels, pike, and carp, just as much as for its own supply.[244] This finds a close parallel in the management of deer parks and dedicated wild bird farms.

Towards the end of the Middle Ages, the growing popularity of hunting, falconry, and fishing as recreational sports is indicated by the emergence of practical manuals or how-to guides written in the vernacular, such as the *Boke of St. Albans*, first printed in 1486 in England (which in its second edition printed ten years later included a *Treatyse of Fysshynge wyth an Angle*), and the German booklet, *How to Catch Fish and Birds by Hand*, first printed in Heidelburg in 1493 but only surviving in a later Strasbourg edition of 1498.[245] Fishing manuals especially, with their advice on the best times and places to fish, entailed a close observation of the natural environment, and they also showed a commendable concern with maintaining a sustainable fish stock, such as by throwing back small catches and not overfishing. Comparisons of birds and fishes to various classes and professions of medieval society, such as we find in the *Boke of St. Albans* and *How to Catch Fish and Birds by Hand*, betokens just as much a deep appreciation and intimate knowledge of the unique qualities of each individual fish or bird as it does a social commentary on the human condition. Here humans are imagined as not so different from animals, with various representatives of each species almost interchangeable with the other.[246]

Finally, one has to ask if there was any kind of protest or resistance to the hunting of animals during the Middle Ages? The medieval cultural acceptance of hunting may seem overwhelming, but isolated examples can be found of those who bravely swam against the tide. Far more than St. Francis of Assisi (who was actually quite conventional in his relationship with animals), the twelfth-century English hermit, Godric of Finchale, should perhaps be adopted as the patron saint of wildlife rehabbers, for even in the depths of winter he would go out barefoot in order to feel for any ground-dwelling creatures frozen from the cold, whereupon he would warm them in his armpits or on his bosom, and he would spy out the dense, woody undergrowth in search of any stray animals who were "weighed down, fatigued, or half-dead from the importune weather of the season" and try to "restore them insofar as he could with his medicinal art." In what his servants must have regarded as an annoying habit, Godric would snatch out of their hands any small bird or animal that he perceived they had caught in a trap or snare and release it in some woodland glade or meadow; if they tried to hide their catch from him, he would ferret out their hiding place (like a hunter of hunters), no matter how well hidden or how much they tried to keep it from him. He also provided refuge to hares and any other wild beasts that, guided by "divine instinct," sought protection in his hermitage as they fled from the hands of hunters or avoided their ambushes, which he later released into their "customary haunts in the woods" where presumably they were safe, while the forest "plunderers" went away, their "hopes frustrated." In one instance, Godric gave asylum to a stag while sending away the hunters, retainers of Bishop Ranulf Flambard of Durham, with a curt reply to their enquiries. Unlike the usual heroes of the "hermit and the hunter" topos, Godric has minimal interaction with human hunters in need of religious conversion to a hermit's life, and no recompense is expected of the stag beyond its companionship. Preparatory to embarking upon his solitary life in the wilderness, Godric experimented with vegetarianism, forsaking the company of "fowlers and hunters" and living off plant roots, becoming, in the words of his biographer, Reginald of Durham, "not a man but a worm, not rational, but more a brutish beast." Perhaps most non-conformist of all, Godric befriended even snakes, usually portrayed in bestiaries as the most vile and venomous of creatures, who were allowed to warm themselves in a great "flock" or heap by Godric's fire and to slither in between his feet and legs, offering the saint "their companionship" and consenting to be stroked and handled.[247]

Another example of a medieval protest to hunting is provided by the twelfth-century Syrian aristocrat, Usama ibn Munqidh, who in his memoirs relates a second-hand story of a Qur'anic scholar who tried to save an exhausted partridge that hid under a large rock upon which he was sitting and reciting the Qur'an. The story is almost exactly analogous to the "hunter and the hermit" version in the West, except that the scholar was unable to save the partridge. Usama's hunter came up and questioned the scholar about the partridge, who tried to dissemble like Godric and say that "it didn't come through here," yet the hunter spied it under the rock and proceeded to drag it out and break its legs, throwing it to the goshawk that pursued it. The story is not very flattering to the scholar (his ruse was unsuccessful and the hunter berated him for his lie), but the empathetic feelings that he had for the partridge, his heart "breaking into pieces" for the death of the bird, seem all the more genuine for that.[248] An early medieval flyting, or exchange of insults conducted in verse, has a bear being baited by dogs turn on his tormenters with the taunt, "If you do not yield, then prepare to meet your maker!" This was part of a whole genre of Latin beast poetry in which animals talk as naturally as humans, allowing the authors to imagine the world from their perspective.[249] Then there is a whole series of marginal manuscript illustrations in which a man is hunted by a hare, instead of the other way around, in some cases with the human all trussed up and slung over the rabbit's shoulder on a stick, as was usually done to hares after being hunted.[250]

Was this simply a medieval jest, a version of the "world turned upside down" theme, or was the illuminator trying to say something deeper about the cruelty of animal slaughter? If there were the equivalents of "animal rights activists" in the Middle Ages, they surely did not think in the same terms as we do today. Conscientious objectors to medieval hunting had to cloak their protest in ways that would not be seen as threatening to the social order, presenting it in the guise of religious piety or a carnivalesque game. They were rare, but they did exist.

## Animals and disease

Humans and animals have intimately shared and transmitted to each other various diseases for millennia, going back to prehistoric times. As Paleolithic hunters, humans undoubtedly acquired some diseases through eating the poorly cooked or raw flesh of their quarry: taenia and trichinosis from wild cattle and boar; anthrax and brucellosis from wild cattle, sheep, and goats; tularemia from rabbits and squirrels; and glanders from horses. With the domestication of animals and the accompanying rise of urban centers in the Neolithic age, opportunities for disease transmission vastly increased. New reservoirs were created for disease microbes as large numbers of different animal species lived side by side with each other and their human masters, and as a critical mass was reached of human and animal populations to sustain endemics. In this way we probably got whooping cough, smallpox, tuberculosis, influenza, rabies, and measles (related to rinderpest and distemper in animals). By means of domestication new routes of transmission were also opened in addition to the usual method of ingesting animal flesh: the consumption of milk and other dairy products; contamination of drinking supplies by animal feces used as manure; sharing of blood-sucking parasites such as mosquitoes, flies, ticks, and fleas; even sexual contact between humans and the newly tame species, which is sometimes depicted in the art of the time. As new routes of travel were opened for trade or warfare that connected formerly localized regions, which were usually achieved with the aid of animals such as camels and horses, other diseases came into play, such as bubonic plague that had an ancient endemic focus in the marmot population of central Asia and which then spread to Europe by means of the black rat that infested both human homes and shipping.[251] Nor should we forget that domestication of animals produced a less varied, more restrictive diet for early man, even if it was perhaps more ample and reliable than the hunt, which could have resulted in vitamin deficiencies that are the cause of such diseases as beriberi, rickets, and pellagra.[252]

When did our ancestors themselves become aware of a disease connection between animals and humans, and how did they conceive of it? The eleventh-century Muslim Persian physician, Avicenna, drew one such connection, which was that certain ground-dwelling animals such as frogs, snakes, and mice could herald a pestilence emanating from the earth when they issued from their homes in great numbers, or else birds fled their nests, likewise a sign that they were attempting to evade terrestrial corruption (see Part I). There is also the possibility that such a connection had its antecedents in ancient traditions of augury, like the belief in Romano-Celtic cultures that certain birds, namely crows and ravens, could portend death. Throughout the Middle Ages, farmers would have witnessed periodical animal murrains that killed off large numbers of their livestock and which reminded them that animals themselves contracted disease, sometimes independently of human epidemics, sometimes concurrently. The Anglo-Saxon Leech-Book of c.1000 C.E. provided herbal as well as magical/religious remedies for cattle that seemed to be in a mortal state, including those afflicted with a "lung disorder." This was apparently a continuation of Roman practice.[253] During the High Middle Ages, from about 1066 to 1300, English chroniclers record a total of nine animal murrains, four of which were described as a general "mortality of animals," but the other five specifically targeting sheep. By and large, however, manorial records show local flocks recovering quite well from any losses sustained in these outbreaks.[254]

Perhaps the most serious outbreak of animal murrain in medieval history occurred during the Great Famine of 1315–22, which began with very heavy and extended downpours of rain in much of northern Europe during the summer months of 1315–16; these in turn probably triggered an explosion in the pasture parasite population that was ingested by ruminants, and they seem to have led to a severe shortage of dry, well-cured, un-moldy hay. (The effects of this worsening climate may also have been exacerbated by overgrazing of livestock on existing pastures.) Consequently, it was sheep and cattle that were most affected by the Great Famine murrains: The fact that both species of livestock were together wiped out at once and that this occurred on a national scale created what has been described as "the worst agrarian crisis faced by England as a whole since the aftermath of the Norman invasion."[255] The losses sustained by some manors during the famine years are very impressive indeed: At Bolton Priory, located in the north of England that was normally a stronghold of the wool industry, the sheep flock fell from over

3000 before the famine to 913 in 1317, a decline of nearly 70 percent. This was coupled with a devastating drop in head of cattle during the great cattle plague of 1319–21: By 1321 numbers had fallen to a total of 84, including oxen, from a high of 750 before the famine, a loss of nearly 90 percent. The estates of the abbey of Crowland in eastern England saw their total sheep flock fall from nearly 11,000 before the famine to less than 2000 in 1321, a drop of 82 percent, while the royal stock-farm at Clipstone in Sherwood Forest, Nottinghamshire, had its sheep flock decline by 50 percent and its goat herd by 82 percent in one year, 1316–17. Inevitably, such losses severely affected wool exports from the country, a mainstay of the English economy, which can be gleaned from custom records kept by various ports. The sufferings of cattle are also recorded by three manors in Huntingdonshire owned by Ramsey Abbey, where mortalities ranged from 84 to 96 percent in 1319–20, and by the estates of the bishop of Winchester in the south of England, where the total oxen population declined by 54 percent or more in one year between 1320 and 1321. Many more such examples for both sheep and cattle can be cited from surviving records from individual manors scattered throughout the country. Chroniclers from France and Germany also point to a "great pestilence of oxen and cattle" there, and eventually the murrains seem to have come to Wales and Ireland. The diseases responsible for these epizootics probably included rinderpest and liver-fluke, and possibly anthrax.[256]

When the Black Death struck in 1348–49, animal murrains once again stalked the land. The most famous testimony here comes from the English chronicler, Henry Knighton, who records that "there was a great murrain of sheep throughout the realm, so much so that in one place more than 5000 sheep died in a single pasture, and their bodies were so corrupt that no animal or bird would touch them." Livestock also suffered from a lack of herders due to the human mortality from the plague, and according to Knighton, "for want of watching animals died in uncountable numbers in the fields and in bye-ways and hedges throughout the whole country."[257] The fact that animals were dying as well as men during the Black Death seems to have led to some speculation about the relationship between the two species with regard to disease. This was explored in some plague treatises that were written during the late Middle Ages in response to the Black Death. Among the more academically minded of these treatises, there is a separate section, usually at the end, devoted to various "doubts" or "problems" arising from the plague. Partly this is intended as an academic disputation "in order to exercise the minds of the young," i.e., students at a university, but partly it is also in response to questions from ordinary people, since one author, the mid-fifteenth century Italian physician, Saladin Ferro de Esculo, complains that they came "from foolish and common folk [*ydeotis et vulgaribus*]," which he did not care to address until now for the sake of brevity and to avoid "murmurings" against him from other doctors and skilled physicians.[258]

Two talking points of debate are particularly relevant with respect to animals and disease. These are discussed in several treatises, extending from the

*Long Consilium* (Long Casebook) of the Perugian physician, Gentile da Foligno, written in 1348, to the *Quaestiones de Peste* (Questions on the Plague) by Dr. Johann Vinck of Sulzfeld from the second half of the fifteenth century.[259] All these are similar enough to each other that there obviously was a standard format or common pool of academic discussion on the subject, or else later authors simply copied from earlier ones; however, there are some minor differences and points of clarification among the treatises that, although subtle, are nonetheless significant.

One of the doubts or problems elaborates on Avicenna's observation in the fourth book of the *Canon*, that one of the signs of a coming pestilence was that birds flee their nests or that frogs, snakes, and mice suddenly emerge from underground, signifying a rotting "putridity" in the earth. The Strasbourg physician, John of Saxony, writing in the early fifteenth century, apparently witnessed this phenomenon first hand in the house of a neighbor, who had eight of ten children die within eight days from boils on the throat (*bocium*). On the doctor's advice, the household transferred itself to another dwelling and was free of the disease within a fortnight, but those who stayed behind, including servants and a "vagabond," all died as "mice also scurried into the same house so that the rest of the household that stayed put could not fend them off," leaving a "lamp burning that is [still] there to the present [day]."[260] From a modern point of view and knowledge of the plague, this obviously speaks to rats as carriers of the disease in the form of bacteria-infected fleas that seem to have bitten the children on or near the head as they lay in bed.

To Avicenna's scenario our medieval authors add the opposite case, whereby birds accustomed to living in the mountains or flying up on high suddenly come down to the plain and fly near the earth in order to avoid a "pestilential putridity" in the air. (At the same time, according to Vinck, toads, mice, locusts, and other ground-dwelling creatures "burrow themselves into the earth" to escape the "infected" air.) The explanation given is that either the air, a more subtle or finer substance and one "more lightly altered" than the earth, putrefies first, in which case birds will flee from it to down below, or else it sometimes happens that the corruption first arises from within the bowels of the earth, when birds and other creatures used to dwelling on the ground flee to up on high. (Saladin Ferro de Esculo adds the further caveat that air is more prone to putridity than earth because it contains two qualities especially conducive of corruption, namely, heat and moisture, whereas earth is a substance usually characterized as cold and dry.) A progression can also occur in which birds dwelling on high will flee to the plain and then, after "a lapse of time," will return to the mountains, which signifies that the putridity that first was "kindled" in the air has now migrated and been kindled in the earth. In the former case the pestilence arises more quickly and is more virulent, since air is a finer substance and presumably burns hotter and more readily, but in the latter case the disease lasts longer when it originates in the "crasser" earth.[261]

Johann Vinck adds still another possibility, which is that an "infection" occurs in the third element, the "waters," which is signified by the fact that "fish rise up [to the surface] and lie there dead."[262] The Moorish physician from Almería in Spain, Ibn Khatima, writing during the first outbreak of the Black Death in February 1349, heard from some Christian merchants who had arrived in his city that this very phenomenon happened on the coast of Turkey, where the plague was "rampant" and where "there were broken up and rotten fish on top of the water," so that "a great amount of them had piled up which caused a heavy stench and other foul smells." Although the merchants claimed that the fish had all been struck by lightning, Khatima averred that this opinion was wrong, since "lightning is extinguished by water" and that, in any case, only a small number of fish "directly hit" by the lightning would be affected. Instead, just as air can change and become rotten "so that land animals perish," so can water "change and turn foul, so that sea animals are destroyed, because the water attains a different mixture and character." This frequently happens in stagnant pools and shallow lakes, where "pestilent and rotten air settles upon them," while at the same time they are hit by "heavenly rays which encourage the process." In Khatima's opinion, therefore, a change could occur in the air or water that was "especially dangerous to a specific kind of animal," which presumably would mean that beasts were affected by a plague more than men, instead of the other way around. Nor were plants excluded from plague's impact, for in an appendix Khatima states that "just as the change of air is noxious and dangerous with animals, we see the same thing with plants"; the evidence for this can be seen when plants "decay and wither" prematurely, the leaves and fruit falling off before their time, which happens "because of the surrounding rotten atmosphere that no longer suffices to make the plants bloom."[263]

Khatima's contemporary Spanish colleague from the Christian north, Jacme d'Agramont of Lérida, concurred with his opinion, declaring that a pestilence could break out among fishes "through change in the water in its quality or in its substance," such that a mortality would be seen among fishes but not among men, birds, or other beasts. However, a universal pestilence of the kind that struck in 1348, caused by a substantial change in the air, "is common to all beasts and to all living things," as can be seen from reptiles fleeing their holes in the ground and birds fleeing their nests, and from grain and fruit in regions affected by the plague being infected to the point that "they are like poisons to all those who eat them."[264] Other plague doctors writing during the late fourteenth and fifteenth centuries quote yet another Spanish medical authority, Arnold of Villa Nova, to the effect that "a pestilence of brute beasts does not pass on to rational [creatures, i.e., men], nor vice-versa," or that "a pestilence of brute beasts does not cease and that of rational creatures on the contrary does." Such species-specific illnesses explained why "sometimes men die and animals do not, and sometimes cows [die] and other [animals] do not, and sometimes sheep or pigs [die] and other [animals] do not."[265] However, Villa Nova was writing more than a quarter of a century before the universal experience of the Black Death. Nonetheless, the fifteenth-century physicians, Primus of Görlitz and John of Speyer, both gave as an example of how "something is a poison to one individual species and not to others" the one cited by Avicenna in his *Canon*, that the *napellus* or monkshood plant "is a poison to men and is food to thrushes." Albert the Great explained that this was due to the fact that the thrush has "a cold heart and narrow pathways," so that the poison has time to become digested and changed to food before it reaches the heart.[266] Another example concerned fishes: Görlitz said that a fish can live in the water and not in the air, since the latter is contrary to its nature and "is not proportionate to its life nor to its complexion, but a body of water is"; in the same way, men cannot live in the water but only in the air. Speyer mentioned that according to Aristotle's *De Animalibus*, "fishes do not suffer from the plague" because they live in water and consequently their blood, unlike that of humans, is not "easily kindled by an external heat."[267] Poisonous animals such as serpents and toads, which some chroniclers portrayed as raining from the sky during the Black Death and actually communicating plague through their venomous bites, were also obviously immune to their own poison.[268]

Towards the end of the Middle Ages, in 1481, the Florentine humanist, Marsilio Ficino, in his *Consiglio* against the plague, openly disagreed with those who asserted that a sign of a coming pestilence was birds fleeing the air or worms and snakes leaving the ground, because "what is poisonous to men is not necessarily poisonous to other animals." Nonetheless, he testified to some common beliefs among the populace that were more open to a shared experience of disease between animals and humans, such as that "where hawks fly there is no bad air" because "hawks flee poisoned air," or that "pestilence passes from men to pigs," which may be "because of some commonalities, not of [vital] spirits, but of complexion." Ficino himself noted a similarity between men and beasts in terms of "absorbing" the plague vapor and yet not being "made sick by it": For example, there were two instances where "a cat and a dog have brought the plague from home to home, without being sick themselves," and at the same time in Florence, "a two-year-old child remained for a whole day tightly embraced with a sick seven-year-old and many times ate the same pieces chewed by the other child, and the plague didn't stick to him."[269] Wholesale slaughter of cats and dogs were, in fact, sometimes undertaken by towns, such as Geneva in the sixteenth century, in the mistaken belief that they were spreading the plague, when ironically they would have helped keep down the rat population that was truly responsible for the disease.

As for the underlying reason why a pestilence should arise in one element or another favored by different beasts (i.e., birds in the air, fishes in water, serpents and mice in the earth), some doctors give the explanation that this happens "by reason of the forms or images of the heavens that [make] it necessary [to be so]," or in other words, as a result of the conjunctions or "influences" of the planets, a common enough explanation of the plague in many medieval treatises. But then they add the caveat that such an event "is unknowable by a physician, inasmuch as he is [only] a physician, [whereas] the flying birds that are accustomed to dwell in the mountains and also to fly up on high [know when] to flee the putridity [i.e., corruption of the air] and [thus] they come down towards the plain and fly near the earth."[270] This seems to imply that animals know more than even learned men like doctors when it comes to forecasting disease.

The other question that was asked with respect to disease and animals is, "Why do men die and not cows and other animals?" or why does the "Plague infect man more than any other animal?", a species-specific conundrum that had already been raised before the plague by Arnold of Villa Nova and the thirteenth-century Florentine physician, Taddeo Alderotti. Pietro di Tussignano and Sigmund Albich, two plague doctors writing respectively in 1398 and 1406, give this query a sense of immediacy by observing that, although now it is wondered why "men die and not cows," nevertheless "in the [past] year that has [just] elapsed, it was the opposite [i.e., cows died and not men]." The fact that these physicians make the same observation in two separate years separated by almost a decade leads to the suspicion that it was a formulaic one and not in response to an actual event. Foligno also observes that the opposite can be the case, i.e., that cows and other animals can die during an epidemic more than men, but Saladin Ferro de Esculo and Johann Vinck admit of only the first possibility, that men are affected by the plague more than beasts. Foligno, Tussignano, Albich, and Vinck give an all-too brief answer to this question: either this is "by reason of the influence of the heavens" (Vinck), or because there is a "singular property" that is "inherent in the air" and "by means of which [property] a putridity is generated in [the bodies of] men and not cows," but the contrary can just as easily be the case.[271] This attributes the difference to outside influence and not to any inherent distinctions to be made between animals and humans themselves.

If this explanation is unsatisfactory, then Saladin Ferro de Esculo provides a much more detailed answer, in which indeed he almost seems to be casting around for a suitable response that will appease his reader. (We have already seen that he professed himself to be very self-conscious of what other physicians thought of him.) Thus, his possible solutions to the problem include: that "men are hotter and more moist than other animals" and therefore their internal matter is more prone to putridity or corruption, an opinion he backs up with a quote from Galen's *De Simplici Medicina* (Concerning Simple Medicine); that humans have a higher respiration rate than other animals and thus breathe in more of the corrupt air, which is bolstered by an appeal to Aristotle's *Problems*; that men are more "porous" than beasts and so the air enters more quickly through their skin; and finally, that humans have more subtle or "inflammable" vital spirits which can more easily be kindled by the plague, in the same way as air is more readily putrefied than earth, or else that man is of a more "noble" and "temperate" complexion, so that being in the middle it is more easily changed. This last notion obviously owes much to Aristotle's principle of the "golden mean" in the *Ethics*, but the authority that Saladin appeals to here is the *Colliget* by Ibn Rushd, or Averroes (1126–98), which pointed out that man was the last to be fashioned by God and therefore the most complex in terms of his bodily organs. Saladin adds another proof of his position from the *Colliget*, namely, that "barbarous men" who dwell in the desert, such as Berbers, Arabs, and those having a "feral" nature, i.e., who live as nomads, "sometimes escape very difficult diseases without a doctor and medicine because they assimilate somewhat with brute beasts." This is not owing to a healthier lifestyle out in the open air, but because animals and their ilk have a "rusticity and crassness" to their complexion, which allows them to "fall ill more rarely." Thus, what would seem to be a disadvantage that men have when compared to beasts—that they get sick more easily and succumb to diseases—is turned into a positive attribute in that the underlying cause of this propensity is attributed to the human's higher, more refined make-up, at least among his more civilized kind.

Saladin's human-privileging solution to the problem of why men are more susceptible to plague than animals is echoed by several other physicians. Primus of Görlitz considers not only the possibility that "the pestilential air must needs corrupt the humor and [vital] spirits of men [as compared] to the humors and spirits of beasts because they are finer and clearer," which makes men "more prone to receiving the aforesaid infection of the air," he also considers the opposite case, i.e., that the humors and spirits of beasts are crasser and more opaque, which means that when animals "expire during a pestilence," the disease is "stronger" or more virulent, and then men have more to fear, because they will die "in greater numbers."[272] An anonymous medical practitioner from Montpellier writing in 1349 said that "corruption of the air makes its stamp on human bodies" more quickly than on anything else because "of the primary soft matter out of which [we] are composed," and a Lübeck doctor writing in 1411 stated that "a politic, civilized man ought to take more care than beasts" during a plague, since wild animals like lions, bears, and boars eat "raw, corrupt meats" and have no care for what they do in nature, whereas medicine was invented for men "who are able to govern their body and be virtuous in spirit."[273] For his final answer to this question, Saladin resorts to the brief solution favored by other doctors, namely, that a differential susceptibility among men and beasts to disease may simply be due to the "diverse proportion" of the corruption to be found in the air, which in turn can be put down to the "form" of the heavens.[274]

Although he does not address the question directly, the Lérida physician, Jacme d'Agramont, who, like Foligno, was writing during the Great Mortality of 1348, provides yet another explanation for why humans suffer from the plague more than animals. Arguing along Aristotlean lines, Agramont distinguishes "three degrees of life" that are "subject to the laws of nature": the first being "herbs and plants of all kinds," which "take nutriment and grow"; the second are "fishes and birds and other beasts," who in addition have "feeling"; and the third belongs to man who alone "has reason and understanding."

Even though the pestilence can fall "upon all these degrees of living things," Agramont implies that it is does not do so equally, since man, being at the top of the food chain, consuming both plants and animals, can imbibe the corruptions arising in all other life forms, which during a plague "acquire the property of poisoning and destroying" what eats them in the same proportion as they convey benefit to our bodies in normal times.[275] One can object, however, that by following a preservative regimen of the very type that Agramont himself prescribes, man can easily avoid suspect plants or meats that may have been corrupted by the pestilential air. In any case, the point is rendered moot if we adhere to the opinion of most medieval physicians that it was corrupt air, not food and drink, that posed the most danger to humans during a time of plague, which brings us back to the starting point of our query.

In and of themselves, each problem or question as stated above makes sense enough, but when taken together the two are, well, problematic. The main reason why men observed animal behavior during a plague, such as birds fleeing their nests or snakes, frogs, and mice issuing from the ground, was to imitate their actions. To flee the plague is a piece of advice that is very commonly found in medieval plague treatises: An oft-quoted turn of phrase (apparently derived from Galen) was "to start early, go far, and return late." However, some doctors acknowledged that, by dispensing this advice, they inadvertently caused patients to suffer when abandoned by friends and relatives, so that they were left with no one to tend to their necessities, such as nourishment and evacuation.[276] Indeed, Foligno complained that this was an inhuman practice "that is usually accorded to brute beasts," and he prescribed preventative measures so that humans could live up to their better natures.[277] And yet, fleeing the plague doesn't make sense if the only thing separating men from animals in terms of their susceptibility to disease is their respective complexions, because this then assumes that plague, which strikes equally both men and animals, is a "universal" event from which no living thing can escape, no matter who or where they are. Indeed, this is the very definition that medieval doctors gave to true epidemics, as opposed to "vintages," of pestilence. According to Heinrich Ribbenitz of Prague, the signs of an epidemic are that animals become infected when eating the "fruits of the earth," which signifies that a foul stench corrupting water and air is emanating from the "interior parts of the earth."[278] An anonymous south German treatise from the mid-fifteenth century also notes that when "the water which flows through the veins of the earth is poisoned," it then "poisons people who drink it ... and then also the animals."[279]

However, animals and humans were believed to be equally susceptible to plague even when there was a localized outbreak arising from a near cause, such as the (false) belief that Jews poisoned wells to give the disease to Christians. This is why the city councilors of Zofingen in present-day Switzerland decided in December 1348 to test poison they allegedly found "in the house of a Jew called Trostli" on "a dog, a pig, and a chicken," which they did "by mixing it with something edible, namely bread or meat," whereupon "the animals succumbed to a most swift death and their life was snuffed out as in a moment."[280] The German natural history scholar, Konrad of Megenberg, writing about the plague in Germany in c.1350, repeats this story, but he also relates how that "if there had been such poison that could infect brute animals, as our adversaries say they have tested on them, then without doubt horses, cows, and sheep and livestock that drink the water [from suspect wells or springs] ought to have been infected and died in great numbers like humans, which has not been seen."[281] The authors of fourteenth-century treatises on poisons, such as Cristoforo de Honestis, explained that poisonous substances—which most plague tracts identified as the agent responsible for the Black Death—acted in a universal way, afflicting "man and beast alike."[282] Animals were also recommended by some plague doctors for a medieval form of drug testing: The Lübeck doctor, Heinrich Lamme, writing in c.1400, said that the plague medicine, theriac, should be tested for its efficacy by first feeding monkshood (*napellus*) or else a poisonous mushroom to a dog and then giving the dog theriac. If the dog vomits up the medicine it is no good, but if he keeps it down, it is good to use.[283] The fifteenth-century physician and humanist, Michele Savonarola, said that some men, in order to determine whether a plague was in the air, collected "dew before sunrise" and then gave it to a dog to drink, "and if the dog dies it is a clear sign of bad air and incoming plague." However, Savonarola also was of the opinion that it would be better to conduct this experiment on a man, "because what is poison for the man is not poison for the dog, and vice-versa," but obviously one would have difficulty finding willing volunteers for such trials![284]

A similar principle of compatibility between human and animal complexions was apparently behind animal treatments for plague. We have already seen that one "cure" for plague boils involved plucking the anus of a live, black hen or rooster and holding or tying it down over the bubo: It was believed that the hen would then suck up the plague poison, which would go straight to the animal's heart, killing it instantly, just like in a human. Anyone who then ate the chicken would die as if he had been poisoned, and a Lübeck doctor writing in 1411 said he saw proof of this one day when he "observed how a certain domestic raven ate of such a chicken who had died in this way, and in three days [the raven] also died."[285] John of Saxony, a physician practicing in the early fifteenth century in Strasbourg, recommended to a neighbor complaining of cattle murrain that he administer barberry juice to his cows by "forcefully" pouring it down the cow's throat after lifting up its head and neck. He and all the other farmers in the village accordingly did so and reported back that "they felt that [their cows] had been greatly preserved from the plague" by the remedy. Barberries were one of several sour or acidic fruits recommended by physicians for human consumption during a plague. John of Saxony's lesson that he took away from this experience was that remedies like barberries "that act by way of the first and second virtues can be indifferently applied to brute beasts and to men," although he was careful to amend this by saying that not all "medicaments for humankind can be recovered from cures for cows."[286]

On the other hand, if humans were more susceptible to plague owing to their more refined and "civilized" natures, then that was all the more reason why they should flee so that they not suffer the consequences of their own superiority, and presumably their higher faculties would have endowed them with the ability to recognize when they should do so. And yet, the urge to flee, even to the point of abandoning one's own children and close relatives, implied, as Gentile da Foligno said, that men could behave no better than "brute beasts."[287] Even so, the tradition whereby animal behavior was acknowledged to be one of the signs that predict when pestilences were about to occur, which should be acted upon by humans, was a long and ingrained one. As already explained above, this could include birds fleeing their nests from on high, serpents, mice, and other ground-dwelling animals emerging from the ground, and fishes washing up dead on the seashore as indicative of corruption in the air, earth, or water; but medieval doctors also believed that corruption of the air could spontaneously generate a "multitude" of flies, worms, spiders, locusts, mice, frogs, toads, lizards, snakes, and other "vermin," just like worms could be spontaneously generated in the human body as a result of excess heat and moisture when infected by the plague. Presumably because of such an abundance of potential prey, predators were also believed to be more on the prowl and hence more visible during a time of plague, such as wolves and other "bad" or "raging" beasts coming out of their dens and entering cities or towns or else ravens, owls, and other "strange birds" or "creatures of the night" now having cause for much "rejoicing"; by the same token, "choice" birds such as partridges, quail, and pheasants, whose meat was also in demand by humans seeking to adhere to a good regimen with regard to food and drink, took flight and became scarce.[288] A Bohemian plague treatise from c.1450 mentions that a pestilence among men will follow "when there is a mortality among living things and among beasts, either among birds or fishes or among animals that dwell in caverns in the earth or that go along the ground." But if "this mortality comes upon animals that dwell amongst you, such as cats, dogs, and the like that have an affinity to the human condition," this was a sign that a plague among humans would happen soon and then "there is no better [thing to do] than to flee."[289] All this implies that men were dependent upon animals and their precocious foresight to know the right time to flee: too early or too late could be costly and even fatal.

While medieval schoolmen like Thomas Aquinas explained the special facility of animals to divine natural catastrophes like plague as owing to instinct (*estimativa*), this then poses the question of whether man could match brute instinct with his more ponderous faculty of reason. A little-known treatise hidden away in the Vatican Archives, *Quod liceat pestilentiam fugere* ("That it should be permitted to flee the pestilence"), by the fifteenth-century bishop of Brescia, Dominico Amanti, actually raises this question indirectly in the course of his efforts to justify flight from the plague, even by members of the Church having a duty to their flocks. Amanti claims that he who would deny to men the right to flee the plague, on the grounds that it is presumptuous to attempt to evade the will of God or that it is against charity, is "an enemy of nature who does not wish men to do what animals do by instinct," such as when dolphins flee a storm or kites, storks, and other birds flee from corrupt air. Moreover, this is what revered authorities like Avicenna mean for us to do when they teach us "how to recognize pestilential air from its qualities." To do otherwise would be to reduce man to a level below that of beasts despite his divinely endowed rational qualities, "in that a beast accommodates itself to that which affects it by sense alone, but a man, who partakes of reason by means of which he comprehends the consequences [of his actions], perceives the causes of things and their progress." Man has the unique perspective of being able to look into the past, present, and future all at once and to make connections among them in order to draw the appropriate lessons, such as "that corrupt air harms a man and that the disease of the pestilence is contagious." Therefore, man should at least be able to match the instincts of animals, which Amanti points out was a defense also employed by the Roman statesman, Lucullus (117–56 B.C.E.), who responded to critics complaining about the number of lavish houses he occupied at various seasons of the year by claiming that he was doing no more than what migrating birds such as swallows and cranes do as they change their nests.[290] The obvious objection to Amanti's argument is that, by the same token of man's superior status and abilities accorded to him by God, which gives him the capacity to reason that he should flee the plague, what distinguishes men from beasts, indeed the very essence of what it is to be human, also entails a self-sacrificing obligation to resign oneself to God's will in the form of the plague, as well as to do one's duty by one's fellow men and stay behind to help them through the disease. This is, in fact, one of the core beliefs of the Prophetic tradition of medieval Christianity's rival religion, Islam, with respect to the plague.[291]

The interdependence of animals and humans in the face of disease, which is an underlying principle behind the modern-day laboratory testing of animals, is actually the more logical conclusion to draw when the two problems of why animals flee the plague and why men are infected and die more from the disease are considered together. In the end, explaining why there should be a difference between animals and men in terms of suffering from disease proved to be a conundrum that medieval doctors and scholastics were never able to satisfactorily resolve.

## Animals on trial

To modern minds, one of the more bizarre and incomprehensible aspects of the Middle Ages are the animal trials that reached their apogee towards the end of the medieval period during the fifteenth and sixteenth centuries and which have received popular treatment in films such as *The Advocate* (1993), also known by its alternate title, *The Hour of the Pig*. One cannot help but be bemused by the spectacle of domestic animals such as pigs (by far the most frequent defendants in such trials, perhaps owing to the fact that more and more pigs roamed the streets of towns by the late Middle Ages), cows, oxen, goats, horses, donkeys, dogs, and roosters being put in the dock and solemnly sentenced to death; how much more risible to us is the entirely separate category of Church proceedings that targeted a host of lowly "pests" and "vermin" such as rats, mice, moles, eels, serpents, locusts or grasshoppers, worms, flies, caterpillars, beetles, leeches, snails, and weevils, in which it was hardly likely that the animals could be brought to the bar of judgment but where the legal argumentation back and forth between prosecution and defense was even more elaborate, and which often ended in excommunication and exorcism of the accused. The nineteenth-century antiquarians who first brought these trials to light typically saw them as evidence of a "primitive" or "superstitious" society, to be explained either as products of a culturally "childish disposition to punish irrational creatures" or of an equally foolish presumption to attribute rational or demonic attributes to animals.[292] But, as we will see, such trials are in fact the logical terminus of a long history of judgment and punishment of animals throughout Western civilization, and they appear, albeit in less elaborate form, in other cultures as well.[293] What is more, when we actually examine the content and context of the trials in depth, they reveal some attitudes towards animals that accord quite well with how most of us today in the West regard our fellow creatures and which we would find eminently balanced and reasonable.

Laws that address animals that commit acts of violence against other animals or against humans date back to the earliest civilizations. In ancient Mesopotamia, law codes including that of Hammurabi from the second millennium B.C.E. make a distinction between an ox that gored as a first offense and one that was a "habitual gorer"; in the latter case a fine was payable to the victim's family, but in the former no claims were allowed or else an equal distribution of the gored ox's remains was mandated. But in neither case was the goring ox punished, except to be sold so that both parties could be equally compensated. This changes in the biblical law under the Hebrews: Even an ox that gored a man as a first offense was ordered to be stoned to death and its flesh to be allowed to rot uneaten; otherwise, an ox goring another ox was treated the same as under the Mesopotamian laws (Exodus 21:28). In the case of an ox that "had the propensity to gore," the human owner shared in the culpability and was put to death, unless the victim's family agreed to his ransom. The stoning of the ox seems to go beyond God's promise to Noah that he would hold animals as well as humans accountable for the shedding of human blood (Genesis 9:5); rather, this punishment is interpreted as directed not so much against the animal itself, which may have done the deed inadvertently, but as a reassertion of the "cosmic order" in that the ox had committed "treason" or insurrection against God who had ordained a hierarchy that placed man above animals, as stated in Genesis 1:26. Thus, the ox was punished by the entire community in the same way as was a human who had been judged to have rebelled against God by committing idolatry or unnatural acts such as bestiality or incest. Further emphasizing the ox's offense against God was that even in death its flesh was considered unclean and a source of contamination to a believer's purity.[294] The Hebrews' hierarchical view of justice is seen to be an innovation, one that contrasts with that of earlier societies such as Mesopotamia's and non-Western ones that place animals on a more linear and equal continuum with humans.[295]

The next comparable set of laws that judged animals for "crimes" were the Germanic codes of the early Middle Ages (see "Animals on the farm" and "Animals as pets"). During the classical period in Greece, Athens apparently had a ritualistic "trial" and punishment of animals and even of inanimate objects that killed a man and where no other, human guilty party could be found, but there is no evidence that actual trials were ever held.[296] We have seen that the Germanic laws generally provided compensation in the form of wergild to be paid by the owner to the victim's family if an animal happened to kill a human being, but that otherwise no blame was attached to either the animal or its owner since the incident was considered the product of accident or chance; indeed, the Burgundian laws allowed for no compensation whatsoever for this very reason, the same principle as was applied in the Hammurabi code. Otherwise some codes provided for an exchange of animals, which was also applied in the case of animals killing other animals and had its precedent in Roman law; this same principle of "noxal surrender" of the offending animal, in this case to the crown, was used in early English laws, known as *deodands*, that extended up into the high Middle Ages.[297] On the whole, the Germanic codes seem far more concerned with injuries, thefts, or killings committed *against* animals by humans rather than the other way around.

When we come to animal trials proper during the Middle Ages, the first example we know of is said to be the execution by burning in 1266 of a pig for having eaten a child at Fontenay-aux-Roses, near Paris, in France. This sentence was evidently the result of a civil procedure approximating that used for humans accused of secular crimes. However, there was a parallel form of justice employed by separate Church courts under canon law that tried crimes of a moral nature (i.e., that offended God) which were likewise applied to both animals and humans. Three such cases against moles, locusts, and serpents are known from the ninth century, three against flies, mice, and caterpillars in 1120–21, and one against eels in 1225, although it is not until the fifteenth century that we have the full record of such trials that tell us the exact circumstances of the indictments and the judicial procedures used. After these early and high medieval examples, 12 trials, both secular and ecclesiastical, are recorded for the fourteenth century, 36 for the fifteenth century, 57 for the sixteenth century, 56 for the seventeenth century, 12 for the eighteenth century, and 9 for the nineteenth century. In geographical terms, the animal trials during the Middle Ages seem to be concentrated in northern and eastern France and in Switzerland, although not too much can be read into this fact since this may merely be an accident of record survival, and the pattern changes from century to century.[298]

Historically, two main schools of thought have grown up among modern scholars concerning the medieval animal trials. One is perhaps the rather obvious point that putting animals on trial as if they were human beings elides the basic distinctions between men and beasts. This has been called the "positivist" interpretation, which views the milieu of the animal trials in the late Middle Ages as representing a shift in attitudes from earlier times, whereby animals were now accorded anthropomorphic and rational attributes to the point that they were held accountable for their actions.[299] A more counterintuitive interpretation, however, sees the trials as actually a reaffirmation of the natural order of a God-created world, in which humans ranked above all other beasts by virtue of their unique rationality and endowment with a soul. In this view, the trials were intended, not as a punishment of animals per se, but rather as public spectacles with a message directed exclusively at their human spectators, namely, that human justice was reasserting its primacy over the animal world and that the divinely ordained hierarchy was being re-established, which had temporarily been overturned or challenged by the animal crime.[300]

Evidence can be marshaled in support of both interpretations. On the one hand, the very fact that medieval people went through the motions of trying beasts at their bar of judgment speaks to some semblance of equality. For example, a now lost fresco in the church of Sainte-Trinité in Falaise, France, apparently depicted the execution of a sow dressed in human clothing in 1386, and receipts for hangman bills to execute infanticidal pigs in three French towns in 1386, 1394, and 1403 read very much like those for human executions, where the condemned animals were either dragged to their appointed place or brought bound in a cart.[301] On the other hand, those who would argue that the trials actually reinforced human superiority over animals point out that the jurists who conducted the trials never expected the animals they put in the dock to treat with them on terms of equality, i.e., that they could make a rational argument in their own defense. That would be regarded as just as foolish to a medieval observer as to a modern one. This is why humans were always appointed to act on the animals' behalf, in effect treating them as minors who were incapable of representing themselves except through a guardian. Moreover, since the trials reached their peak at the very time that Europe was entering the Renaissance, the Scientific Revolution, and the Enlightenment, the notion that contemporaries viewed the trials as anything more than symbolic acts with special significance for humans alone would fly in the face of a linear, progressive narrative to modern history.[302] Instead, the lineage of a hierarchical interpretation to the trials can be traced all the way back to biblical precedents that prescribed public stonings of oxen that gored human victims, and which were cited by the fourteenth-century French jurist,

Jean Boutillier, in his justification for bringing due process of law against "the beast that kills a man."[303]

One weakness of the hierarchial interpretation is that it relies almost exclusively on the testimony of a learned, elite culture; the popular mindset of ordinary country folk, on the other hand, seems to have been far more receptive to the idea that animals could have their own, special knowledge, particularly in terms of their ability to portend natural events.[304] A good example of popular culture attributing human attributes to animals, particularly with regard to systems of justice, are the fabliaux or folk tales concerning Reynard the Fox, one of the more famous and widely known examples of the medieval animal epic whose traditions stretched back to classical origins, such as works by the legendary Greek fabulist, Aesop.[305] The French *Roman de Renart* (Romance of Reynard the Fox) seems to have first arisen towards the end of the twelfth century, and subsequently it became popularized in other languages, including German, Dutch, and English, and was printed in a prose version by William Caxton in 1481. The feudal judicial system of the time plays a role in several "branches" or variant tales of the original French romance, beginning with Isengrin the Wolf's lodging of a formal charge of seduction and rape of his wife, Hersent, against Reynard before the court of King Noble the Lion, whose main concern is to maintain peace in his realm and avoid a private war between his subjects. The main object of the Reynard romance is to parody through its animal characters human institutions and customs, such as the legal system with which the author shows himself to be thoroughly conversant, but it does also show that medieval culture could easily envision animals as equally subject to the law, even if only in a literary way.[306] The message was further reinforced on a popular level by plastic media, such as stained glass, manuscript and woodcut illustrations, and church woodcarvings and sculptures, namely on choir stalls and misericord seats, that depicted various aspects or stages of Reynard's trial.[307] That message couldn't have been clearer to ordinary listeners, readers, or visual observers of the tale: If animals could put themselves on trial in their own world, why couldn't they then appear in a human court to face human judgment?

There is a third interpretation of the animal trials, which argues that they are best understood within the specific, socio-legal context of the late medieval and early modern periods, which is when the trials had their heyday. Central to this interpretation is that the animal trials have much in common with the witch-hunts and beliefs in demonic magic that emerged at much the same time, and which were the product of an inextricable blend of learned and popular culture. Also key to this point of view is a close study of the ecclesiastical animal trials, such as of vermin, which are largely ignored by both the positivist and hierarchical schools but which forge a common link between the witch and animal trials through their mutual use of exorcisms.[308]

One of the more illuminating and fully documented of the ecclesiastical animal trials took place in 1587, when the community and parish of St. Julien

*Figure 3.9* THE TALE OF REYNARD THE FOX, misericord carving from Bristol Cathedral, England, early sixteenth century. Here Reynard is about to be hanged for his crimes, with the execution presided over by the lions, King Noble and his queen, while the bear, wolf, goose, cat, and squirrel all prepare the gallows. © Dominic Strange www.misericords.co.uk http://www.misericords.co.uk

in the Savoy region of France sued the "green flies" (perhaps cherry weevils) that destroyed their vineyards before the bishop's court at Maurienne. Central to the argument for the defence of the weevils by their advocates, Anthony Filliol and Peter Rembaud, was a principle already enunciated throughout the Middle Ages from St. Augustine to St. Thomas Aquinas and going even further back to Aristotle: that brute beasts are judged to live and act "by their natural instinct alone," since they "are lacking in the sense and use of reason." Therefore, to excommunicate the flies for their supposed crime of devouring the vineyards, as the syndics of St. Julien wished, would be an "unaccustomed and unusual mode and form of proceeding," since according to canon law as enunciated by Gratian's *Decretum*, such a sentence can only be promulgated "by reason of the contumacy" of the accused, but "it is certain that the said animals cannot be established to be in contempt [of the Church], since reason and justice do not rule the said animals." Filliol and Rembaud were, in fact, making a very modern-sounding argument that animals should not be tried before the bar of human justice at all, since they could not be expected to follow its rational procedures and obey our laws. To the prosecution's argument that "man was created and established that he might rule other creatures and dispose of the terrestrial orb [as he sees fit] in equity and justice," the defence responded that both divine and natural law, which were eternal and could not be changed, as well as the dictates of reason "demand that the said animals live as well as they can off of these plants, which seem to have been created for the use of the said animals."[309] Apparently, the syndics and community of St. Julien were wary enough of their chances of a favorable judgment to attempt an out-of-court settlement with the flies. On the 29th of June, at an assembly in the town square where a settlement was recorded and read out in French (as opposed to the formal Latin of the rest of the trial record), the townspeople agreed to set aside a piece of land, called *Le Grand Feisse*, especially for the weevils "for their sustenance and for them to live in," which lay on the outer edge of the village and whose exact boundaries and species of plants were specified. Nevertheless, the counsel for the defense of the weevils still contested the settlement on the grounds that the assigned land "is not sufficient nor suitable for the nourishment of the said animals since it is an infertile place and renders nothing."[310] What the end result was we will never know, for this is the last record that we have on a case that had dragged on for nearly six months: In a sublime piece of irony, or perhaps justice, insects or vermin had the last word on the whole affair by eating and gnawing through the last page of the document recording the trial!

What is remarkable about all this legal back and forth just for the sake of some weevils is how it contrasts with a series of witch trials in nearly the same region and on almost the same charge, destroying vineyards by an act of nature, from a century before (see Part I). In that case, the accused witches were burned at the stake for destroying the wine harvest through a killing frost, heavy rains, and storms, whereas the flies who were actually caught in the act of eating the crop were treated with almost kid gloves, being provided with a nature sanctuary all their very own (perhaps the first one of its kind for insects, which is now quite common in the modern world)[311]. One could say that the witches were associated with the devil in their trials, while the weevils were portrayed by the defense as God's agents, acting out his natural laws, although witches were not even allowed defense representation, on the basis of a precedent established in heresy cases.[312] This discrepancy has never been satisfactorily explained, namely, why animals should be put on trial with such elaborate protocol, when the same was barely accorded to some humans, such as witches or Jews, whose "trials" were merely the foregone conclusion of a wave of panic and persecution.[313] One possible explanation has to do with the plague, or Black Death, that swept Europe beginning in the mid-fourteenth century, and which also played a role in the witch trials. Although some late medieval plague doctors, such as Saladin Ferro de Esculo, did argue that humans and animals differed in their susceptibility to disease, such differences were virtually obliterated when even beasts with their "crasser" constitutions fell ill, which is when man really had to worry (see "Animals and disease"). As noted at the beginning of Part I, plague was viewed in the Middle Ages as

God's judgment upon human beings for their past and present wickedness and wrongdoing; if then animals also succumbed to the plague, as they assuredly did during the Black Death, wouldn't the natural conclusion be that they, too, must necessarily fall within the purview of God's judgment, and by implication, also man's?

This interpretation is supported by a discussion of animal trials by the seventeenth-century jurist, Gaspar Bailly, in his *Traité des Monitoires* (Treatise concerning Ecclesiastical Monitions, or Warnings), published at Lyons in 1668. In this work, Bailly provides a hypothetical, and allegedly typical for his day, trial against "gnawing insects" or caterpillars (*bronchos seu erucas*) that destroy vineyards, complete with argumentation back-and-forth between prosecution and defense, much like we've seen with the weevils of St. Julien in 1587. Only this time, a sentence of excommunication, or "malediction and anathematization," is solemnly pronounced by the court, which is to take effect if the bugs do not depart from the vines within six days of the admonition. What carries the day for the prosecution is its argument that, despite acknowledging that these insects are indeed "deprived of the use of reason," nevertheless they are overturning the natural order of creation, in which God ordained that lesser creatures exist for the benefit of man, and that even though the animals may not be in conscious contempt of canon law, still one cannot call into question "the power that God has given to the Church … [as the Mistress] over all of creation," charging her with "the conservation of His most perfect handiwork, that is to say, man, who was made in His image and likeness." This argument was virtually unassailable, in that it appealed to the vanity and authority of the very Church court that was trying the case. Bailly follows this by enumerating a whole host of examples, both from his own day and from the recent and distant past, where "vermin and insects" had merited anathema due to "the damage to the fruits of the earth" that they had brought upon the inhabitants of various places.[314] Plague doctors from the late Middle Ages would fully agree that even insects were subject to God's judgment, namely in the form of the plague, for in their treatises they often cite the sudden appearance of a swarm of flies, locusts, worms, and other vermin as one of the "signs" heralding an imminent plague, since these creatures were believed to be spontaneously generated by a great corruption of the air, which was held to be the true cause of the disease (see Part I). So while even insects, among other animals, were to be given every consideration under the law, their association with an environmental catastrophe on the order of the plague clearly merited the just judgment, and now condemnation, of men no less than of God.

## Animals in the bed

Prohibitions against bestiality, or sexual intercourse between humans and animals, go back to biblical times, where passages from the Old Testament forbid inter-species breeding or "defilement" in both general and specific terms and prescribe the death penalty for both the human and animal participants (Exodus 22:19; Leviticus 18:23, 19:19, 20:15–16). Despite the fact that animals almost certainly were the unwilling or innocent partners of such unions, Hebrew law still demanded their death in order to remove the taint of such associations that were seen to upset the natural hierarchy of man's primacy over his fellow creatures, in much the same way as animal executions were justified in the case of accidental gorings. However, the biblical aversion to bestiality, which is seen as the basis for attitudes towards the practice in the West during the Middle Ages, was rather unique in the ancient culture of the Near East; the nearest analogous laws were those of the Hittite culture in present-day Turkey, which forbade intercourse with pigs under pain of death, yet both actors could be pardoned by the king, and sex with horses and donkeys was specifically allowed.[315]

The history of subsequent approaches to bestiality down through Western culture seems to be a story of progression from complete acceptance of it by pagan Greek, Roman, and Germanic societies to complete intolerance in Christian Europe by the late Middle Ages. Some apparently real-life examples of bestiality are discussed by the Roman author, Aelian (who wrote in Greek), in his influential treatise, *On the Characteristics of Animals*, from c.170 C.E. In these examples, humans treat the animal objects of their desire no differently from human ones, falling in love with them and giving them sweetheart gifts, while the animals in turn evince human emotions such as jealousy and love-sickness and are often the initiators of relationships, especially baboons and goats that were believed to be lecherous. If we are to believe the author, the line between humans and beasts was blurred to the extent that half human, half animal hybrids were born of such unions. Aelian seems to draw the lesson from all this that animals are not so very different from people; indeed they, like humans, can at times rise above the ordinary emotions to fall in love with their paramours for their own sake, regardless of their physical appearance, as in the case of the seal that gave its affections to an ugly diver. Bestiality therefore posed no threat to man's essential humanity, such as his capacity to reason, since love itself was by its very nature an irrational exercise.[316] Even the gods condescended in Greek and Roman mythology to mate with humans in animal form, such as Zeus' appearance to both male and female lovers in the form of an eagle, swan, bull, or snake. (Alexander the Great's mother, Olympias of Epirus, claimed that her famous son was the product of her union with Zeus in the form of a serpent.[317]) Early medieval law codes enacted by the various Germanic tribes make no mention of bestiality as a crime deserving of penalty, and Germanic and Celtic mythology actually celebrates heroes who acquire superhuman strength owing to unions between their human and animal parents, such as a bear or horse.[318] All this indicates that no stigma was attached to bestiality in ancient pagan cultures.

As Europe made the transition from the late Roman Empire to the early Middle Ages, Church fathers such as St. Ambrose and St. Augustine resurrected the old biblical hierarchy that placed humans over animals, not only on the grounds that God made man in his image but also using the Aristotlean notion that human beings were uniquely rational creatures. To this they added the interpretation that bestiality or intimate intercourse with animals somehow "brutalized" or degraded man's uncreated and immortal soul (a concept that was definitely *not* part of Aristotle's philosophy, as later Muslim authors who attempted to synthesize pagan thought with their religion were to discover). Nevertheless, early medieval Church penitentials in the West, which prescribed condign punishments for certain categories of sin, tended to treat bestiality rather leniently, as akin to masturbation, with penances of a year or less, depending on the circumstances surrounding the human participant such as his age, marital status, or membership in the ecclesiastical or secular community. This seems to bespeak of a confident assurance of the separation between the human and animal kingdoms, along the lines of the classical pagan approach, as well as a view of the animal partner as simply one that did not participate or figure in the act. But as we near the millennium, Western penitentials, perhaps influenced by ones from the Eastern orthodox tradition, adopted a more severe attitude towards bestiality, equating it with more serious offenses than masturbation, such as homosexuality or sodomy, and therefore they began prescribing penances lasting years rather than mere months. Basing themselves on the book of Leviticus, churchmen also began turning their attention to punishment of the animal partner, insisting on their deaths but still allowing for their hypothetical "offspring" to be used by man on practical grounds.[319]

The advent of the year 1000 and the dawn of the high Middle Ages is taken as marking a dramatic shift in attitudes towards bestiality on the part of both elite and popular medieval culture. Beginning in the eleventh century, learned churchmen such as Ivo of Chartres expounded on the necessity of executing the animal partner of the bestial act in order to erase all memory of the shameful deed. Paralleling the reasoning behind the animal trials, this was done not so much as a punishment upon the animal, since no guilt or blame could be attached to a creature that lacked the ability to reason and thus make the conscious choice or possess the intent to commit a crime, but rather was for the benefit of humans, in order to erase the memory of a sin that violated the natural separation between the species. This justification was continued and further developed by thirteenth-century scholastics such as Thomas of Chobham, Alexander of Hales, and, above all, by St. Thomas Aquinas, who classified bestiality as among the "sins against nature," or "unnatural vices" of a sexual kind that did not lead to procreation, namely, masturbation, homosexuality, and lecherous acts that did not "observe the due mode of intercourse," i.e., oral and anal sex between a man and a woman. Aquinas ranked bestiality as the greatest of these sins because not only did it not produce more of the human species, but it also crossed the species boundary, causing the human perpetrator to forget his essential humanity (the exact opposite of the lesson drawn by Aelian).[320]

Bestiality also began to be legislated against at this time. Norwegian canon law from the twelfth century gave a man accused of "having carnal dealings with cattle of any sort" the choice of either undergoing trial by ordeal of the "hot iron," i.e., holding a piece of red-hot metal in his hand and then determining his guilt or innocent based on how the wound healed, or else becoming an outlaw and going into exile.[321] Although in general Church courts were more focused on prosecuting homosexuality, secular laws came on the books that made bestiality a capital crime, and the first executions of humans as well as animals for the practice appear in the fifteenth century in Mallorca, Spain, and Corbeil, France.[322] Such prosecutions continued well into the early modern period: The seventeenth-century Saxon jurist, Benedict Carpov, declared that the punishment by burning of animals caught in the act of sodomy, i.e., intercourse with a human being, was for the purpose of being "an example or terror" to the human onlookers witnessing the execution, and he justified the practice, which was on the German law books since the Carolina code enacted in 1532, by quoting Leviticus.[323]

Ironically, the greater severity of such legislation to maintain the separation between animals and humans is taken by modern scholars as a sign that people of the high and late Middle Ages "had become more uncertain about the distance between human and animal natures."[324] Meanwhile, medieval popular culture seems to have concurred with the schoolmen that sexual intercourse between an animal and a human was a dangerous thing. Clear evidence for this comes from the travelogue by the twelfth-century courtier, Gerald of Wales, who recounts a cautionary tale about a "wretched woman" who had "guilty relations" with the royal white goat of Connacht in Ireland. This leads Gerald to reflect on the unspeakable nature of the crime of bestiality, which is due to the fact that it causes humans who were established by God as lords over all other animals to lose the privileges of their "high estate" and "descend to the level of the brutes, when the rational submits itself to such uses with an animal!" Gerald further comments that although animals are less culpable than humans because they are "subject to rational beings in all things" and only obey their natural urgings, nevertheless they are punishable for the crime by virtue of the fact that "nature makes known her indignation and repudiation of the act," since the animal was "created not for abuse but for proper use."[325]

Gerald's tales are the sort of things one would expect to read in a book like Sir John Mandeville's *Travels* that retail the exotic marvels of the East; the fact that monstrosities show up so close to home either speaks volumes of how Gerald and other Britons viewed the Irish—even at this early date it seems that Ireland was viewed as a sub-human backwater—or else (and possibly both) that bestiality was now popularly seen as a crime that could blur "the lines that separated human from animal." It is significant that Gerald himself speculates on whether the offspring of cattle that had been abused by human beings could legitimately be killed, since they could be considered half human and their deaths an act of murder or homicide, a question that did not even factor into earlier penitentials that had freely sanctioned their use. His bestiality stories are exactly the sort of tall, old wives' tales one would expect him to pick up on his travels from members of the local populace; that they are barely credible is not surprising, but at the very least they do illustrate and reflect what was going on in the "popular imagination" of the time.

Closely related are contemporary beliefs in unions between humans and demons that likewise could produce hybrids: Learned churchmen like Aquinas explained this was possible, despite the fact that demons did not have physical bodies, as due to the fact that a demon, assuming the form of a corporeal manifestation of the air (i.e., a spirit), could act as succubus to a man, taking his seed, and then transferring it when acting as incubus to a woman. Later demonological treatises, such as the *Malleus Maleficarum* (Witches' Hammer) of 1487, greatly elaborated on such theories.[326] Already by the twelfth century, medieval visions of hell imagined both male and female sinners being impregnated by the devil and then giving birth to monstrous beasts that tear apart their body as they come out with their "iron heads," sharp beaks, and venomous stings. Also at this time, authors began retailing tales of both legendary figures, such as Merlin, and of real-life contemporaries who were the products of unions between a demon and a human mother, and by the thirteenth century, women were being burnt at the stake on the basis of such accusations.[327] Because demons could appear to their victims as not only fellow humans but also in the guise of animals, usually as goats, snakes, or dogs, bestiality now became a form of heresy and therefore even more serious a crime by the time of the witch trials beginning in the late Middle Ages.[328]

With beliefs in hybrid offspring and supernatural couplings, late medieval interpretations of bestiality could be said to have come full circle to what had been considered the norm in classical times. Only now, instead of being tolerated or even admired, man's intimacy with animals was viewed with fear, disgust, and horror. To a large degree, this was undoubtedly because the unbridgeable divide or clear boundary that formerly separated animal from human could no longer be assumed to be secure in the face of bestiality, but was all too easily crossed with results that threatened man's very definition of what it was to be human. And arguably, this is where attitudes towards bestiality have remained up to the present day.

## Animals and magic

Since earliest times, ancient civilizations have used animals in their magical practices, the most common of which was divination, or the foretelling of the future through the reading of certain signs, or auguries. One of the specialized forms of divination was extispicy, which claimed to predict future events through the interpretation of patterns in the viscera, including the intestines, liver, and lungs, of animals that had been ritually slaughtered as sacrifice to the gods. This was the preferred practice of the Mesopotamians, as well as of the Romans. The soothsayer Spurrina, for example, used this method to forecast the impending doom of Julius Caesar. The Romans also read significance in the observed behaviors of live creatures: Thus, according to the Roman historian, Suetonius, Caesar's assassination in 44 B.C.E. was also forecast by the fact that a few days earlier horses that Caesar had settled along the Rubicon River refused to graze and shed buckets of tears. Likewise, a wren with a sprig of laurel in its beak was pursued into the Senate house by various birds of prey and torn to pieces.[329] Other omens of death in Romano-Celtic societies included the appearance of certain dark birds that fed on carrion, such as ravens and crows.

Such uses of animals for divination continued into the early Middle Ages. In his *Etymologies* from the seventh century, Isidore of Seville lists among contemporary magical practices that of augury by means of understanding "the cries and the flight of birds," which was probably a continuation of Roman traditions. This was especially used "by those making a journey," in which case it was called "auspices," which were of two kinds, one by observing the flight of birds with the eyes, the other by listening to the "chatter ... that is, the voices and tongues of birds" with the ears. Isidore's information on augury and auspices using birds was repeated almost word for word in the early twelfth century by Hugh of St. Victor in his discussion of magic in an appendix to the *Didascalicon*. Likewise, from the beginning of the eleventh century, Burchard of Worms in his penitential, the *Corrector*, which comprises book 19 of his *Decretum*, provides the following formula in his interrogation of parishioners concerning magical practices:

> Have you believed what some are wont to believe? When they make any journey, if a crow croaks from their left side to their right, they hope on this account to have a prosperous journey. And when they are worried about a lodging place, if then that bird which is called the mouse-catcher, for the reason that it catches mice and is named from what it feeds on, flies in front of them, across the road on which they go, they trust more to this augury and omen than to God.

Burchard prescribed penance of fasting for five days on bread and water for indulging in such beliefs.[330] However, some predictive behaviors of animals, such as that birds, serpents, mice, etc. flee corruption in the air or earth and thus can warn men of a coming plague, were believed to be entirely within the realm of naturally occurring phenomena and therefore were endorsed by the Church and the scholastic establishment. Animals also played an important role in demonstrating a saint's miraculous powers, which bordered on the magical.[331]

Much more information on magical practices with animals becomes available with the witch trials and demonological treatises that are recorded and written during the late Middle Ages. Like with the animal trials and bestiality, these give evidence of a much closer relationship and a blurring of the boundaries between humans and animals, which in the case of magic occurs in one of three ways: use of animal parts or blood in magical incantations; the infliction of harmful magic upon animals as well as humans; and association with and even transformation into animals by witches during the practice of magic. With regard to the first category, one of the more common uses of animal parts was in erotic or love magic. While the "Eye of newt and toe of frog / Wool of bat and tongue of dog / Adder's fork and blind-worm's sting / Lizard's leg and owlet's wing" may be familiar lines from Shakespeare's play, *Macbeth* (IV, i, 14–15), the witch's arsenal in love magic was alleged to be far more varied and innovative than these stereotypes suggest. One of the techniques involved natural sympathetic magic, in which the sexual organs obtained from a lustful animal were believed to act as an aphrodisiac in humans. The principle thus elided the difference between animals and humans: What produced sexual arousal in a beast would automatically do the same in a man, with no need for resort to the invocation of demons. Some thirteenth-century Church authorities on magic, such as William of Auvergne and Albert the Great, claimed that sorcerers attempted to increase the power of these aphrodisiacs by seizing the sexual organs while the animals were in the actual act of making love; how a magician would actually do this in the case of wild animals, such as prescriptions to obtain a wolf's penis or the semen and testicles of a stag, is anyone's guess.[332] What were believed to be particularly lusty animals, such as sparrows, were also ideal for the purposes of inducing arousal. Among the more bizarre prescriptions were to arouse a woman's passion by putting ant eggs in her bath, which was alleged to have the power to "so inflame her womb that willy-nilly she would be driven to copulation," or to bite into the heart of a dove and then write the name and naked image of the desired woman using an eagle's quill dipped in the dove's blood on parchment made from the skin of a female dog.[333] By the fourteenth and fifteenth centuries, records of witch trials and necromancer manuals survive, demonstrating the actual practice of such rituals.[334]

Late medieval witches were also accused of causing harm to animals, such as a neighbor's livestock, in much the same way as they were reputed to cause bodily harm to humans, often by means of poison powders or potions. (Image magic, however, seems to have been exclusively reserved for human victims.) In one of the earliest theological treatises to discuss the sect of witches, Johannes Nider's *Formicarius* (The Ant-Colony) from the 1430s, the author describes how an alleged witch called "Stadelin," who was the leader of a sect in the town of Boltingen in the diocese of Lausanne in Switzerland, confessed to causing through his magic "sterility in humans and animals," which included a seven-year long run of aborted fetuses in sheep that coincided with abortions in a human victim, and to driving "horses mad when their wealthy riders mounted them." Such charges were repeated by the *Malleus Maleficarum* of 1487, which added that witches could bewitch man or beast by a "mere look" alone.[335] In a series of eleven trials in Switzerland from 1397 to 1499, a host of farm animals including mostly cattle but also pigs and horses were alleged to have been bewitched, either by themselves or alongside their human owners, with one case at Appenzell in 1427 of a witch supposedly caught in the act by the farmer as she was entering his barn to carry out her poisoning. Earlier in 1420 at the same town, a woman was beheaded for being a poisoner, using the same poison both to harm cattle and kill a neighbor by means of giving him the proverbial "poisoned apple." At Lucerne in 1454 and again in 1480, accused sorceresses were said to have bewitched cows so that they gave blood instead of milk. The motive usually ascribed to these acts was the intent to harm the human owner through his animals, rather than harm the animals on their own account, either because the animals were essential to the intended victim's livelihood or because he had sentimental attachments to them.[336]

Most interesting of all is the third category, whereby witches were said to associate with or turn into animals in order to accomplish their magic. English trials during the sixteenth and seventeenth centuries made a specialty of accusing the witch of keeping a demonic familiar in order to carry out her magic in the form of a domestic or exotic animal pet, such as a cat, dog, toad, hare, fox, badger, etc.[337] The trial and hanging of the Chelmsford witches in 1589 was illustrated by a popular woodcut that featured an accused witch, Joan Prentice, sitting on a stool and suckling her animal familiar with the teat of her breast, while nearby under a gallows on which three witches hang cavort an assortment of other animal familiars, two of which are given names, "Jack" and "Jill."[338] The English chronicler, William of Malmesbury, in his account of the Sorceress of Berkeley from c. 1140, recounts how she daily communed with a jack-daw, "a very great favorite" of hers, through which she learned of her impending death on a day when the bird "chattered a little more loudly than usual."[339] In a cluster of trials in Switzerland, witches were accused of riding on wolves: Since there is no suggestion that the wolves were demon familiars, the very fact that women were able to tame and ride such ferocious beasts was perhaps justification enough for their extraordinary powers as a witch. However, when many of these cases were investigated by the authorities, more plausible explanations of mistaken identity were forthcoming, such as that the alleged witch was actually walking a dog or riding a donkey.[340] Burchard of Worms in his *Corrector* assigns a ten-day fast as penance for believing in werewolves, or that a man can be transformed into a wolf. The position of the Church for much of the Middle Ages, as represented by the authority of St. Thomas Aquinas in the *Summa Theologiae* (Summary of Theology), was that belief in transformations of matter which cannot be effected by the "natural powers," among which Aquinas included "that the human body be changed into the body of a beast," was simply delusion in which "a demon can work on man's imagination." This is also the position of some late medieval theologians, such as Jacopo Passavanti and Bernardino of Siena.[341]

However, by the middle and late fifteenth century, beliefs in animal transformations by witches are taken more seriously and are treated as being quite possibly real. In Nider's *Formicarius*, a witch called "Scavius" or "the scabby man" was said to be able to "transform himself into a mouse and thereby escape from the hands of his enemies." Martin Le Franc's *Defender of Ladies* of 1440 has the Defender make the standard argument that foolish old women are easily deceived by the devil into thinking in their minds that they can fly through the air on broomsticks or transform into animals, but the Adversary makes a strong case based on the trial record of a confessed witch that "ten thousand old women in a troop" assembled at a synagogue to which they flew, where they transformed into "the shapes of cats or goats" and kissed the devil "on his ass as a sign of their obedience." By the time we reach the definitive *Malleus Maleficarum* of 1487 by Heinrich Kramer and Jacob Sprenger, claims by witches to transform into animals or fly through the air are now declared to occur both in the imagination or fantasy as well as in reality, arguing on the grounds of ample quotations from scripture and a far fuller dossier of recorded trials. Witches were also said to act like wolves in devouring human children, something that was otherwise against human nature "and indeed against the nature of all beasts."[342]

As we have seen, the whole phenomenon of the witch-hunt was intimately bound up with animals, whether as ingredients, victims, or as partners and transformed participants in the *maleficia* or harmful magic believed to be performed by necromancers and witches. Since the act of witchcraft, as this was defined by the late Middle Ages, was, above all, a conscious choice to turn away from God and adhere to the devil, rather than simply the practice of magic through the occult forces of nature, animals could become involved in witchcraft only through their close association with humans, who alone possessed the power to make rational decisions. At the same time, those who chose to practice witchcraft were believed to descend to the level of beasts in their rejection of God-given human virtues. Animals' association with human magic and witchcraft was thus a very intricate and complex one, whose various strands made them an inseparable part of this chapter of human history. With the final acceptance by witch-hunters of beliefs in human transformation into animals by the end of the Middle Ages, the identification of witches with beasts was now fully complete.

Overall, Europeans by the end of the Middle Ages seem to have developed a closer relationship with their animals, and to have elided more of the differences between a human and an animal, despite learned efforts to the contrary, in a number of facets of their everyday interactions with their beasts. This can be seen most clearly in medieval uses of animals as pets, as partners in bestiality and magical practices, and in human views of animal susceptibility to disease, especially in the wake of the Black Death. Animal trials, particularly those in ecclesiastical courts, took the position that animals did not have the same capacity for rational culpability as humans, but nonetheless they did have certain rights, as well as a degree of accountability, namely, that even lowly beasts like flies or weevils must try to survive, propagate, and sustain their species as part of a God-created order subject to His laws, a position that even today most of us might find eminently reasonable; any punishments of animals were to be carried out as an example to humans of this demonstrable fact. Meanwhile, the presence of animals in human affairs became more pronounced as more classes of medieval people hunted, as pastoral farming increased, and as more humans ate more meat and fish. Only in the theater of war can we say that animals played rather less of a crucial role in the human endeavor by the late Middle Ages, reflected in the depreciation of the value of the medieval warhorse, as medieval people entered the modern era of the "military revolution" that emphasized artillery and infantry tactics over cavalry charges.[343]

# Afterword

No book can adequately sum up the environmental history of an entire people, let alone of an age. But at the very least, we can chart the course of the most important shifts in attitudes towards the environment that took place in the European mentality during the Middle Ages with regard to the three areas of our study: the elements, forests, and beasts. It is these changes, and the reasons behind them, that hold the most apropos lessons for our own, environmentally turbulent times.

The Middle Ages was the scene of some of the most momentous environmental changes and catastrophes in history: the Dust Veil Event of 536, the Floods of 589, the Medieval Warm Period, the Little Ice Age, the Great Famine of 1315–22, the Black Death of 1348–49, the Great Wind of 1362. Such milestones could hardly happen without some big attitudinal changes in response. Medieval people seem to have become aware that human activity could have far-reaching impacts upon their environment, that their relationship with nature cut both ways, that Mother Earth was not an inexhaustible and all-forgiving source of bounty. Air or water when poisoned would kill when breathed in or drunk. Forests that were poorly managed would no longer be productive. Animals could be just as noble in sentiment and behavior as humans, and humans just as beastly as animals. This especially applied to animals that humans had formed bonds with, such as horses, dogs, and falcons. It is not for nothing that Lady Fortune allows a horse to leave its stable and usurp its human master as head of his household in the fourteenth-century charivari, the *Roman de Fauvel* (Story of the Fawn-Colored Beast).[1] These are all assumptions we now take for granted, but which had to be learned by our medieval ancestors.

Some of our most cherished attitudes towards the environment were thus forged in the crucible of nature that had sorely tested Europe by the end of the Middle Ages. That crucible forced medieval society to make choices as to how to respond to the crisis at hand. Some decisions, such as to persecute Jews and witches on suspicion of poisoning, or "polluting," the environment in the form of water or air, we would regard today as wrong-headed. Others, such as to manage woodland pasture that would benefit wild and domestic animals as well as man, or to elide the rigid demarcations between man and

## 234  Afterword

beast, in spite of the best efforts of medieval thinkers and churchmen to the contrary, we might better agree with. As we make our own choices in response to our environmental crises, let us remember the lesson imparted by our medieval forbears, that some of those choices, even if they are made in acknowledgment of the impact of natural forces such as climate change, may in retrospect be the wrong ones.

# Notes

Preface

1 For a history of all these developments, readers should consult Christopher McGrory Klyza and Stephen C. Trombulak, *The Story of Vermont: A Natural and Cultural History* (Hanover, N.H.: University Press of New England, 1999).
2 Barbara A. Hanawalt and Lisa J. Kiser, "Introduction," in *Engaging with Nature: Essays on the Natural World in Medieval and Early Modern Europe*, eds. Barbara A. Hanawalt and Lisa J. Kiser (Notre Dame, IN: University of Notre Dame Press, 2008), p. 2. R.G. Collingwood's book, *The Idea of Nature* (Oxford: Clarendon Press, 1945), omits the Middle Ages entirely, focusing only on the classical world of ancient Greece, the Renaissance, and then modern times.
3 Keith Thomas, *Man and the Natural World: Changing Attitudes in England, 1500–1800* (Oxford: Oxford University Press, 1983).

Introduction

1 A.W. Henschel, "Document zur Geschichte des Schwarzen Todes," in *Archiv für die gesammte Medicin* ed. Heinrich Haeser, 10 vols. (Jena, 1841–49), 2: 45–46.
2 Richard C. Hoffman, "Homo et Natura, Homo in Natura: Ecological Perspectives on the European Middle Ages," in *Engaging with Nature: Essays on the Natural World in Medieval and Early Modern Europe*, eds. Barbara A. Hanawalt and Lisa J. Kiser (Notre Dame, IN: University of Notre Dame Press, 2008), p. 13. The concept of "new ecology" is further explained in Daniel Botkin's *Discordant Harmonies: A New Ecology for the Twenty-First Century* (New York: Oxford University Press, 1990).
3 Charles R. Bowlus, "Ecological Crises in Fourteenth Century Europe," in *Historical Ecology: Essays on Environment and Social Change*, ed. Lester J. Bilsky (Port Washington, N.Y.: Kennikat Press, 1980), pp. 86–99.
4 For a recent reassessment of the impact of the Black Death, see John Aberth, *From the Brink of the Apocalypse: Confronting Famine, War, Plague, and Death in the Later Middle Ages*, 2nd edn. (London and New York: Routledge, 2010), pp. 79–213. For a reassessment of neo-Malthusian arguments made for the Middle Ages, such as by M.M. Postan, see P.R. Schofield, "Medieval Diet and Demography," in *Food in Medieval England: Diet and Nutrition*, eds. C.M. Woolgar, D. Serjeantson, and T. Waldron (Oxford: Oxford University Press, 2006), pp. 239–53.
5 M.G.L. Baillie, "Putting Abrupt Environmental Change back into Human History," in *Environments and Historical Change: The Linacre Lectures (1998)*, ed. Paul Slack (Oxford: Oxford University Press, 1999), pp. 60–72.
6 *The Years without Summer: Tracing A.D. 536 and Its Aftermath*, ed. J.D. Gunn (Oxford: Archaeopress, 2000).
7 Lynn White, Jr., "The Historical Roots of our Ecologic Crisis," *Science*, 155 (1967):1203–7. Readers also should consult White's further clarifications made in a later essay,

"Continuing the Conversation," in *Western Man and Environmental Ethics: Attitudes Toward Nature and Technology*, ed. Ian G. Barbour (Reading, MA: Addison-Wesley Publishing Co., 1973), pp. 55–64.

8   David Herlihy, "Attitudes Toward the Environment in Medieval Society," in *Historical Ecology: Essays on Environment and Social Change*, ed. Lester J. Bilsky (Port Washington, N.Y.: Kennikat Press, 1980), pp. 100–16.

9   Karen Jolly, "Father God and Mother Earth: Nature-Mysticism in the Anglo-Saxon World," in *The Medieval World of Nature: A Book of Essays*, ed. Joyce E. Salisbury (New York and London: Garland Publishing, 1993), pp. 211–52.

10  René Dubos, *A God Within* (New York: Charles Scribner's Sons, 1972); idem, "Franciscan Conservation versus Benedictine Stewardship," in *Ecology and Religion in History*, eds. David and Eileen Spring (New York: Harper and Row, 1974), pp. 114–36; idem, "A Theology of Earth," in *Western Man and Environmental Ethics: Attitudes Toward Nature and Technology*, ed. Ian G. Barbour (Reading, MA: Addison-Wesley Publishing Co., 1973), pp. 43–54.

11  M.-D. Chenu, *Nature, Man, and Society in the Twelfth Century: Essays on New Theological Perspectives in the Latin West* (Chicago: University of Chicago Press, 1968), pp. 4–5.

12  Alain de Lille, *The Complaint of Nature*, trans. Douglas M. Moffat (New York: Henry Holt and Co., 1908); Veronica Fraser, "The Goddess Natura in the Occitan Lyric," in *The Medieval World of Nature: A Book of Essays*, ed. Joyce E. Salisbury (New York and London: Garland Publishing, 1993), p. 132.

13  Bernard Silvester, *Cosmographia*, ed. Peter Dronke (Leiden: E.J. Brill, 1978), pp. 141, 154; Brian Stock, *Myth and Science in the Twelfth Century: A Study of Bernard Silvester* (Princeton: Princeton University Press, 1972), pp. 197–202, 216–19; Fraser, "Goddess Natura," pp. 130–31.

14  Fraser, "Goddess Natura," pp. 132–42.

15  Paulus Niavis, *Judicium Jovis* (Leipzig: Martin Landsberg, c. 1492–95), available in a modern edition by Paul Krenkel published by the Berlin Akademie in 1953. I have used an English translation by Frank Dawson Adams from c. 1930, preserved in a handwritten manuscript at McGill University Library in Montreal. See also the commentary by Hoffmann, "Ecological Perspectives on the European Middle Ages," pp. 12–13.

16  J. Donald Hughes, "Early Greek and Roman Environmentalists," in *Historical Ecology: Essays on Environment and Social Change*, ed. Lester J. Bilsky (Port Washington, N.Y.: Kennikat Press, 1980), pp. 47–59; Michael Williams, *Deforesting the Earth: From Prehistory to Global Crisis* (Chicago: University of Chicago Press, 2003), p. 76.

17  See Laura A. Smoller, *Of Earthquakes, Hail, Frogs, and Geography: Plague and the Investigation of the Apocalypse*, in *Last Things: Death and the Apocalypse in the Middle Ages*, eds. Caroline Walker Bynum and Paul Freedman (Philadelphia: University of Pennsylvania Press, 2000), pp. 167–87; and the forthcoming book by John Aberth, *Doctoring the Black Death: Europe's Late Medieval Medical Response to Epidemic Disease*, with Rowman and Littlefield.

## Part I

1   A translated text of *On Airs, Waters, and Places* is available online at the Internet Classics Archive: http://classics.mit.edu.

2   Manfred Ullmann, *Islamic Medicine* (Edinburgh: University of Edinburgh Press, 1978), pp. 9–11, 89, 99–100.

3   Karl Sudhoff, "Pestschriften aus den ersten 150 Jahren nach der Epidemie des 'schwarzen Todes' 1348," *Archiv für Geschichte der Medizin* 7 (1914):83–84.

4   Lutfallah Gari, "Arabic Treatises on Environmental Pollution up to the End of the Thirteenth Century," *Environment and History* 8 (2002):475–83.

5   J. Donald Hughes, "Early Greek and Roman Environmentalists," in *Historical Ecology: Essays on Environment and Social Change*, ed. Lester J. Bilsky (Port Washington, N.Y.: Kennikat Press, 1980), p. 57.

6 Avicenna, *Liber Canonis* (Hildesheim: Georg Olms, 1964), fol. 416r. (book 4, fen 1, tractate 4, chapter 1). This is a reprint of the Latin text of the *Canon* printed in Venice in 1507. My personal preference, however, is for the edition with Gentile da Foligno's commentary, printed in Venice by the "Heirs of O. Scotus" in 1520 but which is not available in a modern reprint. All early printed texts of the *Canon* use Gerald of Sabloneta's thirteenth-century Latin translation from the Arabic (often incorrectly attributed to Gerald of Cremona from the twelfth century, owing to the fact that both men are referred to as Gerardus Cremonensis). See also the commentary in Jon Arrizabalaga, "Facing the Black Death: Perceptions and Reactions of University Medical Practitioners," in *Practical Medicine from Salerno to the Black Death*, ed. L. García-Ballester, R. French, J. Arrizabalaga, and A. Cunningham (Cambridge: Cambridge University Press, 1994), pp. 251–52.

7 Avicenna, *Liber Canonis*, fols. 416r.-v. (book 4, fen 1, tractate 4, chapter 3). See also Roger French, *Canonical Medicine: Gentile da Foligno and Scholasticism* (Leiden: Brill, 2001), p. 282.

8 Avicenna, *The Canon of Medicine: First Book*, trans. O. Cameron Gruner (New York: Augustus M. Kelley Publishers, 1970), pp. 34–37, 175–210.

9 Ibid., p. 190; Gentile da Foligno, *Consilium contra Pestilentiam* (Colle di Valdelsa, c. 1479), p. 33. Another plague doctor who was heavily indebted to Avicenna, both in discussing changes to air during autumn time and the relative virtues of mountainous and low-lying air during a plague, was the Florentine physician, Francischino de Collignano, in a treatise from 1382. See Karl Sudhoff, "Pestschriften aus den ersten 150 Jahren nach der Epidemie des 'schwarzen Todes' 1348," *Archiv für Geschichte der Medizin* AGM 5 (1912):368–71.

10 Avicenna, *Canon of Medicine*, pp. 221–29.

11 Gari, "Arabic Treatises on Environmental Pollution," pp. 481–82.

12 Ullmann, *Islamic Medicine*, p. 24.

13 Carole Hillenbrand, *The Crusades: Islamic Perspectives* (New York: Routledge, 2000), pp. 268–71.

14 For a good, modern translation into English of the *Etymologies*, see *The Etymologies of Isidore of Seville*, ed. and trans. Stephen A. Barney, W.J. Lewis, J.A. Beach, and Oliver Berghof (Cambridge: Cambridge University Press, 2006), pp. 270–300, 317–35.

15 For an English translation of De Universo, see Hrabanus Maurus, *De Universo: The Peculiar Properties of Words and their Mystical Significance. The Complete English Translation*, trans. Priscilla Throop, 2 vols. (Charlotte, VT: MedievalMS, 2009), 1: 279–310, 339–62; 2: 1–32, 159–86.

16 Ramsay MacMullen, *Christianity and Paganism in the Fourth to Eighth Centuries* (New Haven, CT: Yale University Press, 1997), p. 65, n.114; J. Toutain, *Les Cultes Païens dans l'Empire Romain*, 3 vols. (Paris: Ernest Leroux, 1907–20), 3: 300–307.

17 Miranda J. Green, *Dictionary of Celtic Myth and Legend* (London: Thames and Hudson, 1992), pp. 102, 129–30, 134, 152, 178, 184, 188–89, 197–98, 200–202, 208–9, 223–24.

18 Bernadette Filotas, *Pagan Survivals, Superstitions and Popular Cultures in Early Medieval Pastoral Literature* (Toronto: Pontifical Institute of Mediaeval Studies, 2005), pp. 147–48, 197; Stephen Wilson, *The Magical Universe: Everyday Ritual and Magic in Pre-Modern Europe* (London and New York: Hambledon and London, 2000), pp. 11, 68.

19 *Registrum Roberti Mascall, episcopi Herefordensis, A.D. 1404–1416*, ed. Joseph Henry Parry (Canterbury and York Society, 21, 1917), pp. 74–75.

20 Filotas, *Pagan Survivals*, pp. 147–48, 150–51, 197–99; Toutain, *Les Cultes Païens*, 3: 293–95, 357–64; Hughes, "Early Greek and Roman Environmentalists," p. 50; Green, *Dictionary of Celtic Myth*, pp. 155–56, 170–71.

21 Filotas, *Pagan Survivals*, pp. 137–38.

22 Paul Edward Dutton, *Charlemagne's Mustache and Other Cultural Clusters of a Dark Age* (New York and Basingstoke, Hampshire: Palgrave Macmillan, 2004), pp. 169–70; idem, "Thunder and Hail over the Carolingian Countryside," in *Agriculture in the Middle Ages: Technology, Practice, and Representation*, ed. Del Sweeney (Philadelphia: University of Pennsylvania Press, 1995), pp. 111–12; idem, *Carolingian Civilization: A Reader* (Peterborough, Ont.: Broadview Press, 1993), pp. 189–91.

23 *Monumenta Germaniae Historic*: Capitularia regum Francorum 1, p. 228.15, available online at www.dmgh.de; Dutton, *Charlemagne's Mustache*, p. 173.
24 Edward Peters, *Torture*, exp. edn. (Philadelphia: University of Pennsylvania Press, 1996), pp. 36–39, 42–43.
26 Dutton, *Charlemagne's Mustache*, pp. 173–74; idem, "Thunder and Hail," pp. 115–16.
26 *Monumenta Germaniae Historia*: Epistolae Karolini aevi 4, p. 304.3–12; Dutton, *Charlemagne's Mustache*, pp. 171–73; idem, "Thunder and Hail," pp. 113–17, 127–29.
27 Dutton, *Charlemagne's Mustache*, pp. 177, 180, n. 72; idem, "Thunder and Hail," pp. 119–24, n. 49; Richard Kieckhefer, *Magic in the Middle Ages* (Cambridge: Cambridge University Press, 1989), p. 86.
28 Richard Kieckhefer, *Magic in the Middle Ages* (Cambridge: Cambridge University Press, 1989), pp. 38–39.
29 MacMullen, *Christianity and Paganism*, pp. 136–37.
30 Karen Jolly, "Father God, Mother Earth: Nature-Mysticism in the Anglo-Saxon World," in *The Medieval World of Nature: A Book of Essays*, ed. Joyce E. Salisbury (New York and London: Garland Publishing, 1993), pp. 221–24, 235; Thomas D. Hill, "The Æcerbot Charm and its Christian User," *Anglo-Saxon England* 6 (1977):213–21.
31 Maureen A. Tilley, "Martyrs, Monks, Insects, and Animals," in *The Medieval World of Nature: A Book of Essays*, ed. Joyce E. Salisbury (New York and London: Garland Publishing, 1993), pp. 93–94, 101.
32 *Lives of the Saints*, trans. J.F. Webb (Harmondsworth, Middlesex: Penguin Books, 1965), pp. 95–96, 98–99. Bede also tells a story of how Cuthbert's successor, Ethelwald, calmed a storm on Farne Island. See Bede, *A History of the English Church and People*, ed. Leo Sherley-Price (Harmondsworth, Middlesex: Penguin Books, 1955), pp. 265–66.
33 Pierre Riché, *Daily Life in the World of Charlemagne*, trans. Jo Ann McNamara (Philadelphia: University of Pennsylvania Press, 1978), pp. 27–28.
34 Henry of Huntingdon, *The History of the English*, ed. Thomas Arnold (London: Longman and Co., 1879), p. 189.
35 Green, *Dictionary of Celtic Myth*, pp. 205–7, 210.
36 Paul C. Buckland, "The North Atlantic Environment," and Thomas H. McGovern, "The Demise of Norse Greenland," in *Vikings: The North Atlantic Saga*, eds. William W. Fitzhugh and Elisabeth I. Ward (Washington D.C.: Smithsonian Institution Press, 2000), pp. 152–53, 330–31, 335.
37 The classic study of the medieval climate on the basis of documentary data is Pierre Alexandre's *Le Climat en Europe au Moyen Age* (L'Ecole des Hautes Etudes en Sciences Sociales, *Recherches d'Histoire et de Sciences Sociales*, 24, 1987), which tracks trends in climate from the years 1000 to 1425 in hundreds of narrative sources from France, Germany, and northern Italy. Alexandre uses decennial indices based on the number of months in which the weather is mentioned in the chronicles as being "mild," "harsh," "rainy," or "dry". For other studies of European climate based on documentary data, see *European Climate Reconstructed from Documentary Data: Methods and Results*, eds. Burkhard Frenzel, Christian Pfister, and Birgit Gläser (Akademie der Wissenschaften unde der Literatur, Paleoclimate Research, 7, 1992).
38 H.H. Lamb, *Climate, History and the Modern World*, 2nd edn. (London and New York: Routledge, 1995), pp. 74–107; idem, *Climate: Present, Past and Future*, 2 vols. (London: Metheun and Co., 1977), 2: 21–279.
39 Brian Fagan, *The Great Warming: Climate Change and the Rise and Fall of Civilizations* (New York: Bloomsbury Press, 2008), pp. 1–45; idem, *The Little Ice Age: How Climate Made History, 1300–1850* (New York: Basic Books, 2000), 3–21; Lamb, *Climate, History, and the Modern World*, pp. 177–82; idem, "An approach to the Study of the Development of Climate and Its Impact in Human Affairs," in *Climate and History: Studies in Past Climates and Their Impact on Man*, eds. T.M.L. Wigley, M.J. Ingram, and G. Farmer (Cambridge: Cambridge University Press, 1981), pp. 294–301; idem, *Climate: Present, Past and Future*, 2: 435–38; *The Climate of Europe: Past, Present and Future. Natural and Man-Induced Climatic*

40 Fagan, *The Little Ice Age*, pp. 55–57; Lamb, *Climate, History and the Modern World*, pp. 327–29.
41 Lamb, *Climate, History and the Modern World*, pp. 177–80; idem, *Climate: Present, Past and Future*, 2:435–38.
42 A good discussion of the interrelationship between history and climate in all these events is available in Neville Brown, *History and Climate Change: A Eurocentric Perspective* (London and New York: Routledge, 2001), pp. 67–200; and idem, "Approaching the Medieval Optimum, 212 to 1000 AD," in *Water, Environment and Society in Times of Climatic Change*, eds. Arie S. Issar and Neville Brown (Dordrecht: Kluwer Academic Publishers, 1998), pp. 69–95.
43 Brown, *History and Climate Change*, p. 198.
44 Historians who both acknowledged and were skeptical of climate's impact upon history include: B.H. Slicher van Bath, *The Agrarian History of Western Europe, A.D. 500–1850* (London: Edward Arnold, 1963), p. 8; Georges Duby, *Rural Economy and Country Life in the Medieval West*, trans. Cynthia Postan (London: Edward Arnold, 1968), pp. 305–6; Emmanuel Le Roy Ladurie, *Times of Feast, Times of Famine: A History of Climate since the Year 1000*, trans. Barbara Bray (Garden City, N.Y.: Doubleday and Co., 1971), p. 255.
45 Helge Salvesen, "The Climate as a Factor of Historical Causation," in *European Climate Reconstructed from Documentary Data: Methods and Results*, eds. Burkhard Frenzel, Christian Pfister, and Birgit Gläser (Akademie der Wissenschaften unde der Literatur, Paleoclimate Research, 7, 1992), pp. 219–33. For a cautionary tale of reading too much into contemporary accounts of the 589 flood in Italy, see Paolo Squatriti, "The Floods of 589 and Climate Change at the Beginning of the Middle Ages: An Italian Microhistory," *Speculum* 85 (2010):799–826.
46 Annie Grant, "Animal Resources," and Grenville Astill and Annie Grant, "Efficiency, Progress and Change," in *The Countryside of Medieval England*, eds. Grenville Astill and Annie Grant (Oxford: Basil Blackwell, 1988), pp. 177, 214, 217.
47 Richard C. Hoffman, "Economic Development and Aquatic Ecosystems in Medieval Europe," *American Historical Review* 101 (1996):638–40.
48 Bruce M.S. Campbell, "Economic Rent and the Intensification of English Agriculture, 1086–1350," in *Medieval Farming and Technology: The Impact of Agricultural Change in Northwest Europe*, eds. Grenville Astill and John Langdon (Leiden: Brill, 1997), pp. 225–27; Christopher Dyer, *Making a Living in the Middle Ages: The People of Britain, 850–1520* (New Haven, CT: Yale University Press, 2002), pp. 246–51.
49 Lynn White, Jr., *Medieval Technology and Social Change* (Oxford: Clarendon Press, 1962), pp. 39–78; Duby, *Rural Economy*, pp. 88–112; Jean Gimpel, *The Medieval Machine: The Industrial Revolution of the Middle Ages* (New York: Holt, Rinehart and Winston, 1976), pp. 29–58; Dyer, *Making a Living in the Middle Ages*, pp. 21–24.
50 For a general overview of the revision of the "agricultural revolution" thesis, see: Adriaan Verhulst, "The 'Agricultural Revolution' of the Middle Ages Reconsidered," in *Law, Custom, and the Social Fabric in Medieval Europe: Essays in Honor of Bryce Lyon*, ed. Bernard S. Bachrach and David Nicholas (Studies in Medieval Culture, 28, 1990), pp. 17–28; and Christopher Dyer, "Medieval Farming and Technology: Conclusion," in *Medieval Farming and Technology: The Impact of Agricultural Change in Northwest Europe*, eds. Grenville Astill and John Langdon (Leiden: Brill, 1997), pp. 293–312.
51 Karl Brunner, "Continuity and Discontinuity of Roman Agricultural Knowledge in the Early Middle Ages," in *Agriculture in the Middle Ages: Technology, Practice, and Representation*, ed. Del Sweeney (Philadelphia: University of Pennsylvania Press, 1995), pp. 21–40; Bjørn Poulsen, "Agricultural Technology in Medieval Denmark," in *Medieval Farming and Technology: The Impact of Agricultural Change in Northwest Europe*, eds. Grenville Astill and John Langdon (Leiden: Brill, 1997), pp. 134–38; John Langdon, "Agricultural Equipment," in *The Countryside of Medieval England*, eds. Grenville Astill and Annie Grant (Oxford: Basil Blackwell, 1988), pp. 95–103.
52 Erik Thoen, "The Birth of 'The Flemish Husbandry': Agricultural Technology in Medieval Flanders," in *Medieval Farming and Technology: The Impact of Agricultural Change in*

*Northwest Europe*, eds. Grenville Astill and John Langdon (Leiden: Brill, 1997), pp. 74–81. For the debate over open field systems as developed in England, see Tom Williamson, *Shaping Medieval Landscapes: Settlement, Society, Environment* (Macclesfield, Cheshire: Windgather Press, 2003), pp. 1–27.

53 John Langdon, *Horses, Oxen and Technological Innovation: The Use of Draught Animals in English Farming from 1066 to 1500* (Cambridge: Cambridge University Press, 1986), pp. 48–62.

54 Poulsen, "Agricultural Technology in Medieval Denmark," pp. 132–33.

55 Campbell, "Economic Rent," pp. 228–29.

56 Janken Myrdal, "The Agricultural Transformation of Sweden, 1000–1300," in *Medieval Farming and Technology: The Impact of Agricultural Change in Northwest Europe*, eds. Grenville Astill and John Langdon (Leiden: Brill, 1997), pp. 151–53; Langdon, *Horses, Oxen and Technological Innovation*, pp. 12–16.

57 Michael Toch, "Agricultural Progress and Agricultural Technology in Medieval Germany: An Alternative Model," in *Technology and Resource Use in Medieval Europe: Cathedrals, Mills, and Mines*, eds. Elizabeth Bradford Smith and Michael Wolfe (Aldershot, UK and Brookfield, VT: Ashgate, 1997), pp. 158–69; Lynn White, Jr., "The Historical Roots of our Ecologic Crisis," in *The Ecocriticism Reader: Landmarks in Literary Ecology*, ed. Cheryll Glotfelty and Harold Fromm (Athens, GA: The University of Georgia Press, 1996), p.8; Mavis Mate, "Agricultural Technology in Southeast England, 1348–1530," in *Medieval Farming and Technology: The Impact of Agricultural Change in Northwest Europe*, eds. Grenville Astill and John Langdon (Leiden: Brill, 1997), pp. 251–74.

58 *The Cambridge Economic History of Europe, volume II: Trade and Industry in the Middle Ages*, eds. M. Postan and E.E. Rich (Cambridge: Cambridge University Press, 1952), pp. 430–41; John Hatcher, *English Tin Production and Trade before 1550* (Oxford: Clarendon Press, 1973), pp. 8–42; Gimpel *The Medieval Machine*, pp. 59–74; Oliver Rackham, *The History of the Countryside* (London: J.M. Dent and Sons, 1986), pp. 369–72; André E. Guillerme, *The Age of Water: The Urban Environment in the North of France, A.D. 300–1800* (College Station, TX: Texas A & M University Press, 1988), pp. 138–40; Patrice Beck, Philippe Braunstein, and Michel Philippe, "Wood, Iron, and Water in the Othe Forest in the Late Middle Ages: New Findings and Perspectives," in *Technology and Resource Use in Medieval Europe: Cathedrals, Mills, and Mines*, eds. Elizabeth Bradford Smith and Michael Wolfe (Aldershot, UK and Brookfield, VT: Ashgate, 1997), pp. 173–84; Dyer, *Making a Living in the Middle Ages*, p. 244.

59 Bryan Ward-Perkins, *From Classical Antiquity to the Middle Ages: Urban Public Building in Northern and Central Italy, A.D. 300–850* (Oxford: Oxford University Press, 1984), pp. 119–54; Guillerme, *The Age of Water*, pp. 105–17; Paolo Squatriti, *Water and Society in Early Medieval Italy, A.D. 400–1000* (Cambridge: Cambridge University Press, 1998), pp. 11–33, 44–65; Klaus Grewe, "Water Technology in Medieval Germany," and Roberta Magnusson and Paolo Squatriti, "The Technologies of Water in Medieval Italy," in *Working with Water in Medieval Europe: Technology and Resource-Use*, ed. Paolo Squatriti (Leiden: Brill, 2000), pp. 130–32, 236–58.

60 Richard Holt, "Medieval England's Water-Related Technologies," Grewe, "Water Technology," and Josephine Rouillard, "Medieval Hydraulics in France," in *Working with Water in Medieval Europe*, pp. 88–94, 137–41, 167–68.

61 Guillerme, *The Age of Water*, pp. 97–98; Holt, "Medieval England's Water-Related Technologies," Grewe, "Water Technology," and Magnsson and Squatriti, "Technologies of Water," in *Working with Water in Medieval Europe*, pp. 97–100, 145–59, 236–44; Hoffmann, "Economic Development and Aquatic Ecosystems," pp. 643–45.

62 William H. Tebrake, "Hydraulic Engineering in the Netherlands," in *Working with Water in Medieval Europe: Technology and Resource-Use*, ed. Paolo Squatriti (Leiden: Brill, 2000), pp. 101–27; idem, *Medieval Frontier: Culture and Ecology in Rijnland* (College Station, TX: Texas A & M University Press, 1985), esp. pp. 185–243; idem, "Ecology of Village Settlement in the Dutch Rijnland," in *Pathways to Medieval Peasants*, ed. J.A. Raftis (Toronto: Pontifical Institute of Mediaeval Studies, 1981), pp. 2–26; Rackham, *History of the Countryside*, p. 379.

63 *The Visigothic Code*, ed. S.P. Scott (Boston: Boston Book Co., 1910), pp. 296–97.

64 Holt, "Medieval England's Water-Related Technologies," Benoit and Rouillard, "Medieval Hydraulics," Magnusson and Squatriti, "Technologies of Water," and Thomas F. Glick and Helena Kirchner, "Hydraulic Systems and Technologies of Islamic Spain: History and Archaeology", in *Working with Water in Medieval Europe*, pp. 53–54, 173–76, 224–27, 231–36, 279–92, 302–9; Guillerme, *The Age of Water*, pp. 51–77; Rackham, *History of the Countryside*, pp. 383–87; Károly Takács, "Medieval Hydraulic Systems in Hungary: Written Sources, Archaeology and Interpretation," in *People and Nature in Historical Perspective*, eds. József Laszlovszky and Péter Szabó (Budapest: Central European University and Archaeolingua, 2003), pp. 289–311.
65 Richard Holt, *The Mills of Medieval England* (Oxford: Basil Blackwell, 1988), pp. 1–4, 119–20, 133–36; Holt, "Medieval England's Water-Related Technologies," in *Working with Water in Medieval Europe*, pp. 1–50, 67–68.
66 Holt, *Mills of Medieval England*, pp. 5–69, 107–28, 145–70; David Crossley, "The Archaeology of Water Power in Britain before Industrial revolution", Paolo Squatriti, "Advent and Conquests of the Water Mill in Italy." and Richard Holt, "Mechanization and the Medieval English Economy," in *Technology and Resource Use in Medieval Europe: Cathedrals, Mills, and Mines*, eds. Elizabeth Bradford Smith and Michael Wolfe (Aldershot, U.K. and Brookfield, VT: Ashgate, 1997), pp. 109–45, 149–56; Squatriti, *Water and Society*, pp. 126–42; Holt, "Medieval England's Water-Related Technologies," Benoit and Rouillard, "Medieval Hydraulics in France," and Magnusson and Squatriti, "Technologies of Water," in *Working with Water in Medieval Europe*, pp. 57–79, 169–73, 179–80, 203–14, 258–65; John Muendel, "Mills in the Florentine Countryside," in *Pathways to Medieval Peasants*, ed. J.A. Raftis (Toronto: Pontifical Institute of Mediaeval Studies, 1981), pp. 83–115.
67 Holt, "Mechanization in Medieval England," p. 145.
68 Holt, *Mills of Medieval England*, pp. 17–35, 136–44; Holt, "Mechanization in Medieval England," pp. 145–48.
69 M.-D. Chenu, *Nature, Man, and Society in the Twelfth Century: Essays on New Theological Perspectives in the Latin West*, ed. and trans. Jerome Taylor and Lester K. Little (Chicago: University of Chicago Press, 1968), pp. 1–48.
70 William of Conches, *Philosophia Mundi*, ed. Gregor Maurach (Pretoria: University of South Africa Press, 1974), pp. 15–16, 18–33.
71 Bernard Silvester, *Cosmographia*, ed. Peter Dronke (Leiden: E.J. Brill, 1978), with a useful summary on pp. 29–49; Brian Stock, *Myth and Science in the Twelfth Century: A Study of Bernard Sylvester* (Princeton: Princeton University Press, 1972).
72 Alain de Lille, *The Complaint of Nature*, trans. Douglas M. Moffat (New York: Henry Holt and Company, 1908), pp. 25, 36–38. For a history of medieval interpretations of Genesis 1:28, see Jeremy Cohen, *"Be Fertile and Increase, Fill the Earth and Master It": The Ancient and Medieval Career of a Biblical Text* (Ithaca, N.Y.: Cornell University Press, 1989), pp. 221–305.
73 Alexander Neckam, *De Naturis Rerum*, ed. Thomas Wright (London: Rolls Series, 34, 1863), pp. 62–68, 127–39, 158–84.
74 Vincent of Beauvais, *Speculum Quadruplex sive Speculum Maius*, 4 vols. (Douai: Balthazar Belleri, 1624), 1: 304–6.
75 Beauvais, *Speculum Maius*, 1: 318–19, 338, 344–46, 350–53.
76 Beauvais, *Speculum Maius*, 1: 386–90, 411.
77 Beauvais, *Speculum Maius*, 1: 425–552.
78 Bartholomaeus Anglicus, *On the Properties of Things: John Trevisa's Translation: A Critical Text*, 2 vols. (Oxford: Clarendon Press, 1975), 1: 566–67.
79 Anglicus, *On the Properties of Things*, 1: 646–87; 2: 693–881.
80 Thomas of Cantimpré, *Liber de Natura Rerum. Volume 1: Text*, ed. Helmut Boese (Berlin: Walter de Gruyter, 1973), p. 405.
81 Cantimpré, *Liber de Natura Rerum*, pp. 351–84, 396–412. The sections on precious stones, metals, and springs are reproduced in books 6, 7, and 8 of Konrad von Megenberg's *Buch der Natur*, available in the edition by Robert Luff and George Steer (Tübingen: Niemeyer Verlag, 2003), pp. 463–522.

82 Albert the Great, *On the Causes of the Properties of the Elements*, trans. Irven M. Resnick (Milwaukee, WI: Marquette University Press, 2010), pp. 71–77; Dagmar Gottschall, "Conrad of Megenberg and the Causes of the Plague: A Latin Treatise on the Black Death Composed ca. 1350 for the Papal Court in Avignon," in *La Vie Culturelle, Intellectuelle et Scientifique a la Cour des Papes d'Avignon*, ed. Jacqueline Hamesse (Brepols, 2006), pp. 319–32.
83 *Godeschalcus und Visio Godeschalci* ed. Erwin Assman (Neumünster: Karl Wachholtz, 1979), p. 48.
84 White, Jr., "The Historical Roots of our Ecologic Crisis," pp. 3–14.
85 Roger D. Sorrell, *St. Francis of Assisi and Nature: Tradition and Innovation in Western Christian Attitudes toward the Environment* (Oxford: Oxford University Press, 1988), p. 101.
86 Ibid, p. 43.
87 Edward A. Armstrong, *Saint Francis: Nature Mystic: The Derivation and Significance of the Nature Stories in the Franciscan Legend* (Berkeley and Los Angeles: University of California Press, 1973), pp. 235–36.
88 Thomas Sieger Derr, "Religion's Responsibility for the Ecological Crisis: An Argument Run Amok," *Worldview* 18 (1975): 39–45; R. Attfield, "Christian Attitudes to Nature," *Journal of the History of Ideas* 44 (1983):369–86.
89 Sorrell, *St. Francis of Assisi*, pp. 45, 63–64, 73–74; Attfield, "Christian Attitudes to Nature," p. 379.
90 Armstrong, *Saint Francis: Nature Mystic*, pp. 242–43.
91 A good, general introduction to The Little Ice Age is Brian Fagan's *The Little Ice Age: How Climate Made History, 1300–1850* (New York: Basic Books, 2000), but I have relied more on: H.H. Lamb, *Climate, History and the Modern World*, pp. 187–207; idem, *Climate: Present, Past and Future*, 2: 449–63; idem, "Climate in the Last Thousand Years," in *The Climate of Europe: Past, Present and Future: Natural and Man-Induced Climatic Changes: A European Perspective*, eds. Hermann Flohn and Roberto Fantechi (Dordrecht: D. Reidel Publishing, 1984), pp. 38–53; Christopher Dyer, *Standards of Living in the Middle Ages: Social Change in England, c. 1200–1520* (Cambridge: Cambridge University Press, 1989), pp. 258–73.
92 Fagan, *The Little Ice Age*, pp. 55–57; Jean M. Grove, *Little Ice Ages: Ancient and Modern*, volume 1, 2nd edn. (London and New York: Routledge, 2004), pp. 564–90.
93 A.T. Grove and Oliver Rackham, *The Nature of Mediterranean Europe: An Ecological History* (New Haven: Yale University Press, 2001), pp. 130–35.
94 See Alexandre, *Le Climat en Europe au Moyen Age*, and critique in John Aberth, *From the Brink of the Apocalypse: Confronting Famine, War, Plague, and Death in the Later Middle Ages*, 2nd edn. (London and New York: Routledge, 2010), pp. 11–12.
95 William Merle, *Consideraciones Temperiei pro 7 Annis*, trans. G.J. Symons (London: Edward Stanford, 1891).
96 Aberth, *From the Brink of the Apocalypse*, pp. 16–19; W.C. Jordan, *The Great Famine: Northern Europe in the Early Fourteenth Century* (Princeton: Princeton University Press, 1996), pp. 7–39; Ian Kershaw, "The Great Famine and Agrarian Crisis in England, 1315–22," in *Past and Present* 59 Hoffmann, "Economic Development," and Aquatic Ecosystems (1973): 3–50; pp. 638–58.
97 Aberth, *From the Brink of the Apocalypse*, p. 27; Jordan, *Great Famine*, pp. 43–60; Dyer, *Making a Living in the Middle Ages*, pp. 228–46; Kershaw, "Great Famine and Agrarian Crisis," pp. 3–50; G. Beresford, "Climatic Change and its Effect upon the Settlement and Desertion of Medieval Villages in Britain," in *Consequences of Climatic Change* eds. C. Delano Smith and M. Parry (Nottingham: University of Nottingham Press, 1981), pp. 30–39; Christopher Dyer, "Documentary Evidence," and Grenville Astill and Annie Grant, "Efficiency, Progress and Change," in *The Countryside of Medieval England*, eds. Grenville Astill and Annie Grant (Oxford: Basil Blackwell, 1988), pp. 33–35, 230–34.
98 Aberth, *From the Brink of the Apocalypse*, pp. 25–27.
99 An opinion with which my former teacher on the Black Death at the University of Cambridge, John Hatcher, concurs. See John Hatcher, *The Black Death: A Personal History* (Boston: Da Capo Press, 2008), p. xi.

100 Jon Arrizabalaga, "Facing the Black Death: Perceptions and Reactions of University Medical Practitioners," in *Practical Medicine from Salerno to the Black Death*, eds. L. García-Ballester, R. French, J. Arrizabalaga, and A. Cunningham (Cambridge: Cambridge University Press, 1994), p. 248.
101 Gentile da Foligno, *Consilium contra Pestilentiam* (Colle di Valdelsa, c. 1479), p. 37; Pietro di Tussignano, *Consilium pro Peste Evitanda*, reprinted in *The Fasciculus Medicinae of Johannes de Ketham, Alemanus: Facsimile of First Edition of 1491*, trans. Charles Singer (Birmingham, Ala.: Classics of Medicine Library, 1988), col. 4; Sudhoff, "Pestschriften," 7 (1913):83; 9 (1916):128; Codex Latin 363, Bayerische Staatsbibliothek, Munich, fols. 124v.; MS 1227, Universitätsbibliothek, Leipzig, fols. 150v.–151r; Maino de Maineri, *Libellus de Preservatione ab Epydimia*, ed. R. Simonini (Modena, 1923), p. 17.
102 H. Émile Rébouis, *Étude Historique et Critique sur la Peste* (Paris: Alphonse Picard, 1888), p. 80.
103 Karl Sudhoff, "Pestschriften," 4 (1911):421, 423; 6 (1913):354; 11 (1919):166.
104 Sudhoff, "Pestschriften," AGM 4 (1911):423.
105 Sudhoff, "Pestschriften," AGM 9 (1914):264.
106 Sudhoff, "Pestschriften," AGM 8 (1915):188, 194. In modern German, the verb, *verunreinigen*, can mean to pollute or else to soil or dirty something.
107 Sudhoff, "Pestschriften," AGM 8 (1915):284, 287.
108 Arrizabalaga, "Facing the Black Death," pp. 251–56.
109 See, for example, an anonymous south German treatise from the early fifteenth century, which mentions the triple conjunction of 1345 as a remote cause of plague even though "it is passed over by astrologers." See Sudhoff, "Pestschriften," AGM 11 (1919):166. The Paris masters were perhaps most indebted to a French Jewish astrologer, Levi ben Gerson, as their specific source for the planetary conjunction of 1345: Gerson prognosticated it in an astronomical work penned for Pope Clement VI in 1344. See B.R. Goldstein and D. Pingree, *Levi ben Gerson's Prognostication for the Conjunction of 1345* (Transactions of the American Philosophical Society, 80, 1990).
110 Rébouis, *Étude Historique*, pp. 76–78. For later treatises that quote the Paris masters' explanation of the causes of plague almost word-for-word, see: Sudhoff, "Pestschriften," AGM 11 (1919):167; 16 (1924–25):81; 16 (1924–25):79–81.
111 Jacme d'Agramont, "Regiment de Preservacio a Epidimia o Pestilencia e Mortaldats," trans. M.L. Duran-Reynals and C.-E.A. Winslow, *Bulletin of the History of Medicine*, 23 (1949):59–63. Agramont's explanation of a "contra-natural" change in the air was repeated decades later by a fellow Spanish physician, Juan de Aviñón, a physician practicing in Seville between 1353 and 1382, who wrote a *Regimen against the Pestilence*. A similar idea seems to be expressed by the Jewish doctor, Rabbi Isaac ben Todros, who practiced in Avignon during the 1370s and described the cause of the plague as a change in the air "that deviates from the natural order and opposes nature." See M.V. Amasuno Sárraga, *La Peste en la Corona de Castilla Durante la Segunda Mitad del Siglo XIV* (Estudios de historia de la ciencia y de la técnica, no. 12, 1996), p. 220; L. Zunz, *Jubelschrift zum Neunzigsten Geburtstag* (Berlin, 1884), p. 109.
112 Taha Dinanah, "Die Schrift von Abi G'far Ahmed ibn 'Ali ibn Mohammed ibn 'Ali ibn Hatimah aus Almeriah über die Pest," *Archiv für Geschichte der Medizin* 19 (1927):34–35.
113 Dinanah, "Die Schrift," p. 36; Foligno, *Consilium*, p. 33. Foligno's third *dubium* is repeated by Pietro di Tussignano and Sigmund Albich of Prague. See Sudhoff, "Pestschriften," AGM 9 (1916):127; and Tussignano, *Consilium*, col. 4.
114 Rébouis, *Étude Historique*, pp. 80–84. The opinion that near causes were ultimately indebted to higher ones was likewise repeated by other plague treatises: Sudhoff, "Pestschriften," AGM 11 (1919):166; 16 (1924–25):38, 170–71.
115 Dinanah, "Die Schrift," pp. 38–39.
116 Foligno's attitude was that the cause of the plague was irrelevant, since it was more important to focus on how to prevent or cure it, in the same way as one didn't care how a fire was set or how a poisonous asp came into being: One simply wanted to save one's house from burning down or check the spread of the asp's poisonous bite. This was directly contrary to Raymond Chalin de Vinario's approach, who criticized physicians for

being unable to treat or cure diseases since they did not know exactly how they were caused. See Foligno, *Consilium*, pp. 2–3; Robert Hoeniger, *Der Schwarze Tod in Deutshland* (Berlin, 1882), p. 160. Foligno's ambivalence was also evident in his half-hearted answer to his seventh doubt, about how the higher planets could be the cause of "corruption and death" in the lower regions if they were meant to give us nourishment and life, and in his short *consilium* dedicated to the college of physicians at the University of Genoa: See Foligno, *Consilium*, pp. 31, 36; Sudhoff, "Pestschriften," AGM 5 (1912):333.

117 Agramont did consider the case of a pestilence particular to a given locality, such as his own city of Lérida, but here he considered the cause not to be bad air, but bad regimen, especially the eating of too many fruits. Agramont, "Regiment de Preservacio a Epidimia," pp. 60, 64–68.

118 An exception to the Paris orthodoxy is perhaps John of Burgundy, writing in 1365, who gave as the cause of plague air corrupted by the "influence of the heavenly bodies," which he explained did not mean that "the air is corrupted in its substance, because it is an uncompounded substance and that would be impossible, but it is corrupted by reason of evil vapors mixed with it." But how this is different from substantial corruption by mixing bad vapors with the air is not explained. See *The Black Death*, trans. and ed. Rosemary Horrox (Manchester: Manchester University Press, 1994) p. 185.

119 Samuel K. Cohn, Jr., *The Black Death Transformed: Disease and Culture in Early Renaissance Europe* (London and New York: Arnold and Oxford University Press, 2003), pp. 235–36.

120 *The Black Death*, trans. and ed. Horrox, pp. 185, 192. Burgundy's rant against physicians ignorant of astrology as being incapable of curing plague is echoed by a fourteenth-century treatise that discusses plague from a purely astrological point of view: "A physician who is ignorant of astronomy is one into whose hands no one, indeed, ought to commit himself, because such a physician is, not undeservedly, reckoned to be imperfect, that is, unproven." See Sudhoff, "Pestschriften," AGM 16 (1924–25):96.

121 Robert Hoeniger, *Der Schwarze Tod in Deutschland* (Berlin, 1882), pp. 160–69. Vinario may in fact refer to Burgundy as the "one master beyond the Alps who was preeminent among those whom I knew [who] subtly touched upon this, albeit in few words," i.e., on the influence of the heavens during a plague. The only plague doctor who seems to defend Avicenna from a charge of laying too little emphasis on celestial causes is the Florentine, Francischino de Collignano, who asserts that Avicenna, in the first fen of the fourth book of the *Canon*, in the chapter on pestilence, declares that "every [cause] comes from one of the heavenly forms, which make that cause inevitable". But since he doesn't mention near causes here, it is not clear that Collignano has properly understood Avicenna. See Sudhoff, "Pestschriften," AGM 5 (1912):367.

122 Ole Benedictow, *The Black Death, 1346–1353: The Complete History* (Woodbridge, Suffolk: Boydell Press, 2004).

123 A very similar account of the dangers posed by wells is given by the fourteenth-century Italian doctor, Mariano di Ser Jacopo, who quotes Avicenna's *Canon* as the source but does not connect it with earthquakes. According to Jacopo, Avicenna's version went like this: "Once, having opened a water well that had remained closed for a long time, a vapor came out of it that was so putrid and so poisonous that it immediately killed those who were near, and faster and more powerfully than any poison in the world." See R. Simonini, "Il Codice di Mariano di Ser Jacopo spora 'Rimedi Abili nel Tempo di Pestilenza'," *Bollettino dell'Istituto Storico Italiano dell'Arte Sanitaria* 9 (1929):165.

124 Sudhoff, "Pestschriften," AGM 11 (1919):44–51, 53; Sabine Krüger, "Krise der Zeit als Ursache der Pest? Der Traktat De mortalitate in Alamannia des Konrad von Megenberg," in *Festschrift für Hermann Heimpel zum 70. Geburtstag am 19. September 1971* (Göttingen: Vandenhoeck and Ruprecht, 1971), pp. 862–83; Gottschall, "Conrad of Megenberg," pp. 319–32. The treatise discovered by Sudhoff in an Erfurt manuscript is incorporated into Megenberg's better-known work, *De Mortalitate in Alamannia* (Concerning the Mortality in Germany), but which has substantial unrelated digressions, such as on the accusation that Jews poisoned wells to give the plague to Christians, and on the perversion of university

Notes   245

studies. My quotations are all from the Sufhoff version. A partial translation of this is also available in *Black Death*, trans. and ed. Horrox, pp. 177–82, although neither Sudhoff nor Horrox seems aware that this treatise is by Megenberg.

125  Albertus Magnus also seems to have made the connection between pestilence and earthquakes. See Gottschall, "Conrad of Megenberg," p. 329, n. 27.

126  Cantimpré's examples of earthquakes go completely unmentioned by Megenberg, such as one in the north Syrian town of Tyre and an eruption of Mount Etna in Sicily. See Cantimpré, *Liber de Natura Rerum*, pp. 408–9.

127  Konrad von Megenberg, *Der Buch der Natur*, vol. 2: *critical text*, eds. Robert Luff and Georg Steer (Tübingen: Max Nienmeyer Verlag, 2003), pp. 131–37.

128  Sudhoff, "Pestschriften," AGM 8 (1915):244–46. This passage may have been inspired by Cantimpré's description of the earth as encircled by the ocean that penetrates "interior passageways like the veins of the body with blood," and by which the earth's "aridity is everywhere irrigated." See Cantimpré, *Liber de Natura Rerum*, p. 406. This passage is, in fact, completely omitted in Megenberg's corresponding section on the earth in book 2, chapter 32, of the *Buch der Natur*.

129  Sudhoff, "Pestschriften," AGM 9 (1916):126. A German plague *Consilium* of 1481 follows Albich's view that physicians should restrict themselves to "secondary and immediate" causes and leave "universal and remote" causes to astrologers, although it clearly gives preference to the latter in that all causes of plague are ultimately derived from higher ones. See Sudhoff, "Pestschriften," AGM 16 (1924–25):37–38.

130  Sudhoff, "Pestschriften," AGM 17 (1925):106–7, 117–18.

131  Sudhoff, "Pestschriften," AGM 11 (1919):123–24.

132  Sudhoff, "Pestschriften," AGM 16 (1924–25):22.

133  Sudhoff, "Pestschriften," AGM 5 (1912):48; 17 (1925):23–24.

134  Sudhoff, "Pestschriften," AGM 4 (1911):213–14; 7 (1913):83–85. See also the closely related treatise of Johann of Glogau, in Sudhoff, "Pestschriften," AGM 9 (1916):68–69. John of Saxony likewise gives three categories of a plague epidemic, which he defines as "major, minor, and medium," caused respectively by air altered in both its qualities and substance, or in its qualities only, either in terms of those "particular" to the nature of air or "opposite" to it. See Sudhoff, "Pestschriften," AGM 16 (1924–25):21.

135  For treaties that devote a special chapter or section to signs of the plague, see: Rébouis, *Étude Historique*, pp. 84–92; Agramont, "Regiment de Preservacio a Epidimia," pp. 71–73; Hoeniger, *Der Schwarze Tod*, pp. 170–71; *I Trattati in Volgare della Peste e dell'acqua Ardente di Michele Savonarola*, ed. Luigi Belloni (Rome: Congresso Nazionale della Società Italiana de Medicina Interna, 1953), pp. 7–8; Amasuno Sárraga, *La Peste en la Corona de Castilla*, pp. 225–26; Zunz, *Jubelschrift zum Neunzigsten Geburtstag*, pp. 108–13; H. Pinkhof, *Abraham Kashlari, over Pestachtige Koorsten* (Amsterdam, 1891), pp. 22–23; Sudhoff, "Pestschriften," AGM 4 (1911):210–13; 5 (1912):367; 6 (1913):354; 7 (1913):82–85; 8 (1915):188–89, 245; 9 (1916):120, 124, 140–42; 11 (1919):126, 133–34, 151, 167–68; 16 (1924–25):24–25, 81–82, 108–10, 178; 17 (1925):45–46.

136  Rébouis, *Étude Historique*, p. 84; Sudhoff, "Pestschriften," AGM 11 (1919):167; 16 (1924–25):81.

137  Foligno, *Consilium*, pp. 30, 33; Tussignano, *Consilium*, col. 4; Codex Latin 363, Bayerische Staatsbibliothek, Munich, fols. 123r.-v.; MS 1227, Universitätsbibliothek, Leipzig, fols. 146v., 148r.; Zunz, *Jubelschrift zum Neunzigsten Geburtstag*, p. 111; Sudhoff, "Pestschriften," AGM 4 (1911):210–11; 5 (1912):369–70; 6 (1913):354; 7(1913):82–83; 9 (1916):127–28, 140–41; 11 (1919):57, 73, 126; 16 (1924–25):5, 21, 25, 178; 17 (1925):40–41. One exception who bucked this trend was Giovanni della Penna of Naples, who maintained that the "cause of this plague greatly threatens during the summer time of the year." Penna's observation, however, seems based not on direct observation but rather on the medical theory that summer was traditionally the season for prevalence of hot and dry choleric humors, which Penna saw as the root cause of plague's occurrence in the human body.

138 Cohn, *The Black Death Transformed*, pp. 155–73; Aberth, *From the Brink of the Apocalypse*, p. 92.
139 For doctors' precautions against the air, see: Foligno, *Consilium*, pp. 5–8; Agramont, "Regiment de Preservacio a Epidimia," pp. 76–80; Rébouis, *Étude Historique*, pp. 94–100; Dinanah, "Die Schrift," p. 52; Zunz, *Jubelschrift zum Neunzigsten Geburtstag*, pp. 112–13; Pinkhof, *Abraham Kashlari*, pp. 24–25; Tussignano, *Consilium*, col. 2; *I Trattati in Volgare della Peste*, pp. 8–10; Amasuno Sárraga, *La Peste en la Corona de Castilla*, pp. 229–30; *Consiglio contro a Pistolenza per Maestro Tommaso del Garbo*, ed. Pietro Perrato (Bologna: Presso Gaetano Romagnoli, 1866), pp. 13–20; Simonini, "Il Codice di Mariano di Ser Jacopo," pp. 167–68; Maineri, *Libellus de Preservatione*, pp. 21–23; Marsilio Ficino, *Contro alla Peste* (Florence: Appresso i Giunti, 1576), pp. 4, 13–14; K. Sudhoff, "Ein anderer pestretetet," *Studien zur Geschichte der Medizin* 8 (1909):200, 205; K. Sudhoff, "Ein weiteres deutsches Pest-Regiment aus dem 14. Jahrhundert und seine lateinische Vorlage, das Prager Sendschreiben 'Missum Imperatori' vom Jahre 1371," and "Ein Pestregimen aus dem Anfange des 15. Jahrhunderts," *Archiv für Geschichte der Medizin* 3 (1910):145, 147–48, 151, 408; Sudhoff, "Pestschriften," AGM 4 (1911):197–98, 200, 204, 213, 217, 390, 392, 396–403, 406–9, 411–12, 418–23; 5 (1912):37–43, 368–74, 81, 338–40, 343, 349, 352–56, 361, 368–75, 386, 393–94; 6 (1913):315–19, 352–55, 360–63, 367, 371–75; 7 (1913):57–63, 67, 70, 73–80, 91–95, 98–100, 106–8; 8 (1915):175–76, 182, 194, 197–203, 207, 210, 237–43, 248, 253–54, 258–59, 264, 267–68, 273–75, 279–83; 9 (1916):57–59, 72, 75, 123–25, 131–32, 148–49, 152–54, 158–60; 11 (1919):53–54, 60–61, 67–69, 73–77, 80–82, 90, 125–28, 131–32, 135–36, 140–41, 157–59, 163–64, 168–69, 175; 14 (1922–23):2, 7–9, 12, 20–21, 23–24, 82–88, 92–94, 102, 130–31, 139–41, 144–45, 154, 156, 159–61, 163, 167; 16 (1924–25):7–8, 11, 13–15, 17, 27–29, 35, 39–40, 49–50, 56–59, 61–63, 65–66, 82–85, 90–91, 110–11, 132–35, 158, 161, 164, 173, 177–78; 17 (1925):13–15, 25–26, 31, 47, 50–51, 81–84, 96–97, 107–8, 122, 127–28, 134–35, 261.
140 Sudhoff, "Pestschriften," AGM 9 (1916):146.
141 Foligno, *Consilium*, pp. 31, 37–38; Codex Latin 363, Bayerische Staatsbibliothek, Munich, fols. 124r.-v.; MS 1227, Universitätsbibliothek, Leipzig, fols. 147r., 149v; Sudhoff, "Pestschriften," AGM 16 (1924–25):61; 17 (1925):24, 42, 79, 96.
142 Ficino, *Contro alla Peste*, p. 9; Sudhoff, "Pestschriften," AGM 5 (1912):77, 338, 377; 6 (1913):320, 363–64; 7 (1913):84–85; 8 (1915):184, 191, 237, 264–65; 11 (1919):61–62, 88, 124, 150, 171; 14 (1922–23):88; 16 (1924–25):8, 40–41, 50, 87, 107–8; 17 (1925):85, 96, 125–26. Plague doctors' prescriptions on air and water were also very much in line with what contemporary health regimens had to say on the subject: See: Pietro de Crescenzi, *Liber ruralia commodorum* (Augsburg: Johann Schüssler, 1471), pp. 10–15; Maino de Maineri, *Regimen Sanitatis*, 3 vols., ed. Séamus Ó Ceithearnaigh (Baile Átha Cliath, 1942–44), 2:142–50; Ken Albala, *Eating Right in the Renaissance* (Berkeley and Los Angeles: University of California Press, 2002), pp. 116–30.
143 Foligno, *Consilium*, p. 5; Hoeniger, *Der Schwarze Tod*, pp. 168–69; Dinanah, "Die Schrift," p. 37; Maineri, *Libellus de Preservatione*, p. 21; Sudhoff, "Pestschriften," AGM 4 (1911):421, 423; 5 (1912):48–49; 11 (1919):134, 166; 17 (1925):137.
144 Foligno, *Consilium*, p. 4.
145 Hoeniger, *Der Schwarze Tod*, p. 164.
146 Sudhoff, "Pestschriften," AGM 7 (1913):84. See also the treatise of John of Glogau: Sudhoff, "Pestschriften," AGM 9 (1916):68.
147 Sudhoff, "Pestschriften," AGM 11 (1919):148. Given physicians' prejudice against retting flax, one is then puzzled by a recipe for a smelling apple in a plague regimen dedicated to the town of Erfurt in Germany in 1405 that calls for "moldy flax"! See Sudhoff, "Pestschriften," AGM 11 (1919):82. For other doctors who mention retting linen or hemp as a near cause of plague, see Sudhoff, "Pestschriften," AGM 16 (1924–25):37, 49.
148 Sudhoff, "Pestschriften," AGM 16 (1924–25):24.
149 Agramont, "Regiment de Preservacio a Epidimia," p. 69. Sudhoff, "Pestschriften," AGM 7 (1913):85; 8 (1915):108; 11 (1919):149; 16 (1924–25):49; 17 (1925):96.

150 Carlo M. Cipolla, *Public Health and the Medical Profession in the Renaissance* (Cambridge: Cambridge University Press, 1976), p. 11; Ann G. Carmichael, *Plague and the Poor in Renaissance Florence* (Cambridge: Cambridge University Press, 1986), pp. 96–99. John Henderson, "The Black Death in Florence: Medical and Communal Responses," in *Death in Towns: Urban Responses to the Dying and the Dead, 100–1600*, ed. Steven Basseh (London and New York: Leicester University Press, 1992), p. 143.
151 *The Black Death*, trans. and ed. Horrox, pp. 198–200.
152 C. De Backer, "Maatregelen Tegen de Pest te Diest in de Vijftiende en Zestiende Eeuw," *Verhandelingen-Koninklijke Academie voor Geneeskunde van Belgie* 61 (1999):275–77.
153 *Agrarian History of England and Wales. Volume 3: 1348–1500*, ed. Edward Miller (Cambridge: Cambridge University Press, 1991), pp. 388–89; *The Black Death*, trans. and ed. Horrox, pp. 203–5.
154 Madeleine Pelner Cosman, *Fabulous Feasts: Medieval Cookery and Ceremony* (New York: George Braziller, 1976), pp. 98–99.
155 *The Black Death*, trans. and ed. Horrox, pp. 205–6.
156 An Lentacker, Wim Van Neer, and Jean Plumier, "Historical and Archaeozoological Data on Water Management and Fishing during Medieval and Post-Medieval Times at Namur (Belgium)," in *Environment and Subsistence in Medieval Europe*, eds. Guy de Boe and Frans Verhaeghe (Zellik: I.A.P. Rapporten, 9, 1997), pp. 84–86; Arrizabalaga, "Facing the Black Death," p. 276; *Agrarian History of England and Wales, Volume 3*, pp. 389–91; Gimpel, *The Medieval Machine*, pp. 82–86; Cosman, *Fabulous Feasts*, p. 94; Hoffmann, "Economic Development and Aquatic Ecosystems," p. 645; Richard C. Trexlar, "Measures against Water Pollution in Fifteenth-Century Florence," *Viator* 5 (1974):460–67.
157 Sudhoff, "Pestschriften," AGM 3 (1910):223–25.
158 Sudhoff, "Pestschriften," AGM 3 (1910):224. For a more detailed treatment of the well-poisoning accusation against the Jews during the Black Death, see Aberth, *From the Brink of the Apocalypse*, pp. 156–88.
159 Christian Guilleré, "La Peste Noire a Gérone (1348)," *Annals de Institut d'Estudis Gironins* 27 (1984):141–42.
160 Frederick W. Gibbs, "Medical Understandings of Poison circa 1250–1600," (Ph.D. dissertation, University of Wisconsin—Madison, 2009), pp. 13–52, 105–39.
161 Sudhoff, "Pestschriften," AGM 11 (1919):150–51; 14 (1922–23):14.
162 *The Black Death*, trans. and ed. Horrox, p. 56.
163 Sudhoff, "Pestschriften," AGM 16 (1924–25):84–85.
164 Agramont, "Regiment de Preservacio a Epidimia," p. 65.
165 Sudhoff, "Pestschriften," AGM 4 (1911):215.
166 Sudhoff, "Pestschriften," AGM 11 (1919):149.
167 John Aberth, *The Black Death: The Great Mortality of 1348–1350. A Brief History with Documents* (Boston and New York: Bedford/St. Martin's 2005), pp. 64–65.
168 Aberth, *The Black Death*, pp. 155–58; Megenberg, *Buch der Natur*, 2:136. Megenberg also mentions in the *Buch der Natur* that the Jews of Vienna had died in such "large numbers" from the plague that "they had to expand their cemetery and bought two houses for the purpose," which would have been a "foolishness" if they "had poisoned themselves."
169 Aberth, *The Black Death*, pp. 126–31, 151–55.
170 Cohn, *The Black Death Transformed*, p. 232.
171 The qualitative aspects of poison are especially emphasized by Albert the Great in his discussion of serpents' venom, which is heavily dependent on Avicenna. See: Albertus Magnus, *On Animals: A Medieval Summa Zoologica*, trans. Kenneth F. Kitchell, Jr. and Irven Michael Resnick, 2 vols. (Baltimore: Johns Hopkins University Press, 1999), 2:1711–13.
172 Gibbs, "Medical Understandings of Poison," pp. 38, 45–47, 55–56, 77–85. See also Lynn Thorndike's discussion of Abano's treatise on poisons in *A History of Magic and Experimental Science. Volumes 3 and 4: Fourteenth and Fifteenth Centuries* (New York: Columbia University Press, 1934), pp. 905–10.
173 Gibbs, "Medical Understandings of Poison," pp. 107–15, 133–34.

174 Avicenna, *Liber Canonis*, fols. 416r., 470r.-v. (book 4, fen 1, tractate 4, chapter 1 and book 4, fen 5, tractate 1, chapter 2).
175 Gibbs, "Medical Understandings of Poison," pp. 107–10, 116.
176 Gibbs, "Medical Understandings of Poison," pp. 110–11, 115–22.
177 Plague treatises that attempt to reconcile the poison and humoral theories mainly do so by explaining that poison, once it is drawn towards the heart, creates a heat and moisture imbalance that alters the humors or vital spirits. See: Hoeniger, *Der Schwarze Tod in Deutschland*, pp. 168–69; Michon, *Documents Inédits sur la Grande Peste de 1348*, ed. L.-A. Joseph Michon (Paris, 1860), pp. 74–75, 77–78; Ficino, *Consiglio*, pp. 1–3, 7–8; Sudhoff, "Pestschriften," AGM 5 (1912):361–62; 9 (1916):147; 11 (1919):124–25, 166; 14 (1922–24):144, 154, 159; 16 (1924–25):16, 23, 36–37, 61–62, 68, 165; 17 (1925):24–25, 55–60, 78. Yet one fourteenth-century English doctor from Oxford absolutely refused to make any concessions to competing theories of humoral or complexional imbalances caused by poison: According to him, the disease can "be compared to poisonous matter, which kills the body by its entire property [of being poisonous] and by its whole substance, not from its heat nor from its coolness," as it seeks out the "mine shafts (*minas*) of the bodily virtues, namely the heart, liver, and the brain." See: Sudhoff, "Pestschriften," AGM 17 (1925):123–24.
178 Foligno, *Consilium*, pp. 3, 31, 39–40. See also Foligno's short *Consilia*, such as in Sudhoff, "Pestschriften," AGM 5 (1912):84. Foligno's statement on poisonous matter is repeated word-for-word by the fifteenth-century physician, Johannes Widman, thus revealing its influence. See Sudhoff, "Pestschriften," AGM 16 (1924–25):6.
179 Sudhoff, "Ein anderer pestretetet," p. 194; Sudhoff, "Pestschriften," AGM 4 (1911):421; 6 (1913):376; 8 (1915):284, 287.
180 Sudhoff, "Pestschriften," AGM 9 (1916):132, 151.
181 Sudhoff, "Pestschriften," AGM 8 (1915):246.
182 Sudhoff, "Pestschriften," AGM 5 (1912):341–42. Another treatise discovered by Sudhoff in a Wiesbaden manuscript is also attributed to Penna, but I believe there are reasons to doubt this attribution. At one point, for example, it is stated that plague "kills by its insidiousness and ensnaring poisonousness and, for instance, by means of its specific form [i.e., as poison]." This would represent a complete reversal of Penna's earlier opinion that plague was *not* caused by the unique quality of poison. See Sudhoff, "Pestschriften," AGM 16 (1924–25):165.
183 Rébouis, *Étude Historique*, p. 83; Agramont, "Regiment de Preservacio a Epidimia," p. 62; Sudhoff, "Pestschriften," AGM 11 (1919):47–48; *Documents Inédits sur la Grande Peste de 1348* ed. L.-A. Joseph Michon (Paris, 1860), pp. 74–75. Other plague doctors mention that if buboes form on the neck, this signifies that the "poisonous matter" has afflicted the brain rather than the heart. See Sudhoff, "Pestschriften," AGM 16 (1924–25):68; 17 (1925):28.
184 I refer readers to my forthcoming book, *Doctoring the Black Death: Europe's Late Medieval Medical Response to Epidemic Disease*, to be published by Rowman and Littlefield. In this work, I will try to explain further why poison proved to be such an attractive explanation for plague and assess the implications that the poison orthodoxy had for plague treatment and whether it constituted a revolutionary or radical new approach to disease. Gibbs makes a similar observation on the universality of poison references in fourteenth-century plague tracts in "Medical Understandings of Poison," p. 116.
185 Aside from treatises otherwise cited in my discussion on poison, readers should consult: *The Black Death*, trans. and ed. Horrox, pp. 186–94; Simonini, "Il Codice di Mariano di Ser Jacopo," pp. 164–67; Maineri, *Libellus de Preservatione*, p. 17; Sudhoff, "Ein anderer pestretetet," p. 194; K. Sudhoff, "Ein deutsches Pest-Regiment aus dem 14. Jahrhundert," *Archiv für Geschichte der Medizin* 2 (1908–9):380, 383; Sudhoff, "Pestschriften," AGM 4 (1911):201, 393–94, 399, 413–17, 421–22; 5 (1912):48, 51, 53, 73–75, 78, 82–84, 333, 342, 349, 361–64, 384, 395; 6 (1913):314, 352–54, 376–77; 7 (1913):65, 72–75, 79, 82, 84, 87, 101, 105–10, 113; 8 (1915):178–81, 187–92, 196–201, 203, 207–15, 237, 245–46, 253–54, 257, 259–60, 267–68, 273, 277–82, 284–87; 9 (1916):57–59,

75, 121, 132, 134, 139, 146–48, 150, 152–54, 160–65; 11 (1919):53, 58, 64–68, 71, 76, 80, 86, 91, 124–25, 128–30, 133–35, 138–39, 141, 144–46, 149, 151, 154–56, 163, 166, 173–75; 14 (1922–23):2–5, 7, 13–14, 21–22, 80, 84–85, 92, 95–97, 101–2, 105, 131, 142, 144, 149, 153–54, 159–60, 163, 166; 16 (1924–25):6–7, 11, 13–14, 16, 23–29, 32–33, 37–39, 50, 61, 65–66, 68, 81, 84, 87–90, 93–94, 103, 125, 127, 152, 163–67, 175, 177, 180; 17 (1925):24–25, 28, 31, 33–34, 38, 55, 59–62, 78–79, 86–89, 98, 112–15, 123–24, 128–31, 134, 137–38, 261. See also Gibbs, "Medical Understandings of Poison," pp. 115–22.
186 Sudhoff, "Pestschriften," AGM 17 (1925):78. Nonetheless, Görlitz had to explain why "this poisonous air afflicts one household and all die and not another," which he did using the standard Galenic explanation that humans have different bodily complexions that dispose them in varying degrees to disease. This answer is by far the most extensive and rambling of the seven "curious problems" posed by the plague that he considers.
187 Zunz, *Jubelschrift zum Neunzigsten Geburtstag*, p. 109–10, 120–23; Dinanah, "Die Schrift," pp. 34–37. Nonetheless, Todros does mention that plague fever comes from a "foreign heat" that "has decayed and been changed to poisonousness," and his fourteenth-century Jewish colleague, Abraham Kashlari, who also wrote a treatise on "pestilential fevers" in Hebrew, describes plague as spread by a "poisonous corruption" of the air or else a poison radiating out from the stars and given off by poisonous plants and trees, which, once entered into the body, poisoned the vital spirits. See Pinkhof, *Abraham Kashlari*, pp. 19–25.
188 See, for example, Ibn Khatima's explanation for "why treatment shows no results, even when ... a prescribed treatment—such as is described by me and outlined by the greatest medical specialists—is done without fail": Aberth, *The Black Death*, p. 61.
189 In his answer to the first doubt of the fourth chapter of the *Long Consilium*, Gentile da Foligno explains that theriac confers benefits during a plague despite its heat-generating properties due to "its specific property and form," by which it can be administered in much smaller amounts than drugs that work by their qualities only. See Foligno, *Consilium*, pp. 30–32; Gibbs, "Medical Understandings of Poison," pp. 122–25. Nonetheless, bleeding and scarification were still prescribed as a means to evacuate poison, especially when present as poisonous, "coagulated" blood or as poisonous matter concentrated in the glands. Thus, Blasius of Barcelona, in a treatise written in 1406 for his patron, King Martin of Aragon, cited his own experience of curing himself of plague for recommending a direct evacuation by surgery of the "poisonous matter imprisoned" within the bubo, accompanied by a "sweating treatment" or drug therapy, the latter especially in the case of "silly" persons who "faint away at the sight alone of the scalpel." See Sudhoff, "Pestschriften," AGM 17 (1925):113–14.
190 Gibbs, "Medical Understandings of Poison," pp. 125–219; Sudhoff, "Pestschriften," AGM 16 (1924–25):117–18, 182.
191 Kieckhefer, *Magic in the Middle Ages*, pp. 81–83; Thorndike, *A History of Magic and Experimental Science*, pp. 904–5.
192 *The Black Death*, trans. and ed. Horrox, pp. 193–94.
193 *Witchcraft in Europe, 400–1700: A Documentary History*, eds. Alan Charles Kors and Edward Peters, 2nd edn. (Philadelphia: University of Pennsylvania Press, 2001), p. 161.
194 See Richard Kieckhefer, *European Witch Trials: Their Foundations in Popular and Learned Culture, 1300–1500* (Berkeley and Los Angeles: University of California Press, 1976), pp. 61–63; Wolfgang Behringer, "Weather, Hunger and Fear," *German History* 13 (1995):1–27.
195 *Witchcraft in Europe*, pp. 158–59, 161.
196 Joseph Hansen, *Quellen und Untersuchungen zur Geschichte der Hexenwahns und der Hexenverfolgung im Mittelalter* (Bonn, 1901), pp. 565–69, 581–89.
197 Richard Kieckhefer, *Forbidden Rites: A Necromancer's Manual of the Fifteenth Century* (University Park, PA: Pennsylvania State University Press, 1997), pp. 54–59, 216–21, 232–34.
198 Hieronymus Mengus, *Flagellum Daemonum* (Bologna: Johannes Rossium, 1578), pp. 299–302.

## Part II

1. J. Hansman, "Gilgamesh, Humbaba and the Land of the Erin-Trees," *Iraq* 38 (1976):23–35; Russell Meiggs, *Trees and Timber in the Ancient Mediterranean World* (Oxford: Clarendon Press, 1982), pp. 49–53.
2. I have used the Penguin translation of the epic by N.K. Sanders, first published in 1960.
3. Pliny the Elder, *Natural History*, book 12, chapter 2. For a discussion of the value of Pliny's work for information about the ancient Mediterranean forest, see Meiggs, *Trees and Timber*, pp. 19–29.
4. J. Donald Hughes, "Early Greek and Roman Environmentalists," in *Historical Ecology: Essays on Environment and Social Change*, ed. Lester J. Bilsky (1980), pp. 47–50; Meiggs, *Trees and Timber*, p. 378.
5. Hughes, "Early Greek and Roman Environmentalists," pp. 50–52.
6. Hughes, "Early Greek and Roman Environmentalists," ed. Lester J. Bilsky (1980), pp. 55–56; Meiggs, *Trees and Timber*, pp. 17–19; Michael Williams, *Deforesting the Earth: From Prehistory to Global Crisis* (Chicago: University of Chicago Press, 2003), pp. 77–79.
7. Oliver Rackham, *The History of the Countryside* (London: J.M. Dent and Sons, 1986), p. 74; idem, "Ecology and Pseudo-Ecology: The Example of Ancient Greece," in *Human Landscapes in Classical Antiquity: Environment and Culture*, eds. Graham Shipley and John Salmon (London and New York: Routledge Press, 1996), p. 38.
8. J. Toutain, *Les Cultes Païens dans l'Empire Romain*, 3 vols. (Paris: Ernest Leroux, 1907–20), 3:295–96; Miranda J. Green, *Dictionary of Celtic Myth and Legend* (London: Thames and Hudson, 1992), pp. 28, 86, 108, 160, 190–91.
9. Pliny, *Natural History*, XVI.95; Green, *Dictionary*, pp. 86, 164, 212–14.
10. Tacitus, *Annales*, XIV.30; Green, *Dictionary*, pp. 86, 213.
11. Toutain, *Les Cultes Païens*, 3:296–99; Green, *Dictionary*, pp. 102, 164, 170–71.
12. Ramsay MacMullen, *Christianity and Paganism in the Fourth to Eighth Centuries* (New Haven, CT: Yale University Press, 1997), p. 65, n.112.
13. *Soldiers of Christ: Saints and Saints Lives from Late Antiquity and the Early Middle Ages*, eds. Thomas F.X. Noble and Thomas Head (University Park, PA: Pennsylvania State University Press, 1995), pp. 15–16.
14. *The Laws of the Earliest English Kings*, trans. and ed. F.L. Attenborough (Cambridge: Cambridge University Press, 1922), p. 71; *The Lombard Laws*, trans. Katherine Fischer Drew (Philadelphia: University of Pennsylvania Press, 1973), pp. 73–74; *The Visigothic Code*, ed. S.P. Scott (Boston: Boston Book Co., 1910), pp. 275–76; William Linnard, *Welsh Woods and Forests: History and Utilization* (Cardiff: National Museum of Wales, 1982), pp. 18–19; *The Earliest Norwegian Laws, being the Gulathing Law and the Frostathing Law*, trans. Laurence M. Larson (New York: Columbia University Press, 1935), p. 269.
15. MacMullen, *Christianity and Paganism*, p. 65, n. 116.
16. Toutain, *Les Cultes Païens*, 3:299–300.
17. Meiggs, *Trees and Timber*, p. 378.
18. Ronald Hutton, *The Pagan Religions of the Ancient British Isles: Their Nature and Legacy* (Oxford: Blackwell, 1991), pp. 252–53.
19. Rackham, *History of the Countryside*, pp. 68–71; idem, *Ancient Woodland: Its History, Vegetation and Uses in England* (London: Edward Arnold, 1980), pp. 97–104; idem, *Trees and Woodland in the British Landscape* (London: J.M. Dent and Sons, 1976), pp. 39–45.
20. Rackham, *History of the Countryside*, pp. 71–73; idem, *Ancient Woodland*, pp. 104–5; idem, *Trees and Woodland*, pp. 45–48; *The Environment in British Prehistory*, eds. I.G. Simmons and M.J. Tooley (Ithaca, N.Y.: Cornell University Press, 1981), pp. 102–10, 152–91; Stephen Budiansky, *The Covenant of the Wild: Why Animals Chose Domestication* (New York: William Morrow and Company, 1992), pp. 116–17.
21. *Environment in British Prehistory*, pp. 231–47, 264–77.
22. Williams, *Deforesting the Earth*, pp. 79–95; Meiggs, *Trees and Timber*, pp. 116–53, 325–70.

23 Christopher Dyer, *Hanbury: Settlement and Society in a Woodland Landscape* (Leicester: Leicester University Press, 1991), pp. 15–18.
24 Rackham, *History of the Countryside*, pp. 74–75; idem, *Ancient Woodland*, pp. 104–9; idem, *Trees and Woodland*, pp. 49–51.
25 Williams, *Deforesting the Earth*, pp. 95–101; Meiggs, *Trees and Timber*, p. 377.
26 Catherine Delano Smith, "Where was the 'wilderness' in Roman times?" in *Human Landscapes in Classical Antiquity: Environment and Culture*, eds. Graham Shipley and John Salmon (London and New York: Routledge, 1996), pp. 154–79.
27 F.W.M. Vera, *Grazing Ecology and Forest History* (Wallingford, U.K. and New York: CABI Publishing, 2000), pp. 119–23.
28 Rackham, *History of the Countryside*, p. 72.
29 Meiggs, *Trees and Timber*, p. 379, is rather pessimistic about historians' ability to assess the extent of Roman deforestation, commenting: "From the available ancient evidence it is impossible to say how seriously Mediterranean forests were depleted by the end of the Roman Empire."
30 Charles Higounet, "Les forêts de l'Europe occidentale du Ve au XIe siècle," *Settimane di Studio*, 13 (1965): 343, 398.
31 Rackham, *History of the Countryside*, p. 130; Chris Wickham, *Land and Power: Studies in Italian and European Social History, 400–1200* (London: British School at Rome, 1994), pp. 155–99; Vera, *Grazing Ecology*, pp. 102–23; D.P. Kirby, "The Old English Forest: Its Natural Flora and Fauna," in *Anglo-Saxon Settlement and Landscape: Papers presented to a Symposium, Oxford 1973*, ed. Trevor Rowley (British Archaeological Reports, 6, 1974), p. 120.
32 Wickham, *Land and Power*, pp. 159–60.
33 I know this from my own personal experience enrolling our family's tree farm in the U.S. Forest Service's "Wildlife Habitat Incentive Program" (WHIP).
34 Wickham, *Land and Power*, pp. 162–70.
35 Christopher Taylor, "The Anglo-Saxon Countryside," in *Anglo-Saxon Settlement and Landscape: Papers presented to a Symposium, Oxford 1973*, ed. Trevor Rowley (British Archaeological Reports, 6, 1974), pp. 5–7; Michael Aston, *Interpreting the Landscape: Landscape Archaeology in Local Studies* (London: B.T. Batsford, 1985), pp. 86–90; M. Bell, "Environmental Archaeology as an Index of Continuity and Change in the Medieval Landscape," in *The Rural Settlements of Medieval England: Studies Dedicated to Maurice Beresford and John Hurst*, eds. M.W. Beresford, John G. Hurst, Michael Aston, David Austin, Christopher Dyer (Oxford: Basil Blackwell, 1989), pp. 269–86; Dyer, *Hanbury*, p. 27; Wickham, *Land and Power*, pp. 168–83; idem, *Framing the Early Middle Ages: Europe and the Mediterranean, 400–800* (Oxford: Oxford University Press, 2005), pp. 507–8; idem, *The Inheritance of Rome: Illuminating the Dark Ages, 400–1000* (New York: Viking Penguin, 2009), pp. 216–17. There is some disagreement with this assessment, such as Della Hooke, "Pre-Conquest Woodland: its Distribution and Usage," *Agricultural History Review*, 37 (1989):114, who argues that in some regions of Anglo-Saxon England there was regeneration of woodland as the result of the collapse of the Roman Empire.
36 *Monumenta Germaniae Historica*, Capitularia regum Francorum I, Karoli Magni Capitularia, n. 32, c. 36, available online at www.dmgh.de.
37 *Laws of the Salian and Ripuarian Franks*, trans. Theodore John Rivers (New York: AMS Press, 1986), p. 50; *The Laws of the Salian Franks*, trans. Katherine Fischer Drew (Philadelphia: University of Pennsylvania Press, 1991), p. 72; *The Burgundian Code*, trans. Katherine Fischer Drew (Philadelphia: University of Pennsylvania Press, 1972), p. 32; *The Lombard Laws*, trans. Katherine Fischer Drew (Philadelphia: University of Pennsylvania Press, 1973), pp. 100, 111–12; *Visigothic Code*, pp. 272–75; *Laws of the Alamans and Bavarians*, trans. Theodore John Rivers (Philadelphia: University of Pennsylvania Press, 1977), pp. 170–71.
38 Rackham, *Ancient Woodland*, pp. 134–35; Wickham, *Land and Power*, p. 185–87; Hooke, "Pre-Conquest Woodland," p. 117, 119–20.
39 *The Laws of the Earliest English Kings*, ed. and trans. F.L. Attenborough (Cambridge: Cambridge University Press, 1922), pp. 51, 71. The exception was felling a tree that

could shelter thirty pigs, i.e., that was a mast-producing tree, in which case the fine was the same as for burning it, since the tree was deemed worth more in its "live" state than as firewood or timber.
40 *Monumenta Germaniae Historica*, Capitularia regum Francorum I, Karoli Magni Capitularia, n. 77, c. 19, available online at www.dmgh.de.
41 Wickham, *Land and Power*, pp. 156–58; Ian Wood, "Before or After Mission: Social Relations Across the Middle and Lower Rhine in the Seventh and Eighth Centuries," in *The Long Eighth Century: Production, Distribution and Demand*, eds. Inge Lyse Hansen and Chris Wickham (Leiden: Brill, 2000), pp. 152–53. Eigil's *Life of Sturmi* is translated in *The Medieval Sourcebook*, available online at www.fordham.edu.
42 Rackham, *Ancient Woodland*, pp. 107–34; H.C. Darby, *Domesday England* (Cambridge: Cambridge University Press, 1977), pp. 171–94; Paul Stamper, "Woods and Parks," in *The Countryside of Medieval England*, eds. Grenville Astill and Annie Grant (Oxford: Basil Blackwell, 1988), pp. 129–32. Measuring woodland in Domesday based on pannage dues is unreliable since oak trees did not always yield a mast crop from year to year.
43 Hooke, "Pre-Conquest Woodland," pp. 121–22; idem, "Early Medieval Estate and Settlement Patterns: The Documentary Evidence," in *The Rural Settlements of Medieval England: Studies dedicated to Maurice Beresford and John Hurst*, eds. Michael Aston, David Austin and Christopher Dyer (Oxford: Basil Blackwell, 1989), pp. 27–30; Christopher Taylor, "The Anglo-Saxon Countryside," and Peter Sawyer, "Anglo-Saxon Settlement: The Documentary Evidence," in *Anglo-Saxon Settlement and Landscape: Papers presented to a Symposium, Oxford 1973*, ed. Trevor Rowley (British Archaeological Reports, 6, 1974), pp. 7–8, 108–12.
44 Williams, *Deforesting the Earth*, p. 102.
45 Rackham, *Ancient Woodland*, pp. 133–34.
46 *The Agrarian History of England and Wales, Volume II: 1042–1350*, ed. H.E. Hallam (Cambridge: Cambridge University Press, 1988), pp. 137–259; Paul Stamper, "Woods and Parks," in *The Countryside of Medieval England*, eds. Grenville Astill and Annie Grant (Oxford: Basil Blackwell, 1988), p. 129.
47 P.T.H. Unwin, "The Changing Identity of the Frontier in Medieval Nottinghamshire and Derbyshire," in *Villages, Fields and Frontiers: Studies in European Rural Settlement in the Medieval and Early Modern Periods* (Oxford: B.A.R. international series, 185, 1983), pp. 339–51; Stamper, "Woods and Parks," pp. 128–29.
48 Leonard Cantor, "Forests, Chases, Parks and Warrens," in *The English Medieval Landscape*, ed. Leonard Cantor (Philadelphia: University of Pennsylvania Press, 1982), pp. 58–59.
49 I consulted the edition translated by Cynthia Postan and published by Edward Arnold in 1968.
50 Duby's continuing influence is attested by Williams, *Deforesting the Earth*, pp. 118, and Wickham, *Land and Power*, p. 196. I will also note that the 2008 updated version of volume one of the *Cambridge Economic History of Europe* on the *Agrarian Life of the Middle Ages* reproduces Richard Koebner's chapter on "The Settlement and Colonization of Europe" from 1966 without any changes.
51 Duby, *Rural Economy*, pp. 67, 70. On the subject of medieval monks as "the land-clearers of Europe," see chapter 9 of Robert K. Winters, *The Forest and Man* (New York: Vantage Press, 1974), pp. 192–242.
52 Duby, *Rural Economy*, p. 84.
53 Duby, *Rural Economy*, p. 67.
54 Constance Hoffman Berman, "Medieval Agriculture, the Southern French Countryside, and the Early Cistercians. A Study of Forty-three Monasteries," *Transactions of the American Philosophical Society* 76 (1986):11–12; Constance Brittain Bouchard, *Holy Entrepreneurs: Cistercians, Knights, and Economic Exchange in Twelfth-Century Burgundy* (Ithaca, N.Y.: Cornell University Press, 1991), pp. 101–4, 110–11.
55 *Records of Feckenham Forest, Worcestershire, c.1236–1377*, ed. Jean Birrell (Worcestershire Historical Society, new series, 21, 2006), pp. xx–xxi, 1–11, 19, 33–37.

Notes    253

56  *A Calendar of New Forest Documents, 1244–1334*, ed. D.J. Stagg (Hampshire Record Series, 3, 1979), pp. 47–51; *The Sherwood Forest Book*, ed. Helen E. Boulton (Thoroton Society Record Series, 23, 1965), pp. 89–92.
57  *Collections for a History of Staffordshire. The Forests of Cannock and Kinver: Select Documents, 1235–1372*, ed. Jean Birrell (Staffordshire Record Society, 4th series, 18, 1999), pp. 71–73, 85–93; Cyril E. Hart, *Royal Forest: A History of Dean's Woods as Producers of Timber*, (Oxford: Clarendon Press, 1966), p. 40.
58  Dyer, *Hanbury*, pp. 27–32.
59  See the chapter by Hermann Aubin, "Medieval Agrarian Society in its Prime: The Lands East of the Elbe and German Colonization Eastwards," in *The Cambridge Economic History of Europe, volume I: The Agrarian Life of the Middle Ages*, ed. M.M. Postan (Cambridge: Cambridge University Press, 1966), pp. 449–86.
60  Duby, *Rural Economy*, pp. 391–400.
61  Williams, *Deforesting the Earth*, p. 123.
62  Charles Young, "Conservation Policies in the Royal Forests of Medieval England," *Albion* 10 (1978):97.
63  Charles Young, *The Royal Forests of Medieval England* (Philadelphia: University of Pennsylvania Press, 1979), pp. 4–8; Raymond Grant, *The Royal Forests of England* (Stroud, Gloucestershire: Alan Sutton, 1991), pp. 3–6; Rackham, *Ancient Woodland*, pp. 133–34; Darby, *Domesday England*, pp. 195–201; Cantor, "Forests, Chases, Parks and Warrens," pp. 60–61.
64  Jean Birrell, "Common Rights in the Medieval Forest," *Past and Present* 117 (1987):27–28; Rackham, *Trees and Woodland in the British Landscape*, pp. 83–84; Dolores Wilson, "Multi-Use Management of the Medieval Anglo-Norman Forest," *Journal of the Oxford University History Society*, 1 (2004): "Woods and Parks," pp. 133–35.
65  *The Honor and Forest of Pickering*, ed. Robert Bell Turton, 4 vols. (North Riding Record Society, new series, 1894–97), 3:234–35.
66  R. Cunliffe Shaw, *The Royal Forest of Lancaster* (Preston, 1956), pp. 64–65, 148–51.
67  Grant, *Royal Forests of England*, pp. 8–10; Wilson, "Multi-Use Management of the Medieval Anglo-Norman Forest," p. 7.
68  Young, *Royal Forests of Medieval England*, pp. 11–17; Grant, *Royal Forests of England*, pp. 12–16; Wilson, "Multi-Use Management of the Medieval Anglo-Norman Forest," p. 8.
69  Young, *Royal Forests of Medieval England*, pp. 18–29; Grant, *Royal Forests of England*, pp. 16–19. The complete Latin text of the Assize of Woodstock is available in *The Sherwood Forest Book*, pp. 59–62.
70  Young, *Royal Forests of Medieval England*, pp. 33–59.
71  Young, *Royal Forests of Medieval England*, pp. 22, 35–36; *A Calendar of New Forest Documents, 1244–1334*, p. 6.
72  Young, *Royal Forests of Medieval England*, pp. 23–24, 36–51.
73  Young, *Royal Forests of Medieval England*, pp. 60–66; Grant, *Royal Forests of England*, pp. 137–38; Hart, *Royal Forest*, p. 7.
74  Young, *Royal Forests of Medieval England*, pp. 67–73; Grant, *Royal Forests of England*, pp. 138–45.
75  Young, *Royal Forests of Medieval England*, pp. 135–42; Grant, *Royal Forests of England*, pp. 145–59.
76  Young, *Royal Forests of Medieval England*, pp. 142–72; Grant, *Royal Forests of England*, pp. 159–72.
77  Young, *Royal Forests of Medieval England*, pp. 74–113; Grant, *Royal Forests of England*, pp. 35–129. Also see the introductions to local forest eyre records, such as: Collections for a History of Staffordshire, pp. 3–16; *Select Pleas of the Forest*, ed. G.J. Turner (London: Selden Society, 1901), pp. xiv–lxxix; *A Calendar of New Forest Documents, 1244–1334*, pp. 20–30; *Records of Feckenham Forest*, pp. xiv–xxiii.
78  *Sherwood Forest Book*, pp. 67–77.
79  Grant, *Royal Forests of England*, pp. 35–40, 49–52, 76–80.

80 *Records of Feckenham Forest*, pp. 21–26, 29. In 1280 it was complained in the New Forest pleas held at Winchester that the pasture of the king's beasts as well as the entire forest and district was "much surcharged" by all the horses, cattle, sheep and other animals too many to count that the abbot of Beaulieu and the priors of Christchurch and Bromhore allowed to roam in the forest (*Calendar of New Forest Documents, 1244–1334*, p. 93). Also see below, n. 107, for the case of Scalby Hay in the Forest of Pickering.
81 Hart, *Royal Forest*, p. 42.
82 *Records of Feckenham Forest*, pp. 54–60, 85–88, 119–30, 140. A similar case to that of Monte Viron is recorded in the Chancery Miscellanea of a forester who tried to hide 600 oak stumps with turf in a vain effort to evade detection. See Young, *Royal Forests of Medieval England*, p. 82.
83 *Calendar of New Forest Documents, 1244–1334*, pp. 65–69, 84–86, 93, 177–94.
84 *Collections for a History of Staffordshire*, pp. 31–39, 74–76, 93–98.
85 *Select Pleas of the Forest*, pp. 44–53.
86 *Sherwood Forest Book*, pp. 143–47.
87 *Sherwood Forest Book*, pp. 104–5.
88 *Honor and Forest of Pickering*, 3:23–43.
89 Cantor, "Forests, Chases, Parks and Warrens," p. 58.
90 Wilson, "Multi-Use Management of the Medieval Anglo-Norman Forest," pp. 1–3.
91 Rackham, *History of the Countryside*, p. 67.
92 The champion of this view has been Massimo Montanari, *L'alimentazione contadina nell'alto Medioevo* (Naples: Liguori, 1979), esp. pp. 469–76.
93 See: *Honor and Forest of Pickering*, 2:175–88, 4:40–41; *Sherwood Forest Book*, pp. 154–55, 160–62, 187–88; *Calendar of New Forest Documents, 1244–1334*, pp. 75–76, 117–18, 205; *Records of Feckenham Forest*, pp. 51, 83–84, 87, 132, 138–39.
94 *Calendar of New Forest Documents, 1244–1334*, p. 123.
95 Rackham, *Trees and Woodland*, pp. 70–73.
96 *Register of Edward the Black Prince*, 4 vols. (London: HMSO, 1930–33), 1:159.
97 Rackham, *Trees and Woodland*, pp. 20–23, 32–33, 69–80; idem, *Ancient Woodland*, pp. 1–8, 137–47; idem, *History of the Countryside*, pp. 65–68; idem, "The Medieval Countryside of England: Botany and Archaeology," in *Inventing Medieval Landscapes: Senses of Place in Western Europe*, eds. John Howe and Michael Wolfe (Gainesville, FL: University Press of Florida, 2002), pp. 15–21; Stamper, "Woods and Parks," pp. 130–32; Péter Szabó, *Woodland and Forests in Medieval Hungary* (Oxford: Archaeopress, 2005), pp. 20–21.
98 Hart, *Royal Forest*, p. 23.
99 *Select Pleas of the Forest*, p. 60.
100 *Honor and Forest of Pickering*, 2:39, 43; *Records of Feckenham Forest*, pp. 17, 32.
101 Vera, *Grazing Ecology and Forest History*, p. 132; Stamper, "Woods and Parks," pp. 136–40.
102 Rackham, *Trees and Woodland*, pp. 80–83; idem, *History of the Countryside*, p. 63.
103 Young, *Royal Forests of Medieval England*, pp. 74–78.
104 Wilson, "Multi-Use Management of the Medieval Anglo-Norman Forest," pp. 8–9; Hart, *Royal Forest*, p. 30.
105 Hart, *Royal Forest*, p. 35.
106 Birrell, "Common Rights in the Medieval Forest," p. 32.
107 *Honor and Forest of Pickering*, 2:44.
108 Stamper, "Woods and Parks," pp. 135–36.
109 Hart, *Royal Forest*, p. 34.
110 Young, *Royal Forests of Medieval England*, pp. 114–34.
111 See, for example, cases from the Northamptonshire eyre of 1209, including that of Henry of Newton, who was wrongfully imprisoned for being found "lying under a certain bush" near where a doe had its throat cut, or of Roger Tock, who was imprisoned because a search of his house found the "ears and bones of wild beasts," but who had lain so long in prison that he was adjudged "nearly dead" and allowed to go free. *Select Pleas of the Forest*, pp. 3–4.

112 Young, *Royal Forests of Medieval England*, pp. 80, 89–91, 102–3.
113 Shaw, *Royal Forest of Lancaster*, p. 134.
114 Birrell, "Common Rights in the Medieval Forest," p. 33.
115 *Honor and Forest of Pickering*, 2:128–29.
116 *Calendar of New Forest Documents, 1244–1334*, pp. 149–64.
117 Shaw, *Royal Forest of Lancaster*, pp. 65–66; *Victoria History of the Counties of England: A History of the County of Buckingham*, 4 vols., ed. William Page (London: A. Constable, 1905–27), 2:133–34; *Victoria History of the Counties of England: A History of Shropshire. Volume I*, ed. William Page (London: A. Constable, 1908), p. 488; *Victoria History of the Counties of England: A History of the County of Lancaster*, 8 vols., eds. William Farrer and J. Brownbill (London: A. Constable, 1906–14), 2:439.
118 Young, *Royal Forests of Medieval England*, p. 82.
119 *Select Pleas of the Forest*, p. 110.
120 *Collections for a History of Staffordshire*, pp. 21–23.
121 *Calendar of New Forest Documents, 1244–1334*, p. 122. In fact, all the communities of the New Forest were in mercy for concealment, "because they had claimed that it was not customary to present pleas of this kind [i.e., offenses against vert and venison], that being manifestly against the custom of the entire kingdom."
122 *Select Pleas of the Forest*, p. 37.
123 *Calendar of New Forest Documents, 1244–1334*, pp. 89–92.
124 *Sherwood Forest Book*, p. 139.
125 *Honor and Forest of Pickering*, 2:125–36.
126 *Select Pleas of the Forest*, p. 101.
127 At the Cannock Forest eyres of 1262 and 1271, 51 and 49 towns, respectively, were fined for non-appearance for a total of £14 and £11. See *Collections for a History of Staffordshire*, pp. 9, 40–42, 100–101.
128 *Select Pleas of the Forest*, p. 88; *Calendar of New Forest Documents, 1244–1334*, pp. 59–60, 73.
129 *Select Pleas of the Forest*, p. 25. A similar case occurred at the Rutland eyre in 1209, when Elias of Lutterworth was fined "for contemptuous speech before the justices." See *Select Pleas of the Forest*, p. 6.
130 *Select Pleas of the Forest*, p. 18.
131 *Select Pleas of the Forest*, p. 20.
132 *Select Pleas of the Forest*, pp. 125–28.
133 Jean Birrell, "Peasant Craftsmen in the Medieval Forest," *Agricultural History Review* 17 (1969):91–107.
134 Shaw, *Royal Forest of Lancaster*, pp. 64–76, 111–67.
135 *Registrum Thome de Cantilupo, episcopi Herefordensis, A.D. 1275–1282*, ed. R.G. Griffiths (Canterbury and York Society, 2, 1907), pp. 46, 159, 211–12; *Registrum Ricardi de Swinfield, episcopi Herefordensis, A.D. 1283–1317*, ed. William W. Capes (Canterbury and York Society, 5, 1909), pp. 19, 50–51, 55, 145–47, 407–9, 431–32; *Registrum Ade de Orleton, episcopi Herefordensis, A.D. 1317–1327*, ed. A.T. Bannister (Canterbury and York Society, 1, 1907), pp. 122–23, 316–18; *Registrum Johannis de Trillek, episcopi Herefordensis, A.D. 1344–1361*, ed. Joseph Henry Parry (Canterbury and York Society, 8, 1911), pp. 33–34, 197–99, 206–7, 216–21, 224; *Registrum Johannis Gilbert, episcopi Herefordensis, A.D. 1375–1389*, ed. Joseph Henry Parry (Canterbury and York Society, 18, 1915), pp. 39–40; *Registrum Johannis Stanbury, episcopi Herefordensis, A.D. 1453–1474*, ed. A.T. Bannister (Canterbury and York Society, 25, 1919), pp. 87–89.
136 Alexander Hamilton Thompson, *Visitations of Religious Houses in the Diocese of Lincoln*, 3 vols. (Publications of the Lincoln Record Society, 7, 14, 21, 1914–20), 2–3:19, 153, 286–87, 290–91, 293–96. The abbot of Selby Abbey in Yorkshire was also cited in 1280 for cutting and selling wood for his own profit: See *The Register of William Wickwane, Lord Archbishop of York, 1279–1285*, ed. William Brown (Surtees Society, 114, 1907), p. 23.

137 Grant, *Royal Forests of England*, pp. 133–35.
138 *Honor and Forest of Pickering*, 3:243.
139 Shaw, *Royal Forest of Lancaster*, pp. 127–28.
140 David Crook, "The Records of Forest Eyres in the Public Record Office, 1179 to 1670," *Journal of the Society of Archivists* 17 (1996):186. Note the discrepancy in dates with Young, *Royal Forests of Medieval England*, p. 158.
141 Cantor, "Forests, Chases, Parks and Warrens," p. 66.
142 *Calendar of the Patent Rolls Preserved in the Public Record Office: Edward III, 1327–1377*, 16 vols. (London, 1891–1916), 7:264.
143 Young, *Royal Forests of Medieval England*, pp. 149–72; Grant, *Royal Forests of England*, pp. 165–72.
144 Young, *Royal Forests of Medieval England*, p. 158.
145 Shaw, *Royal Forest of Lancaster*, p. 68.
146 Hart, *Royal Forest*, pp. 10, 51, 71.
147 Grant, *Royal Forests of England*, pp. 181–83; Crook, "The Records of Forest Eyres in the Public Record Office," p. 188.
148 *Honor and Forest of Pickering*, 1:121, 126–28, 130, 134, 164; 2:201–7. By another count from elsewhere in the rolls, I came up with 280 oaks, 6 ashes, 300 stumps, 200 faggots, and 1235 wagon-loads of wood. See *Honor and Forest of Pickering*, 1:144, 146–47, 151, 161–66, 170, 177–78, 200.
149 Linnard, *Welsh Woods and Forests*, p. 51.
150 Rackham, *History of the Countryside*, p. 138. See also Cantor, "Forests, Chases, Parks and Warrens," pp. 66–70.
151 Dyer, *Hanbury*, p. 28; Christopher Dyer, "'The Retreat from Marginal Land': The Growth and Decline of Medieval Rural Settlements," in *The Rural Settlements of Medieval England: Studies Dedicated to Maurice Beresford and John Hurst*, eds. Michael Aston, David Austin and Christopher Dyer (Oxford: Basil Blackwell, 1989), p. 52; Christopher Dyer, *Making a Living in the Middle Ages: The People of Britain, 850–1520* (New Haven, CT: Yale University Press, 2002), pp. 18, 25.
152 Cantor, "Forests, Chases, Parks and Warrens," p. 58. In 1255 the towns of Ellington, Brampton, Little Stukeley, and Alconbury complained at the Huntingdon eyre that "their lands and meadows are wasted by the deer of the lord king": *Select Pleas of the Forest*, p. 25.
153 Rackham, *Ancient Woodland*, p. 185; idem, *Trees and Woodland in the British Landscape*, p. 155.
154 Hart, *Royal Forest*, pp. 26–30, 43–49.
155 *Agrarian History of England and Wales, volume III: 1348–1500*, ed. Edward Miller (Cambridge: Cambridge University Press, 1991), pp. 408–17.
156 Rackham, *History of the Countryside*, p. 138.
157 David J. Breeze, "The Great Myth of Caledon," in *Scottish Woodland History*, ed. T.C. Smout (Edinburgh: Scottish Cultural Press, 1997), pp. 47–51.
158 Chris Smout, "Highland Land-Use before 1800: Misconceptions, Evidence and Realities," in *Scottish Woodland History*, ed. T.C. Smout (Edinburgh: Scottish Cultural Press, 1997), p. 6.
159 *People and Woods in Scotland: A History*, ed. T.C. Smout (Edinburgh: Edinburgh University Press, 2003), pp. 60–71.
160 Peter R. Quelch, "Ancient Trees in Scotland," and Martin Dougall and Jim Dickson, "Old Managed Oaks in the Glasgow Area," in *Scottish Woodland History*, ed. T.C. Smout (Edinburgh: Scottish Cultural Press, 1997), pp. 30–32, 81–84.
161 Linnard, *Welsh Woods and Forests*, pp. 12–51.
162 Oliver Rackham, "Looking for Ancient Woodland in Ireland," in *Wood, Trees and Forests in Ireland*, eds. Jon R. Pilcher and Seán Mac an tSaoir (Dublin: Royal Irish Academy, 1995), pp. 1–12.
163 Aidan O'Sullivan, "Woodland Management and the Supply of Timber and Underwood to Anglo-Norman Dublin," in *Environment and Subsistence in Medieval Europe*, eds. Guy de Boe and Frans Verhaeghe (Zellik: I.A.P. Rapporten, 9, 1997), pp. 135–41.

164 Michel Devèze, "Forêts françaises et forêts allemandes: Étude historique comparée," *Revue Historique* 235 (1966):362–80; Michel Aubrun, "Droits d'usages forestiers et libertés paysannes (XIe–XIIIe siècle): Leur role dans la formation de la carte foncière," *Revue Historique* 280 (1988):377–86; Bechmann, *Trees and Man*, pp. 225–34.
165 Bechmann, *Trees and Man*, pp. 235–36.
166 Vera, *Grazing Ecology and Forest History: The Forest in the Middle Ages* (New York: Paragon House, 1990), pp. 131–36.
167 Bechmann, *Trees and Man*, p. 236; *Dictionary of the Middle Ages*, ed. Joseph R. Strayer, 13 vols. (New York: Scribner, 1982–89), 5:134.
168 *Ordonnances des Roys de France de la troisième race*, 21 vols. (Paris: Imprimerie royale, 1723–1849), 1: 684–88; 2:244–49; 6:222–37; 8:521–36.
169 *Ordonnances*, 6:226–27. A similar preamble opens the ordinance of 1402, where it says that the waters and forests have been "much trampled, destroyed, and diminished in value by the fault and negligence of some of our officers of the aforesaid waters and forests." See *Ordonnances*, 8:523.
170 Bechmann, *Trees and Man*, pp. 241–51; René de Maulde-la-Clavière, *Ètude sur la condition forestière de l'Orleanais au moyen age et la renaissance* (Orléans: Herluison, 1871).
171 *Ordonnances*, 9:129–36.
172 Bechmann, *Trees and Man*, p. 239; *Dictionary of the Middle Ages*, 5:135–36.
173 *The Oxford Dictionary of the Middle Ages*, ed. Robert E. Bjork, 4 vols. (Oxford: Oxford University Press, 2010), 2:649.
174 Vera, *Grazing Ecology and Forest History*, pp. 245–47.
175 Szabó, *Woodland and Forests in Medieval Hungary*, pp. 47–70, 87–103, 119–33, 147–49.
176 Corinne J. Saunders, *The Forest of Medieval Romance: Avernus, Broceliande, Arden* (Rochester, N.Y.: D.S. Brewer, 1993), p. 17; George H. Williams, *Wilderness and Paradise in Christian Thought: The Biblical Experience of the Desert in the History of Christianity and the Paradise Theme in the Theological Idea of the University* (New York: Harper and Brothers, 1962), pp. 50–64.
177 Gerald of Wales, *The Journey through Wales*, trans. Lewis Thorpe (London: Penguin Books, 1978), p. 106.
178 For a summary of the debate on this issue and a case study, see Szabó, *Woodland and Forests in Medieval Hungary*, pp. 111–15, 142–43.
179 Saunders, *Forest of Medieval Romance*, pp. 44–185.
180 Saunders, *Forest of Medieval Romance*, p. 149–50.
181 J.C. Holt, *Robin Hood*, rev. edn. (London: Thames and Hudson, 1989), pp. 77–79; Stamper, "Woods and Parks," p. 143.
182 In addition to Holt's *Robin Hood*, readers might want to consult: John Bellamy, *Robin Hood: An Historical Enquiry* (London: Croom Helm, 1985); Maurice Keen, *The Outlaws of Medieval Legend*, rev. edn. (London and New York: Routledge, 1987); R.B. Dobson and John Taylor, *Rymes of Robyn Hood: An Introduction to the English Outlaw*, rev. edn. (Stroud, Gloucestershire: Sutton, 1997); Stephen Knight, ed., *Robin Hood: An Anthology of Scholarship and Criticism* (Woodbridge, Suffolk: D.S. Brewer, 1999).
183 Young, *Royal Forests of Medieval England*, p. 105.
184 *Close Rolls of the Reign of Henry III Preserved in the Public Record Office: A.D. 1237–1242* (London: HMSO, 1911), p. 239.
185 *Close Rolls of the Reign of Henry III Preserved in the Public Record Office: A.D. 1234–1237* (London: HMSO, 1908), p. 8.
186 *Close Rolls of the Reign of Henry III Preserved in the Public Record Office: A.D. 1256–1259* (London: HMSO, 1932), pp. 57, 165; *Close Rolls of the Reign of Henry III Preserved in the Public Record Office: A.D. 1261–1264* (London: HMSO, 1936), p. 120; Collections for a History of Staffordshire, p. 21.
187 Young, *Royal Forests of Medieval England*, p. 105; *Calendar of the Patent Rolls Preserved in the Public Record Office: Henry III, A.D. 1232–1247* (London: HMSO, 1906), p. 165; *Close Rolls of the Reign of Henry III: A.D. 1237–1242*, pp. 137–38, 144–45; Hart, *Royal Forest*, pp. 22–23.

188 For English translations of these works, see: *The* Etymologies *of Isidore of Seville*, ed. and trans. Stephen A. Barney, W.J. Lewis, J.A. Beach, and Oliver Berghof (Cambridge: Cambridge University Press, 2006), pp. 341–49; Hrabanus Maurus, *De Universo: The Peculiar Properties of Words and their Mystical Significance. The Complete English Translation*, trans. Priscilla Throop, 2 vols. (Charlotte, VT: MedievalMS, 2009), 2:218–39.
189 Alexander Neckam, *De Naturis Rerum*, ed. Thomas Wright (London: Rolls Series, 34, 1863), pp. 163–77.
190 Vincent of Beauvais, *Speculum Quadruplex sive Speculum Maius*, 4 vols. (Douai: Balthazar Belleri, 1624), 1:873–906.
191 Bartholomaeus Anglicus, *On the Properties of Things: John Trevisa's Translation: A Critical Text*, 2 vols. (Oxford: Clarendon Press, 1975), 2:882–85. I have interpreted Trevisa's adjective, *"gleymy,"* as "sticky," based on the *Oxford English Dictionary*.
192 Thomas of Cantimpré, *Liber de Natura Rerum. Volume 1: Text*, ed. Helmut Boese (Berlin: Walter de Gruyter, 1973), pp. 313–15.
193 Beauvais, *Speculum Maius*, 1:906–1092; Anglicus, *On the Properties of Things*, 2:886–1091; Cantimpré, *Liber de Natura Rerum*, pp. 315–41; Konrad von Megenberg, *Buch der Natur*, eds. Robert Luff and George Steer (Tübingen: Niemeyer Verlag, 2003), pp. 341–409.
194 Albertus Magnus, *De Vegetabilibus*, ed. Karl Jessen (Berolini Reimer, 1867).
195 Magnus, *De Vegetabilibus*, pp. 5–55.
196 Pietro de Crescenzi, *Liber ruralia commodorum* (Augsburg: Johann Schüssler, 1471), pp. 155–212. Crescenzi does describe non-fruit bearing trees, but only in terms of the uses to be derived from their wood products, much in the vein of the thirteenth-century encyclopedias.
197 Master Fitzherbert, *The Book of Husbandry*, ed. Walter W. Skeat (London: English Dialect Society, 1882), pp. 82–90.
198 Ülle Sillasoo, "Plant Depictions in Late Medieval Religious Art," in *People and Nature in Historical Perspective*, eds. József Laszlovszky and Péter Szabó (Budapest: Central European University and Archaeolingua, 2003), pp. 381–93.
199 K. Sudhoff, "Pestschriften aus den ersten 150 Jahren nach der Epidemie des 'schwarzen Todes' von 1348," *Archiv für Geschichte der Medizin* 11 (1919):144.
200 Williams, *Deforesting the Earth*, pp. 136–37.
201 Some of this evidence is presented in *The Black Death*, trans. and ed. Rosemary Horrox (Manchester: Manchester University Press, 1994), pp. 280–84, 296–99.
202 Maurice Beresford and John G. Hurst, *Deserted Medieval Villages* (London: Lutterworth Press, 1971), pp. 3–226; Dyer, *Hanbury*, p. 52; Dyer, *Making a Living in the Middle Ages*, p. 350.
203 Williams, *Deforesting the Earth*, p. 136.
204 Dyer, "'Retreat from Marginal Land'," pp. 48–57.
205 On the management of medieval woodland pastures, see Rackham, *History of the Countryside*, pp. 120–21; idem, *Trees and Woodland in the British Landscape*, pp. 135–41; idem, "Medieval Countryside of England," pp. 21–24.
206 *The Agrarian History of England and Wales, volume III: 1348–1500*, ed. Edward Miller (Cambridge: Cambridge University Press, 1991), pp. 34–173; Dyer, *Making a Living in the Middle Ages*, pp. 350–52.
207 Dyer, *Hanbury*, pp. 52–58.
208 Rackham, *History of the Countryside*, pp. 92–97; Rackham, *Trees and Woodland*, pp. 173–81.

## Part III

1 Juliet Clutton-Brock, *Domesticated Animals from Early Times* (London and Austin, TX: British Museum and University of Texas Press, 1981), pp. 9–25, 46–51; Stephen Budiansky, *The Covenant of the Wild: Why Animals Chose Domestication* (New York: William Morrow and Company, 1992), pp. 19–126; Joanna Swabe, *Animals, Disease and Human*

Society: Human-Animal Relations and the Rise of Veterinary Medicine (London and New York: Routledge, 1999), pp. 31–39.
2  H. te Velde, "A Few Remarks upon the Religious Significance of Animals in Ancient Egypt," *Numen* 27 (1980):76–82.
3  Anthony Preus, *Science and Philosophy in Aristotle's Biological Works* (Hildesheim, Germany, and New York: Georg Olms Verlag, 1975), pp. 21–47.
4  Pierre Pellegrin, *Aristotle's Classification of Animals: Biology and the Conceptual Unity of the Aristotelian Corpus*, trans. Anthony Preus (Berkeley and Los Angeles: University of California Press, 1982).
5  J. Donald Hughes, "Early Greek and Roman Environmentalists," in *Historical Ecology: Essays on Environment and Social Change*, ed. Lester J. Bilsky (Port Washington, N.Y.: Kennikat Press, 1980), pp. 53–55.
6  Hughes, "Early Greek and Roman Environmentalists," pp. 46–47.
7  Hughes, "Early Greek and Roman Environmentalists," pp. 50–52; Debra Hassig, *Medieval Bestiaries: Text, Image, Ideology* (Cambridge: Cambridge University Press, 1995), p. 169.
8  J.M.C. Toynbee, *Animals in Roman Life and Art* (Ithaca, N.Y.: Cornell University Press, 1973), pp. 21–23.
9  Origen, *Contra Celsum*, trans. Henry Chadwick (Cambridge: Cambridge University Press, 1953), pp. 199–200 (IV:23).
10  Origen, *Contra Celsum*, pp. 246–49, 253–54, 261–63 (IV:78–79, 81, 86–88, 98–99).
11  Origen, *Contra Celsum*, pp. 200–202, 246–51, 254–60 (IV:24–26, 79–80, 82–85, 89–97).
12  This is the approach of Joyce Salisbury's *The Beast Within: Animals in the Middle Ages* (New York and London: Routledge, 1994), although my topics differ somewhat from hers.
13  Lynn White, Jr., "The Historical Roots of our Ecologic Crisis," *The Ecocriticism Reader: Landmarks in Literary Ecology*, ed. Cheryll Glotfelty and Harold Fromm (Athens, GA.: University of Georgia Press, 1996), p. 9.
14  Jeremy Cohen, *"Be Fertile and Increase, Fill the Earth and Master It": The Ancient and Medieval Career of a Biblical Text* (Ithaca, N.Y.: Cornell University Press, 1989), p. 268.
15  *Collection Canonum in V Libris*, ed. M. Fornasari (Continuatio Mediaevalis, VI, 1976), p. 496.
16  This is in contrast to modern popular culture in the West, which seems willing to entertain the idea of an afterlife for pets. Two examples that come to mind are the films, *What Dreams May Come* (1998), in which the main character is reunited in heaven not only with his human family but also with his dog, and *Dean Spanley* (2008), where the title character remembers his past life as a Welsh spaniel.
17  Peter G. Sobol, "The Shadow of Reason: Explanations of Intelligent Animal Behavior in the Thirteenth Century," in *The Medieval World of Nature: A Book of Essays*, ed. Joyce E. Salisbury (New York and London: Garland Publishing, 1993), pp. 109–28; Albertus Magnus, *On Animals: A Medieval Summa Zoologica*, trans Kenneth F. Kitchell, Jr. and Irven Michael Resnick, 2 vols. (Baltimore: Johns Hopkins University Press, 1999), 2:1409–39.
18  Salisbury, *The Beast Within*, pp. 4–6.
19  Thomas of Cantimpré, *Liber de Natura Rerum. Volume 1: Text*, ed. Helmut Boese (Berlin: Walter de Gruyter, 1973), pp. 104–5.
20  K. Sudhoff, "Pestschriften aus den ersten 150 Jahren nach der Epidemie des 'schwarzen Todes' von 1348," *Archiv für Geschichte der Medizin* AGM 16 (1924–25):47.
21  Salisbury, *The Beast Within*, pp. 8–9.
22  This topic is pursued in the second half of Salisbury's *The Beast Within*, pp. 103–66. The werewolf theme is also examined in Aleksander Pluskowski, *Wolves and the Wilderness in the Middle Ages* (Woodbridge, Suffolk: Boydell Press, 2006), pp. 172–92.
23  My approach is almost the opposite of that of Salisbury, who sees "imaginary animals" of bestiaries and fables as exerting more of an influence over the attitudes and relationships medieval people had with animals than the real specimens they may have encountered in life. See Salisbury, *The Beast Within*, p. 103; idem, "Human Animals of Medieval Fables," in *Animals in the Middle Ages*, ed. Nona C. Flores (New York and London: Routledge, 1996), p. 49.
24  Clutton-Brock, *Domesticated Animals*, p. 18.

25 Clutton-Brock, *Domesticated Animals*, pp. 15–16, 52–61.
26 Clutton-Brock, *Domesticated Animals*, pp. 62–77.
27 Clutton-Brock, *Domesticated Animals*, pp. 80–101; Juliet Clutton-Brock, *Horse Power: A History of the Horse and the Donkey in Human Societies* (Cambridge, MA: Harvard University Press, 1992), pp. 26–66.
28 Robert Trow-Smith, *A History of British Livestock Husbandry to 1700* (London: Routledge and Kegan Paul, 1957), pp. 34–39.
29 Miranda J. Green, *Dictionary of Celtic Myth and Legend* (London: Thames and Hudson, 1992), pp. 28–29, 44–45, 51–54, 62–63, 88, 90–92, 106–7, 120–23, 173–74, 182–83, 217–18.
30 Studies relevant for the early Middle Ages include: Pam Crabtree, "Animals as Material Culture in Middle Saxon England: The Zooarchaeological Evidence for Wool Production at Brandon," Marco Valenti and Frank Salvadori, "Animal Bones: Synchronous and Diachronic Distribution as Patterns of Socially Determined Meat Consumption in the Early and High Middle Ages in Central and Northern Italy," and Antonietta Buglione, "People and Animals in Northern Apulia from Late Antiquity to the Early Middle Ages: Some Considerations," in *Breaking and Shaping Beastly Bodies: Animals as Material Culture in the Middle Ages*, ed. Aleksander Pluskowski (Oxford: Oxbow Books, 2007), pp. 161–216; N.J. Sykes, "From *Cu* and *Sceap* to *Beffe* and *Motton*: The Management, Distribution, and Consumption of Cattle and Sheep in Medieval England," U. Albarella, "Pig Husbandry and Pork Consumption in Medieval England," D. Serjeantson, "Birds: Food and a Mark of Status," and N.J. Sykes, "The Impact of the Normans on Hunting Practices in England," in *Food in Medieval England: Diet and Nutrition*, eds. C.M. Woolgar, D. Serjeantson, and T. Waldron (Oxford: Oxford University Press, 2006), pp. 56–87, 131–47, 163–64; Helena Hamerow, *Early Medieval Settlements: The Archaeology of Rural Communities in Northwest Europe, 400–900* (Oxford: Oxford University Press, 2002), pp. 126–34; Georges Comet, "Technology and Agricultural Expansion in the Middle Ages: The Example of France north of the Loire," in *Medieval Farming and Technology: The Impact of Agricultural Change in Northwest Europe*, eds. Grenville Astill and John Langdon (Leiden: Brill, 1997), p. 20; Norbert Benecke, "On the Utilization of the Domestic Fowl in Central Europe from the Iron Age up to the Middle Ages," *Archaeofauna* 2 (1993):28–29; Jennie Coy, "The Provision of Fowls and Fish for Towns," in *Diet and Crafts in Towns: The Evidence of Animal Remains from the Roman to the Post-Medieval Periods*, eds. D. Serjeantson and T. Waldron (BAR British Series, 199, 1989), pp. 25–40; Pam Jean Crabtree, "Zooarchaeology at Early Anglo-Saxon West Stow," in *Medieval Archaeology: Papers of the Seventeenth Annual Conference of the Center for Medieval and Early Renaissance Studies*, ed. Charles L. Redman (SUNY Binghamton, Medieval and Renaissance Texts and Studies, 60, 1989), pp. 203–15; *Un Village au Temps de Charlemagne: Moines et Paysans de l'Abbaye de Saint-Denis du VIIe siècle à l'An Mil* (Paris: Ministère de la Culture, 1988), pp. 228–32; Annie Grant, "Animal Resources," in *The Countryside of Medieval England*, eds. Grenville Astill and Annie Grant (Oxford: Basil Blackwell, 1988), pp. 149–87; G.F. IJzereef, "The Animal Remains," in *Farm Life in a Carolingian Village: A Model Based on Botanical and Zoological Data from an Excavated Site*, eds. W. Groenman-van Waateringe and L.H. van Wijngaarden-Bakker (Assen/Maastricht: Van Gorcum, 1987), pp. 39–51.
31 Richard Britnell, *Britain and Ireland, 1050–1530: Economy and Society* (Oxford: Oxford University Press, 2004), pp. 205–11.
32 Salisbury, *The Beast Within*, p. 24.
33 *The Visigothic Code*, ed. S.P. Scott (Boston: Boston Book Co., 1910), pp. 284–86, 288; *Laws of the Salian and Ripuarian Franks*, trans. Theodore John Rivers (New York: AMS Press, 1986), pp. 43–50, 80–82, 177; *The Laws of the Salian Franks*, trans. Katherine Fischer Drew (Philadelphia: University of Pennsylvania Press, 1991), pp. 65–72, 99–101; *The Lombard Laws*, trans. Katherine Fischer Drew (Philadelphia: University of Pennsylvania Press, 1973), pp. 116–18; *Laws of the Alamans and Bavarians*, trans. Theodore John Rivers (Philadelphia University of Pennsylvania Press, 1977), pp. 90–93, 156–57; *The Laws of the Earliest English Kings*, ed. and trans. F.L. Attenborough (Cambridge:

Cambridge University Press, 1922), pp. 51, 73, 133; *The Earliest Norwegian Laws, being the Gulathing Law and the Frostathing Law*, trans. Laurence M. Larson (New York: Columbia University Press, 1935), p. 360; Ann Hyland, *The Medieval Warhorse: From Byzantium to the Crusades* (London: Grange Books, 1994), p. 81.
34 *Laws of the Earliest English Kings*, p. 161.
35 Salisbury, *The Beast Within*, pp. 19–23; *The Visigothic Code*, pp. 172, 287; *Laws of the Salian and Ripuarian Franks*, pp. 43–50, 80–82, 109–10, 211; *Laws of the Salian Franks*, pp. 65–72, 99–101, 125–26.
36 *The Burgundian Code*, trans. Katherine Fischer Drew (Philadelphia University of Pennsylvania Press, 1972), pp. 35, 39–40, 67, 69–70, 80–81.
37 Ann Hyland, *The Horse in the Middle Ages* (Stroud, Gloucestershire: Sutton Publishing, 1999), p. 1.
38 *Agrarian History of England and Wales, volume II: 1042–1350*, ed. H.E. Hallam (Cambridge: Cambridge University Press, 1988), pp. 121–30; John Langdon, *Horses, Oxen and Technological Innovation: The Use of Draught Animals in English Farming from 1066 to 1500* (Cambridge: Cambridge University Press, 1986), pp. 22–38; Robert Trow-Smith, *A History of British Livestock Husbandry to 1700* (London: Routledge and Kegan Paul, 1957), pp. 65–85; Hyland, *The Horse in the Middle Ages*, pp. 13–14; B.H. Slicher Van Bath, *The Agrarian History of Western Europe, A.D. 500–1850*, trans. Olive Ordish (New York: St. Martin's Press, 1963), pp. 67–68; H.C. Darby, *Domesday England* (Cambridge: Cambridge University Press, 1977), pp. 162–70.
39 M.M. Postan, "Village Livestock in the Thirteenth Century," *Economic History Review*, new series, 15 (1962):228–35.
40 D.J. Stone, "Consumption and Supply of Birds in Late Medieval England," in *Food in Medieval England: Diet and Nutrition*, eds. C.M. Woolgar, D. Serjeantson, and T. Waldron (Oxford: Oxford University Press, 2006), p. 153.
41 Langdon, *Horses, Oxen, and Technological Innovation*, pp. 38–62, 80–118, 172–229, 250–53; Bruce M.S. Campbell, *English Seigniorial Agriculture, 1250–1450* (Cambridge: Cambridge University Press, 2000), pp. 123–27.
42 Langdon, *Horses, Oxen, and Technological Innovation*, pp. 4–21, 127–57, 244–50, 254–92; John Langdon, "Was England a Technological Backwater in the Middle Ages?" and Bruce Campbell, "Economic Rent and the Intensification of English Agriculture, 1086–1350," in *Medieval Farming and Technology: The Impact of Agricultural Change in Northwest Europe*, eds. Grenville Astill and John Langdon (Leiden: Brill, 1997), pp. 230–32 and 282–83; Hyland, *The Horse in the Middle Ages*, pp. 39–49; John Langdon, "Agricultural Equipment," in *The Countryside of Medieval England*, eds. Grenville Astill and Annie Grant (Oxford: Basil Blackwell, 1988), pp. 91–94; Campbell, *English Seigniorial Agriculture*, pp. 127–33.
43 This is assuming that 260 fleeces make up a woolsack. See Britnell, *Britain and Ireland*, pp. 216–17.
44 See the records for so many gallons of sulphur and stone or pounds of "ointment," tallow, and "white fat" purchased to make sheep smear for the sheep in Pickering Forest in Yorkshire. Despite this, a total of 70 sheep and 22 lambs died of the pox. *The Honor and Forest of Pickering*, 4 vols., ed. Robert Bell Turton (North Riding Record Society, 1894–97), 4:209, 218, 225–26, 260, 264–65, 269.
45 Comet, "Technology and Agricultural Expansion in the Middle Ages," p. 21; Albarella, "Pig Husbandry," p. 79; Campbell, *English Seigniorial Agriculture*, pp. 165–68.
46 *The Book of Beasts, being a Translation from a Latin Bestiary of the Twelfth Century*, ed. T.H. White (New York: G.P. Putnam's Sons, 1954), pp. 72–91; Alexander Neckam, *De Naturis Rerum*, ed. Thomas Wright (London: Rolls Series, 34, 1863), pp. 258–68; Bartholomaeus Anglicus, *On the Properties of Things: John Trevisa's Translation: A Critical Text*, 2 vols. (Oxford: Clarendon Press, 1975), 2:1111–24; Vincent of Beauvais, *Speculum Quadruplex sive Speculum Maius*, 4 vols. (Douai: Balthazar Belleri, 1624), 1: 1328–82; Cantimpré, *Liber De Natura Rerum*, pp. 108–11, 118–20, 131–33, 148–49, 156–59, 164–66; Albertus Magnus, *On Animals*, 2:1450–54, 1456, 1477–1504, 1528–30, 1536–38.
47 Beauvais, *Speculum Maius*, pp. 1:1329, 1333, 1343, 1372, 1377, 1379.

48 Karl Sudhoff, "Ein chirurgisches Manual des Jean Pitard, Wundarztes König Philippes des Schönen von Frankreich," *Archiv für Geschichte der Medizin* 2 (1908–9):232–33, 245, 252, 254, 259.
49 Sudhoff, "Pestchriften," AGM 5 (1912):364; 6 (1913):344, 347–48; 8 (1915):213, 240, 272, 275, 282; 11 (1919):86, 139, 159, 162; 14 (1923):24, 89; 16 (1924–25):129; 17 (1925):114. On medieval dovecotes, see *Agrarian History of England and Wales*, 2:882–82; 3:881–83.
50 Albertus Magnus, *On Animals*, pp. 1479–1504.
51 R.H.C. Davis, *The Medieval Warhorse: Origin, Development and Redevelopment* (London: Thames and Hudson, 1989), pp. 99–107; Hyland, *The Horse in the Middle Ages*, pp. 53–59; Swabe, *Animals, Disease and Human Society*, pp. 75–77.
52 *Walter of Henley's Husbandry*, trans. Elizabeth Lamond (London: Longmans, Green, and Co., 1890), pp. 11–15, 23–31.
53 Pietro de Crescenzi, *Liber ruralia commodorum* (Augsburg: Johann Schlösser, 1471), pp. 287–362.
54 The *Great Surgery* of the fourteenth-century papal physician, Guy de Chauliac, describes trepanning on humans for head injuries and also provides illustrations of the surgical instruments involved. See Guy de Chauliac, *Inventarium sive Chirurgia Magna. Volume One: Text*, ed. Michael R. McVaugh (Leiden: E.J. Brill, 1997), pp. 177–89.
55 Master Fitzherbert, *The Book of Husbandry*, ed. Walter W. Skeat (London: English Dialect Society, 1882), pp. 47–51, 53–57, 63–74, 76–77. Skeat's argument in the introduction for attribution of authorship to Sir Anthony Fitzherbert rather than to an elder brother, John, is now generally accepted, according to the *Catholic Encyclopedia*, since all of Anthony's elder brothers died young, allowing him to inherit the Norbury estate.
56 *Agrarian History of England and Wales, volume III: 1348–1500*, ed. Edward Miller (Cambridge: Cambridge University Press, 1991), pp. 206, 220, 235–38, 265–67, 282–83, 299–300, 315–21, 377–408; Sykes, "From *Cu* and *Sceap* to *Beffe* and *Motton*," pp. 62–65; Albarella, "Pig Husbandry," pp. 84–86; Bob Wilson, "Mortality Patterns, Animal Husbandry and Marketing in and around Medieval and Post-Medieval Oxford," in *Urban-Rural Connexions: Perspectives from Environmental Archaeology*, eds. A.R. Hall and H.K. Kenward (Oxford: Oxbow Books, 1994), pp. 111–13; Joan Thirsk, *Alternative Agriculture: A History from the Black Death to the Present Day* (Oxford: Oxford University Press, 1997), pp. 8–9; Christopher Dyer, *Making a Living in the Middle Ages: The People of Britain, 850–1520* (New Haven, CT.: Yale University Press, 2002), pp. 340–49, 360; Campbell, *English Seigniorial Agriculture*, pp. 143–51, 154–55.
57 Christopher Dyer, *Standards of Living in the Later Middle Ages: Social Change in England, c. 1200–1520* (Cambridge: Cambridge University Press, 1989), pp. 151–60; Christopher Dyer, *Everyday Life in Medieval England* (London: Hambledon Press, 1994), pp. 77–99; Christopher Dyer, "English Diet in the Later Middle Ages," in *Social Relations and Ideas: Essays in Honour of R.H. Hilton*, eds. T.H. Aston, P.R. Cross, Christopher Dyer, and Joan Thirk (Cambridge: Cambridge University Press, 1983), pp. 197–210; C.M. Woolgar, "Meat and Dairy Products in Late Medieval England," in *Food in Medieval England: Diet and Nutrition*, eds. C.M. Woolgar, D. Serjeantson, and T. Waldron (Oxford: Oxford University Press, 2006), 90–91, 97; *Agrarian History of England and Wales* 2: 825–45; G. Persson, "Consumption, Labour and Leisure in the Late Middle Ages," in *Manger et Boire au Moyen Age*, ed. D. Menjot (Nice: Les Belles Lettres, 1984), pp. 215–18; Richard C. Hoffmann, "Frontier Foods for Late Medieval Consumers: Culture, Economy, Ecology," *Environment and History* 7 (2001), p. 137; Massimo Montanari, "Rural Food in Late Medieval Italy," in *Bäuerliche Sachkultur des Spätmittelalters: Internationaler Kongress, Krems an der Donau 21 bis 24 September 1982* (Vienna: Verlag der Österreichischen Akademie der Wissenschaften, 1984), pp. 317–18.
58 Dyer, *Standards of Living*, pp. 55–61; Dyer, "English Diet," pp. 191–97; Christopher Woolgar, "Fast and Feast: Conspicuous Consumption and the Diet of the Nobility in the Fifteenth Century," in *Revolution and Consumption in Late Medieval England*, ed. Michael Hicks (Woodbridge, Suffolk: Boydell Press, 2001), pp. 8–18; Woolgar, "Meat and Dairy Products," pp. 91–94; C.M. Woolgar, *The Great Household in Late Medieval England* (New

Haven, CT: Yale University Press, 1999), p. 134; C.M. Woolgar, "Diet and Consumption in Gentry and Noble Households: A Case Study from around the Wash," in *Rulers and Ruled in Late Medieval England: Essays Presented to Gerald Harriss*, eds. Rowena E. Archer and Simon Walker (London: Hambledon Press, 1995), pp. 17–31; Richard M. Thomas, "Food and the Maintenance of Social Boundaries in Medieval England," in *The Archaeology of Food and Identity*, ed. Katheryn C. Twiss (Southern Illinois University: Center for Archaeological Investigations, Occasional Paper No. 34, 2007), pp. 141–46; Albarella, "Pig Husbandry," pp. 73–82; Montanari, "Rural Food in Late Medieval Italy," pp. 318–19. On the seasonality of medieval diets, see Christopher Dyer, "Seasonal Patterns in Food Consumption in the Later Middle Ages," in *Food in Medieval England: Diet and Nutrition*, eds. C.M. Woolgar, D. Serjeantson, and T. Waldron (Oxford: Oxford University Press, 2006), pp. 201–14.

59 *Agrarian History of England and Wales* 3:377–408; Stone, "The Consumption and Supply of Birds," p. 153; Umberto Albarella and Simon J.M. Davis, "Mammals and Birds from Launceston Castle, Cornwall: Decline in Status and the Rise of Agriculture," *Circaea: The Journal of the Association for Environmental Archaeology* 12 (1996):12, 16–18, 34–38, 42–58; Elizabeth Shepherd Popescu, et al, *Norwich Castle: Excavations and Historical Survey, 1987–98. Part I: Anglo-Saxon to c. 1345*. (East Anglian Archaeology, 132, 2009), 723–24; Campbell, *English Seigniorial Agriculture*, pp. 120–23, 135–43, 169–87; Britnell, *Britain and Ireland*, pp. 211–13, 409–23; Christopher Dyer, "Documentary Evidence: Problems and Enquiries," and Grenville Astill and Annie Grant, "Efficiency, Progress and Change," in *The Countryside of Medieval England*, eds. Grenville Astill and Annie Grant (Oxford: Basil Blackwell, 1988), pp. 31–33, 227–29.

60 Hoffmann, "Frontier Foods," 137–39.

61 Christopher Dyer, *An Age of Transition? Economy and Society in England in the Later Middle Ages* (Oxford: Clarendon Press, 2005), pp. 135–37, 151–54; Christopher Dyer, "English Peasant Buildings in the Later Middle Ages (1200–1500)," in *Everyday Life in Medieval England* (London: Hambledon Press, 1994), pp. 133–65; Christopher Dyer, "Sheepcotes: Evidence for Medieval Sheepfarming," *Medieval Archaeology* 39 (1995):136–64; Dominique Barthélemy and Philippe Contamine, "The Use of Private Space," in *A History of Private Life. Volume II: Revelations of the Medieval World*, ed. Georges Duby, trans. Arthur Goldhammer (Cambridge, MA.: Harvard University Press, 1988), pp. 444–60; *Agrarian History of England and Wales* 2:882–85, 888–901; 3:875–84; Stone, "Consumption and Supply of Birds," p. 151, 156, 159; Thirsk, *Alternative Agriculture*, pp. 12–13; Grant, "Animal Resources," pp. 166–67.

62 Sudhoff, "Pestschriften," AGM 5 (1912):81; 9 (1916):154; 14 (1922–23):154; 16 (1924–25):40, 49. Sigmund Albich of Prague, writing in 1406, was especially concerned about pigsties.

63 *Agrarian History of England and Wales* 3:390.

64 Sudhoff, "Pestschriften," AGM 5 (1912):364, 395; 7 (1913):63; 8 (1915):178, 202; 11 (1919): 57, 130, 162; 14 (1922–23):4, 98; 16 (1924–25):2, 28, 69, 93, 155; 17 (1925):29, 49, 61, 63.

65 Isidore of Seville, *The Etymologies*, ed. and trans. Stephen A. Barney, W.J. Lewis, J.A. Beach, and Oliver Berghof (Cambridge: Cambridge University Press, 2006), p. 249; *Book of Beasts*, pp. 77–78, 86; Neckham, *De Naturis Rerum*, p. 261; Beauvais, *Speculum Maius*, pp. 1334, 1353; Anglicus, *On the Properties of Things*, pp. 1148, 1187; Cantimpré, *Liber de Natura Rerum*, pp. 133, 164; Konrad von Megenberg, *Der Buch der Natur. Volume 2: Critical Text*, eds. Robert Luff and George Steer (Tübingen: Max Niemeyer Verlag, 2003), pp. 162, 185.

66 Albertus Magnus, *On Animals*, 2:1528.

67 Neckam, *De Naturis Rerum*, pp. 264–65.

68 Fitzherbert, *Book of Husbandry*, 63–65, 73–74.

69 Nigel Longchamp, *A Mirror for Fools: The Book of Burnel the Ass*, trans. J.H. Mozley (Oxford: Blackwell, 1961).

70 Jill Mann, *From Aesop to Reynard: Beast Literature in Medieval Britain* (Oxford: Oxford University Press, 2009), pp. 98–148.

71 *The Minor Poems of John Lydgate. Part II: Secular Poems*, ed. Henry Noble MacCracken (London: Early English Text Society, 192, 1934), pp. 539–66.
72 Trinity College, Cambridge, MS B.11.22, fol. 118v., reproduced in Lilian M.C. Randall, *Images in the Margins of Gothic Manuscripts* (Berkeley and Los Angeles: University of California Press, 1966), plate 20.
73 Clutton-Brock, *Domesticated Animals*, pp. 34–45.
74 Clutton-Brock, *Domesticated Animals*, pp. 106–12; Stephen Budiansky, *The Character of Cats: The Origins, Intelligence, Behavior, and Stratagems of* Felis silvestris catus (New York: Viking Penguin, 2002), pp. 1–35.
75 Budiansky, *Character of Cats*, pp. 26–27.
76 Crabtree, "Zooarchaeology at Early Anglo-Saxon West Stow," p. 204; IJzereef, "The Animal Remains," p. 40; Tove Hatting, "Cats from Viking Age Odense," *Journal of Danish Archaeology*, 9 (1990):179–93.
77 *Un Village au Temps de Charlemagne*, pp. 227–28, 233.
78 Popescu, "*Norwich Castle*," pp. 527, 723, 727–28; Grant, "Animal Resources," p. 160; A. Locker, "Animal and Plant Remains," in *Battle Abbey: The Eastern Range and the Excavations of 1978–80*, ed. J.N. Hare (London: Historic Buildings and Monuments Commission for England, 1985), p. 187; Barbara Noodle, "The Animal Bones," in *Excavations in Medieval Southampton, 1953–1969. Volume 1: The Excavation Reports*, eds. Colin Platt and Richard Coleman-Smith (Leicester: Leicester University Press, 1975), p. 333.
79 *Laws of the Alamans and Bavarians*, pp. 95, 169–70; *Laws of the Salian and Ripuarian Franks*, pp. 48–49; *Laws of the Salian Franks*, pp. 70–71, 207.
80 *Burgundian Code*, p. 84.
81 *Laws of the Alamans and Bavarians*, p. 101.
82 Salisbury, *The Beast Within*, p. 38.
83 *Laws of the Alamans and Bavarians*, p. 168.
84 *Ancient Laws and Institutes of Wales*, ed. Aneurin Owen (London, 184), pp. 426, 495, 743; Hatting, "Cats from Viking Age Odense," pp. 179, 192.
85 *The Lombard Laws*, p. 114.
86 *Laws of the Alamans and Bavarians*, pp. 100–101, 170; *Laws of the Salian and Ripuarian Franks*, pp. 49–50; *Laws of the Salian Franks*, pp. 71–72, 160.
87 Isidore of Seville, *Etymologies*, pp. 253–54.
88 Hrabanus Maurus, *De Universo: The Peculiar Properties of Words and their Mystical Significance. The Complete English Translation*, trans. Priscilla Throop, 2 vols. (Charlotte, VT: MedievalMS, 2009), pp. 240–43.
89 *Book of Beasts*, pp. 61–68; Neckham, *De Naturis Rerum*, pp. 253–54; Anglicus, *On the Properties of Things*, 2:1164–71; Beauvais, *Speculum Maius*, 1:1388–91; Cantimpré, *Liber de Natura Rerum*, pp. 114–16.
90 Beauvais, *Speculum Maius*, 1:1390–95; Anglicus, *On the Properties of Things*, 2:1166–68; Cantimpré, *Liber de Natura Rerum*, pp. 115–16.
91 Albertus Magnus, *On Animals*, 2:1461–63.
92 *Les Livres du Roy Modus et de la Royne Ratio*, 2 vols., ed. Gunnar Tilander (Paris: Société des Anciens Textes Français, 1932), 1:100–103; Wilhelm Schlag, *The Hunting Book of Gaston Phébus: Manuscrit français 616, Paris, Bibliothèque nationale* (London: Harvey Miller Publishers, 1998), pp. 32–33; Edward, Duke of York, *The Master of Game: The Oldest English Book on Hunting*, ed. Wm.A. and F. Baillie-Grohman (London: Chatto and Windus, 1909), pp. 85–104.
93 Beauvais, *Speculum Maius*, 1:1394–98; Megenberg, *Der Buch der Natur*, p. 151; Albertus Magnus, *On Animals*, 2:1462.
94 Neckham, *De Naturis Rerum*, pp. xl–xli, 201–2.
95 Beauvais, *Speculum Maius*, 1:1433; Anglicus, *On the Properties of Things*, 2:1228–29; Cantimpré, *Liber de Natura Rerum*, pp. 151–52; Albertus Magnus, *On Animals*, 2:1523. See also Compton Reeves, *Pleasures and Pastimes in Medieval England* (Oxford: Oxford University Press, 1998), p. 128.

96 Sudhoff, "Ein Chirurgisches Manual des Jean Pitard," p. 254. A skinned cat is also used in a remedy for gout in falcons. See Albertus Magnus, *On Animals*, 2:1615; Rachel Hands, *English Hawking and Hunting in* The Boke of St. Albans (Oxford: Oxford University Press, 1975), p. 38.
 97 Popescu, "Norwich Castle," p. 728; Giovanni De Venuto, "Animals and Economic Patterns in Medieval Apulia (South Italy): Preliminary Findings," in *Breaking and Shaping Beastly Bodies: Animals as Material Culture in the Middle Ages*, ed. Aleksander Pluskowski (Oxford: Oxbow Books, 2007), pp. 217–34; Roberta Gilchrist, "The Animal Bones," in *Further Excavations at the Dominican Priory, Beverly, 1986–89*, ed. Martin Foreman (Sheffield Excavation Reports, 4, 1996), p. 228–31; Hatting, "Cats from Viking Age Odense," p. 192; Rosemary M. Luff and Marta Moreno García, "Killing Cats in the Medieval Period. An Unusual Episode in the History of Cambridge, England," *Archaeofauna* 4 (1995):93–95, 102–11.
 98 Perhaps the 79 cat skeletons found at Haithabu in Germany from the ninth to eleventh centuries are of this type, since they are larger than the Cambridge cats (i.e., were better cared for) and seem to bear no traces of butchery. See Luff and Moreno, "Killing Cats," p. 102.
 99 Esther Cohen, "Law, Folklore and Animal Lore," *Past and Present* 110 (1986):22–23.
100 Etienne de Bourbon, *Anecdotes historiques*, ed. A. Lecoy de Marche (Paris: Librairie Renouard, 1877), pp. 314–29. For a detailed analysis of the Guinefort legend, see Jean-Claude Schmitt, *The Holy Greyhound: Guinefort, Healer of Children since the Thirteenth Century* (Cambridge and Paris: Cambridge University Press and La Maison des Sciences de l'Homme, 1983), esp. pp. 39–87.
101 John Aberth, *From the Brink of the Apocalypse: Confronting Famine, War, Plague, and Death in the Later Middle Ages*, 2nd edn. (London and New York: Routledge, 2010), pp. 133–34.
102 Reeves, *Pleasures and Pastimes*, pp. 125–26; Eileen Power, *Medieval English Nunneries, c. 1275 to 1535* (Cambridge: Cambridge University Press, 1922), pp. 662–63.
103 Reeves, *Pleasures and Pastimes*, pp. 126–28; Grant, "Animal Resources," p. 184.
104 *Honor and Forest of Pickering*, 3:225.
105 Margaret Wade Labarge, *A Small Sound of the Trumpet: Women in Medieval Life* (Boston, MA: Beacon Press, 1986), pp. 124–25.
106 Reeves, *Pleasures and Pastimes*, p. 129.
107 Salisbury, *The Beast Within*, pp. 65–66.
108 Edward, Duke of York, *Master of Game*, p. 75; Crescenzi, *Liber ruralia commodorum*, pp. 336–37.
109 For archaelogical studies of prehistoric hunting, see *Animals and Archaeology: Volume 1. Hunters and their Prey*, eds. Juliet Clutton-Brock and Caroline Grigson (BAR International Series, 163, 1983).
110 Henry Rawlinson, *Babylonian and Assyrian Literature* (New York: P.F. Collier and Son, 1901), pp. 213–29.
111 Text available online at www.reshafim.org.il/ad/egypt/texts/the_pleasures_of_fishing_and_fowling.htm.
112 J.K. Anderson, *Hunting in the Ancient World* (Berkeley: University of California Press, 1985); Robin Lane Fox, "Ancient Hunting: From Homer to Polybios," in *Human Landscapes in Classical Antiquity: Environment and Culture*, eds. Graham Shipley and John Salmon (London and New York: Routledge, 1996), pp. 119–53.
113 Mahesh Rangarajan, *India's Wildlife History: An Introduction* (Delhi: Ranthambhore Foundation, 2001), p. 8.
114 Edward Payson Evans, *Animal Symbolism in Ecclesiastical Architecture* (New York: Henry Holt and Co., 1896), pp. 23–24.
115 Stephen O. Glosecki, "Movable Beasts: The Manifold Implications of Early Germanic Animal Imagery," in *Animals in the Middle Ages*, ed. Nona C. Flores (New York and London: Routledge, 1996), pp. 3–23; Sykes, "Impact of the Normans on Hunting,", p. 164.
116 Sykes, "Impact of the Normans on Hunting," p. 167.
117 Green, *Dictionary of Celtic Myth and Legend*, pp. 28–29, 41, 43–45, 68–69, 84–86, 88–89, 124, 168, 174, 182–85, 194–99, 203.

118 Bernadette Filotas, *Pagan Survivals, Superstitions and Popular Cultures in Early Medieval Pastoral Literature* (Toronto: Pontifical Institute of Mediaeval Studies, 2005), pp. 142–45.
119 *Lombard Laws*, pp. 113–15; *Laws of the Alamans and Bavarians*, pp. 100–101, 170; *The Burgundian Code*, pp. 53, 69, 84–85; *Laws of the Salian Franks*, pp. 71–73, 95–96; *Laws of the Salian and Ripuarian Franks*, pp. 49–51, 76–77, 187–88; *Visigothic Code*, pp. 292–93; *Earliest Norwegian Laws*, pp. 370, 380.
120 *Un Village au temps de Charlemagne*, pp. 226–27.
121 Marco Valenti and Frank Salvadori, "Animal Bones: Synchronous and Diachronic Distribution as Patterns of Socially Determined Meat Consumption in the Early and High Middle Ages in Central and Northern Italy," in *Breaking and Shaping Beastly Bodies: Animals as Material Culture in the Middle Ages*, ed. Aleksander Pluskowski (Oxford: Oxbow Books, 2007), pp. 183–87.
122 Hyland, *Horse in the Middle Ages*, p. 91; Crabtree, "Zooarchaeology at Anglo-Saxon West Stow," p. 206; Sykes, "Impact of the Normans on Hunting," pp. 162–64.
123 *Visigothic Code*, pp. 295–96.
124 *Earliest Norwegian Laws*, pp. 96–97, 103–4, 126–27, 233, 236–37. A medieval description of whales and of whale hunting, some of it based on personal observation from dockside, is given by Albert the Great in *De Animalibus* (see Albertus Magnus, *On Animals*, 2:1666–71). Albert describes baleen, narwhal, and perhaps even right whales and explains that the goal of a whale hunt was to drive the animal to shore where it could be beached; if it fled to the open sea, the cause was lost. But even a beached whale that was staked to the ground and tied to houses could still escape with the return of the tide. Whales were valued just as much for their blubber, used for lighting and obtained by piercing the beached whale through the eye with a lance, as for their meat.
125 Rodulfus Glaber, *The Five Books of the Histories*, ed. and trans. John France (Oxford: Clarendon Press, 1989), pp. 51–53.
126 Charles R. Young, *The Royal Forests of Medieval England* (Philadelphia: University of Pennsylvania Press, 1979), pp. 2–3; Jean Birrell, "Hunting and the Royal Forest," in *L'Uomo e la Foresta, secc. XIII–XVIII*, ed. Simonetta Cavaciocchi (Prato: Istituto Internazionale di Storia Economica "F. Datini," 1996), p. 439.
127 *Honor and Forest of Pickering*, 4:13–20, in a case brought against the abbot of Whitby for using nets to catch deer in his free chase of Hackness, of which he was acquitted.
128 On April 12, 1278, a twelve-man jury at Windiate-upon-Mountains, Malvern, rendered a verdict in favor of the bishop. See *Registrum Thome de Cantilupo, Episcopi Herefordensis, A.D. 1275–1282*, ed. W.W. Capes (Hereford: Cantilupe Society, 1906), pp. 34, 36, 52–56, 59–62.
129 *Victoria History of the Counties of England: A History of Berkshire*, 4 vols., eds. P.H. Ditchfield and William Page (London: A. Constable, 1906–24), 2:342. For other cases of live deer granted by the crown to stock parks, see *Select Pleas of the Forest*, ed. G.J. Turner (London: Selden Society, 1901), pp. 92, 104, 106.
130 *Select Pleas of the Forest*, pp. cix–cxxii; Young, *Royal Forests of Medieval England*, pp. 95–97; John M. Gilbert, *Hunting and Hunting Reserves in Medieval Scotland* (Edinburgh: John Donald Publishers, 1979), pp. 215–24; Raymond Grant, *The Royal Forests of England* (Stroud, Gloucestershire: Alan Sutton, 1991), pp. 27–31; Oliver Rackham, *Ancient Woodland: Its History, Vegetation and Uses in England* (London: Edward Arnold, 1980), pp. 188–95; Oliver Rackham, *The History of the Countryside* (London: J.M. Dent and Sons, 1986), pp. 122–24; Oliver Rackham, *Trees and Woodland in the British Landscape* (London: J.M. Dent and Sons, 1976), pp. 142–51; Leonard Cantor, "Forests, Chases, Parks and Warrens," in *The English Medieval Landscape*, ed. Leonard Cantor (Philadelphia: University of Pennsylvania Press, 1982), pp. 70–82; Paul Stamper, "Woods and Parks," in *The Countryside of Medieval England*, eds. Grenville Astill and Annie Grant (Oxford: Basil Blackwell, 1988), pp. 140–45.
131 Jean Birrell, "Deer and Deer Farming in Medieval England," *Agriculture History Review* 40 (1992):112–26. In addition to the records of deer farming during the fifteenth century in

Needwood Chase studied by Birrell, there are also those for Blansby Park in the forest of Pickering in the early fourteenth century, which was estimated to have a deer population of 1300 at this time: See *Honor and Forest of Pickering*, 2:21, 196, 4:199, 206, 222–24, 227, 244, 251–52, 254.
132 Cambridge University Library, Ely Diocesan Records, D10/2/17.
133 Rackham, *Ancient Woodland*, p. 191; Rackham, *Trees and Woodland*, p. 147; L.M. Cantor and J. Hatherly, "The Medieval Parks of England," *Geography* 64 (1979):71–85; Thirsk, *Alternative Agriculture*, pp. 11–12; Stamper, "Woods and Parks," pp. 145–47.
134 On the intensive management of medieval rabbit warrens, see: Mark Bailey, "The Rabbit and the Medieval East Anglian Economy," *Agriculture History Review* 36 (1988):1–20; Petra J.E.M. van Dam, "New Habitats for the Rabbit in Northern Europe, 1300–1600," in *Inventing Medieval Landscapes: Senses of Place in Western Europe*, eds. John Howe and Michael Wolfe (Gainesville, FL: University Press of Florida, 2002), pp. 57–69; Thirsk, *Alternative Agriculture*, pp. 10–11; Grant, "Animal Resources," p. 166. By the late Middle Ages, rabbit became as popular as fallow deer as a game meat among the English aristocracy, judging from the prevalence of their bones at castle sites: See Sykes, "Impact of the Normans on Hunting," p. 167.
135 *Select Pleas of the Forest*, pp. cxxiii–cxxxiv; Grant, *Royal Forests of England*, pp. 31–34; Cantor, "Forests, Chases, Parks and Warrens," pp. 82–83.
136 *Registrum Johannis de Trillek, Episcopi Herefordensis, A.D. 1344–1361*, ed. Joseph Henry Parry (Canterbury and York Society, 8, 1912), pp. 201–2.
137 In 1441 the owner of Okeover Park in Staffordshire complained that nearly his entire deer population, 100 of 125 animals, was poached. See Birrell, "Deer Farming," p. 115; Jean Birrell, "Procuring, Preparing, and Serving Venison in Late Medieval England," in *Food in Medieval England: Diet and Nutrition*, eds. C.M. Woolgar, D. Serjeantson, and T. Waldron (Oxford: Oxford University Press, 2006), p. 187.
138 *Select Pleas of the Forest*, pp. 137–49; Edward, duke of York, *Master of Game*, pp. 254–55; Rackham, *History of the Countryside*, pp. 39–40; Rackham, *Ancient Woodland*, p. 177; Birrell, "Deer and Deer Farming," p. 116; Sykes, "Impact of the Normans on Hunting," p. 169.
139 Grant, *Royal Forests of England*, p. 31; Rackham, *Ancient Woodland*, p. 177. For other prosecutions in the royal forest, see *Select Pleas of the Forest*, p. 8, 10, 16, 29, 40–43, 54, 58; *Records of Feckenham Forest, Worcestershire, c. 1236–1377*, ed. Jean Birrell (Worcestershire Historical Society, new series, 21, 2006), pp. 51, 71, 104, 108, 114. In 1334–35, the number of hare hunters in Pickering Forest outnumbered deer poachers by 267 to 216. See *Honor and Forest of Pickering*, 3:44–46, 67–70.
140 *Records of Feckenham Forest*, p. 67.
141 Sykes, "Impact of the Normans on Hunting," pp. 166–67; Rackham, *History of the Countryside*, pp. 33–37; Cyril E. Hart, *Royal Forest: A History of Dean's Woods as Producers of Timber* (Oxford: Clarendon Press, 1966), p. 41, n.204; John Charles Cox, *The Royal Forests of England* (London: Methuen and Co., 1905), pp. 30–32.
142 Crabtree, "Zooarchaeology at Early Anglo-Saxon West Stow," P. 206.
143 Rackham, *History of the Countryside*, pp. 33–37; *Victoria History of the Counties of England: A History of the County of Chester*, 3 vols., ed. B.E. Harris (Oxford: Oxford University Press, 1979), 2:179; Cox, *Royal Forests of England*, pp. 32–34; Pluskowski, *Wolves and the Wilderness*, pp. 97, 103–5.
144 Schlag, *Hunting Book of Gaston Phébus*, pp. 66–67; Edward, duke of York, *Master of Game*, pp. 61; *Livres du Roy Modus et de la Royne Ratio*, 1:159–60; John Cummins, *The Hound and the Hawk: The Art of Medieval Hunting* (New York: St. Martin's Press, 1988), p. 241; Richard Almond, *Medieval Hunting* (Stroud, Gloucestershire: Sutton Publishing, 2003), p. 107; Pluskowski, *Wolves and the Wilderness*, pp. 98–101.
145 According to the authors of medieval encyclopedias and hunting manuals, wolves that have once tasted human flesh will stop at nothing to hunt man, even turning aside from a flock of sheep to eat the shepherd, since it tastes sweeter to him "than any other meat." Such legends were repeated by Albert the Great even though he admitted that wolves rarely ate

humans and he had seen them in person. See: Albertus Magnus, *On Animals*, 2:1519; Pluskowski, *Wolves and the Wilderness*, p. 95; Beauvais, *Speculum Maius*, 1:1429; Megenberg, *Buch der Natur*, p. 173; Schlag, *Hunting Book of Gaston Phébus*, pp. 27–28; Richard, duke of York, *Master of Game*, pp. 59–60. Medieval fears of being eaten by another tied into concerns about the resurrection of the body, as evidenced by the legend of St. Magnus of Norway, who miraculously resuscitated a man consumed and regurgitated by wolves. One should also point out that there was a medieval hagiographical tradition that was more friendly to wolves, such as the legend of the head of the ninth-century East Anglian martyr, King Edmund, being guarded by a wolf, or of wolves tamed and fed by saints. See: Pluskowski, *Wolves and the Wilderness*, pp. 79–85, 95–97, 168, 175–85; Salisbury, *The Beast Within*, pp. 62–70, 163–66.

146 *Statutes of the Realm (1225–1713)*, 9 vols. (London: G. Eyre and A. Strahan, 1810–22), 2:655; Rackham, *History of the Countryside*, p. 37; Umberto Albarella and Richard Thomas, "They Dined on Crane: Bird Consumption, Wild Fowling and Status in Medieval England," *Acta Zoologica Cracoviensa* 45 (2002):33; Serjeantson, "Birds," p. 142; Stone, "Consumption and Supply of Birds," pp. 159–60.

147 Pluskowski, *Wolves and the Wilderness*, pp. 22–24, 106–7. Archaeological remains of wolves have been recovered from English sites as late as the thirteenth century and from Scandinavian sites in the fifteenth century, but identification is tentative due to the difficulty of differentiating them from domestic dogs.

148 Richard C. Hoffmann, "Economic Development and Aquatic Ecosystems in Medieval Europe," *American Historical Review* 101 (1996):665–66; Cummins, *Hound and the Hawk*, pp. 121, 270.

149 *Livres du Roy Modus et de la Royne Ratio*, 1:153–56; Cummins, *Hound and the Hawk*, p. 147.

150 For examples, see: *The Sherwood Forest Book*, ed. Helen E. Boulton (Thoroton Society Record Series, 23, 1965), pp. 48–49; *Honor and Forest of Pickering*, 3:99, 118, 136, 149, 164, 168–71, 4:13–20; *A Calendar of New Forest Documents, 1244–1334*, ed. D.J. Stagg (Hampshire Record Series, 3, 1979), p. 94; *Records of Feckenham Forest*, p. 143; Cox, *Royal Forests of England* pp. 34–39; *Victoria History of the Counties of England: Chester*, 2:187. In some cases, the right to hunt these animals in the forest was denied: *Honor and Forest of Pickering*, 3:113, 116.

151 Edward, duke of York, *Master of Game*, pp. 63, 67, 69, 181, 255–57; Pluskowski, *Wolves and the Wilderness*, pp. 102–3, 111–17.

152 Over 200 poachers were regularly listed at eyres for the forests of Cannock and Kinver in Staffordshire, for Rockingham Forest in Northamptonshire, and Pickering Forest in Yorkshire, and over 300 for the Forest of Dean in Gloucestershire. See *Honor and Forest of Pickering*, 3:67–70; Jean Birrell, "Who Poached the King's Deer? A Study in Thirteenth Century Crime," *Midland History* 7 (1982): 10; Stamper, "Woods and Parks," p. 143.

153 *Records of Feckenham Forest*, p. 115. The term used in the text, "engine" (*ingenium*), makes it clear that this was a man-made device or contrivance, and not a real horse.

154 Jean Birrell, "Who Poached the King's Deer," pp. 9–25; idem, "Aristocratic Poachers in the Forest of Dean: Their Methods, their Quarry, and their Companions," *Transactions of the Bristol and Gloucestershire Archaeological Society* 119 (2001):147–54; *Dictionary of the Middle Ages*, 13 vols.) ed. Joseph R. Strayer (New York: Scriber, 1982), 5:360.

155 In one case, at the forest pleas of 1271 for Kinver Forest in Staffordshire, a poacher, Walter Petit, was originally fined ten shillings for stealing away a hind captured by the royal dogs in 1265 but then was pardoned because, as the record piquantly describes him, "he is a miserable wretch." See *Collections for a History of Staffordshire. The Forests of Cannock and Kinver: Select Documents, 1235–1372*, ed. Jean Birrell (Staffordshire Record Society, 4th series, 18, 1999), p. 67.

156 For examples of poachers who acted as guides to hunting parties, see: *Sherwood Forest Book*, p. 128, 135; *Records of Feckenham Forest*, p. 72, 100.

157 Almond, *Medieval Hunting*, p. 102; *Dictionary of the Middle Ages*, 5:360. For other cases of poachers using traps, nets, and snares besides those mentioned in the text, see: *Sherwood Forest Book*, p. 123; *Calendar of New Forest Documents, 1244–1334*, pp. 71, 109–10, 167, 171; *Select Pleas of the Forest*, p. 32, 55–56, 83–84, 90, 107, 112–13; *Records of Feckenham Forest*, pp. 71, 115; *Collections for a History of Staffordshire*, p. 123. See also the case of the Black Prince's bondman, Peter Stonerssh, who was caught with two freshly made nets and eight snares in his house for taking coneys at the prince's warren in Byfleet, Surrey, in 1356. *Register of Edward the Black Prince*, 4 vols. (London: HMSO, 1930–33), 4:182.

158 *Select Pleas of the Forest*, pp. 94–95.

159 *Calendar of New Forest Documents, 1244–1334*, pp. 109–10. In at least one instance, that of John and Martin, sons of Martin le Coupere, their snare had caught a great hart which ran off "carrying the snare through the forest for a great time." The men were fined a lump sum of 20 marks, or a little over £13.

160 *Honor and Forest of Pickering*, 2:78. Slightly less impressive is the feat of Stephen Duraunt of Woodcote, who was able to kill a deer (species unspecified) on October 8, 1279 in "Spretemede" meadow in the forest of Feckenham armed only with a stone, but only after the animal has been worried by two mastiffs. See *Records of Feckenham Forest*, pp. 114–15.

161 For cases of deer being killed with axes, see: *Honor and Forest of Pickering*, 2:77; *Calendar of New Forest Documents, 1244–1334*, p. 112; *Select Pleas of the Forest*, p. 11, 76.

162 Carrying bows, even unstrung ones, in the forest without warrant was established as a punishable offense by the Assize of Woodstock of 1184. For this and examples of prosecutions, see: *Sherwood Forest Book*, p. 61, 66–67; *Calendar of New Forest Documents, 1244–1334*, pp. 73–74; *Select Pleas of the Forest*, p. 2, 6–7, 11–12, 14, 23, 30, 32, 64, 73, 75, 89–90.

163 For examples, see: *Sherwood Forest Book*, pp. 132, 134; *Honor and Forest of Pickering*, 4:51, 60.

164 Jean Birrell, "Peasant Deer Poachers in the Medieval Forest," in *Progress and Problems in Medieval England: Essays in Honour of Edward Miller*, eds. Richard Britnell and John Hatcher (Cambridge: Cambridge University Press, 1996), pp. 68–88; Birrell, "Who Poached the King's Deer?" pp 20–21; Hart, *Royal Forest*, p. 41. For examples of people fined for selling or purchasing venison, see: *Sherwood Forest Book*, p. 113, 132; *Calendar of New Forest Documents, 1244–1334*, pp. 110, 169, 171; *Records of Feckenham Forest*, pp. 15, 27, 111; *Collections for a History of Staffordshire*, p. 117.

165 For dogs confiscated from poachers (who sometimes forcibly rescued them back), see: *Sherwood Forest Book*, pp. 118, 124, 130–31, 134; *Calendar of New Forest Documents, 1244–1334*, pp. 73–74; *Select Pleas of the Forest*, p. 13–15, 17, 56, 75, 79, 112–13; *Records of Feckenham Forest*, pp. 68, 115–16. Not all dog owners paid a fine in lieu of lawing their animals, as the eyre roll for Cannock Forest in Staffordshire in 1272 demonstrates: See *Collections for a History of Staffordshire*, pp. 96–97. For evidence of poachers using packhorses, see: *Sherwood Forest Book*, pp. 135–36, 138–39; *Calendar of New Forest Documents, 1244–1334*, p. 108, 173. I have found a total of just two cases of horses confiscated by foresters: See *Calendar of New Forest Documents, 1244–1334*, pp. 71–72; *Select Pleas of the Forest*, p. 87. For the few cases of poachers seen riding on horseback, see: *Sherwood Forest Book*, pp. 130–31; *Select Pleas of the Forest*, p. 22, 31, 69, 76–77, 90, 95–97, 113–15.

166 Birrell, "Who Poached the King's Deer," pp. 18–19, 21–22; idem, "Peasant Deer Poachers," pp. 79–80. A particularly large cache of 50 fawn hides was uncovered in one eyre in the Forest of Dean. See Hart, *Royal Forest*, p. 42.

167 Beauvais, *Speculum Maius*, 1:1348; Sudhoff, "Pestschriften," AGM 11 (1919):62. Albert the Great even recommended the aborted fetus of a deer killed in the womb of its mother as "very effective against poison": See Albertus Magnus, *On Animals*, 2:1472.

168 Birrell, "Procuring, Preparing, and Serving Venison," p. 179.

169 Cummins, *Hound and the Hawk*, pp. 47–67; "Veneurs s'en vont en Paradis: Medieval Hunting and the 'Natural' Landscape," in *Inventing Medieval Landscapes: Senses of Place in Western Europe*, eds. John Howe and Michael Wolfe (Gainesville, FL: University Press of Florida, 2002), pp. 40–41; Gilbert, *Hunting and Hunting Reserves in Medieval Scotland*, pp. 52–61.

170 Birrell, "Hunting and the Royal Forest," pp. 442–51; Birrell, "Procuring, Preparing, and Serving Venison," p. 179. The earl came to Rockingham Forest for a whole month's worth of hunting, between August 15 and September 14, 1248: *Select Pleas of the Forest*, 91.
171 *Sherwood Forest Book*, pp. 68, 70; *Collections for a History of Staffordshire*, pp. 21–22. An inquest into the forest of Rockingham in Northamptonshire in 1253 also rendered a verdict that the vert and venison were "well preserved" despite accusations of poaching made against the foresters, and the steward of the forest, Hugh of Goldingham, was found not guilty of "many and great trespasses to the venison" at the eyre of 1255. See *Select Pleas of the Forest*, pp. 37, 108, 110.
172 *Honor and Forest of Pickering*, 1:198–99; 2:211.
173 *Victoria History of the Counties of England: A History of the County of Derby*, 2 vols., ed. William Page (London: A. Constable, 1905), 1:401; *Victoria History of the Counties of England; A History of Yorkshire*, ed. P.M. Tillott (Oxford: Oxford University Press, 1961), p. 514.
174 Young, *Royal Forests of Medieval England*, p. 101; *Honor and Forest of Pickering*, 1:123–26, 3:258–59; 2:266–71; *Register of Black Prince*, 3:8–9, 13, 166, 286–87, 407; *Victoria History of the Counties of England: Chester*, 2:176–77, 183. Eventually the abbot of St. Mary's got wise to Edward III's sharp practice with regard to Spaunton Forest and asked for his old tithe back in the Forest of Galtres or else the right to have game in Spaunton Forest forever; he was granted merely a further lease of the game in Spaunton Forest for another five years.
175 My table obviously has a different basis from that of table 4 given in Young, *Royal Forests of Medieval England*, p. 99, which counts only the number of venison "cases" or offenses for 13 forests during the second half of the thirteenth century. Young's table yields an average number of 2–8 cases per year for each forest, but there is no way of knowing how many deer this represents.
176 Glenn Foard, David Hall, and Tracey Britnell, *The Historic Landscape of Rockingham Forest: Its Character and Evolution from the 10th to the 20th Centuries* (Rockingham Forest Trust and Northamptonshire County Council, 2003), p. 19.
177 *Calendar of New Forest Documents, 1244–1334*, pp. 80, 89–92, 102–3, 113. Walter de Kent was accused of suborning 28 jurors in order to convict Walter's own foresters of destroying the vert and venison and of groundlessly removing two verderers, Henry de Ernewde and Robert Ernis; he was also said to have supported "many evildoers in the forest," whose indictments "he put aside," while holding his own pleas of vert and venison and levying fines without the verdict of verderers and foresters. On the other hand, it seems that Ernis and a regarder, Peter le Clek of Brockley, brought a false accusation of murder against Walter's close associates, John de Butestorne and Richard Cole, but the end of the roll recording the result of the enquiry is largely illegible.
178 A perfect example of poachers taking advantage of the habitual feeding behaviors of deer is when some workmen repairing the palings of Guildford Park in Surrey were said to have stretched out snares for taking the deer that came to browse on the little branches of oaks that the workmen had cut down to make the palings. For this and other examples, see: *Select Pleas of the Forest*, pp. 55–56; *Sherwood Forest Book*, pp. 138–39; *Collections for a History of Staffordshire*, p. 123.
179 Rackham, *Ancient Woodland*, pp. 181–84.
180 Birrell, "Hunting and the Royal Forest," p. 440, 446; Young, *Royal Forests of Medieval England*, p. 106; Rackham, *Ancient Woodland*, p. 181.
181 Birrell, "Who Poached the King's Deer," p. 10; Stamper, "Woods and Parks," p. 143; Young, *Royal Forests of Medieval England*, p. 99.
182 At times the record simply states that the number of deer taken is "unknown" or "cannot be discovered," perhaps owing to the fact that the foresters did not dare approach a poaching party of intimidatingly large size. See: *Calendar of New Forest Documents, 1244–1334*, pp. 98, 168, 174–75; *Sherwood Forest Book*, pp. 130; *Collections for a History of Staffordshire*, pp. 83, 115, 121, 127, 131, 133; *Records of Feckenham Forest*, p. 61, 69, 73.

183 For examples, see: *Sherwood Forest Book*, p. 128, 139; *Calendar of New Forest Pleas, 1244–1334*, p. 98; *Calendar of New Forest Documents: The Fifteenth to the Seventeenth Centuries*, ed. D.J. Stagg (Hampshire Record Series, 5, 1983), pp. 3, 22–23.

184 Here in my home state of Vermont, about 80,000 hunting licenses are issued each year, but only 15,000 deer on average are taken in the fall, out of a herd that is currently estimated at 123,000, based on statistics available from the Department of Fish and Wildlife.

185 *Sherwood Forest Book*, pp. 133–34. For cases where it is specified that poachers entered the forest but "took nothing," see *Calendar of New Forest Documents, 1244–1334*, p. 79; *Sherwood Forest Book*, pp. 123–24, 130, 133–34; *Select Pleas of the Forest*, pp. 31, 87–88; *Collections for a History of Staffordshire*, pp. 48, 123, 126, 130.

186 For cases of men arrested and accused of poaching but found not guilty, such as Henry, son of Benselin, who was thrown in prison "for a long time" just for being found "lying under a certain bush" in the wood of Nassington in Northamptonshire hard by "a doe with its throat cut," see: *Select Pleas of the Forest*, p. 3; *Calendar of New Forest Documents, 1244–1334*, p. 87; *Collections for a History of Staffordshire*, p. 130. Poachers summarily killed by foresters without trial include the case of two poachers caught by William Trumwyn in the forest of Cannock on December 20, 1250, who were beheaded after the forester beat them with his sword, and that of John Coxwold, a notorious poacher and thief, who was killed by three foresters in Pickering Forest after refusing to "be peaceably arrested": See *Collections for a History of Staffordshire*, pp. 42–43; *Honor and Forest of Pickering*, 3:233. For especially punitive fines against poachers, such as the £200 levied against Philip Marmion for venison offenses in Cannock Forest in 1272, or the 400 marks, or about £266, assessed upon Roger de Males, a "habitual offender of venison," for taking a buck and other unspecified offenses in the New Forest in 1271–72, see: *Calendar of New Forest Documents, 1244–1334*, p. 107; *Collections for a History of Staffordshire*, p. 121.

187 *Select Pleas of the Forest*, pp. 1–2, 9. For other examples of poachers adopting aliases or attempting to disguise themselves and their dogs with masks or cloths, see: *Select Pleas of the Forest*, pp. 17, 70, 102; *Collections for a History of Staffordshire*, p. 135; *Records of Feckenham Forest*, p. 114.

188 For examples of foresters keeping watch through the night for poachers or of having their persons and animals beaten, trussed, strung up, or shot at in the line of duty, see: *Sherwood Forest Book*, p. 116; *Calendar of New Forest Documents, 1244–1334*, pp. 72, 122; *Honor and Forest of Pickering*, 4:8–9; *Select Pleas of the Forest*, pp. 28, 32, 35, 69, 79–81, 88, 94, 99; *Collections for a History of Staffordshire*, pp. 66–67, 116, 126.

189 Thus, at the eyre of 1280 it was reported that the freemen and villeins of the bishop of Winchester and of the prior of St. Swithin's in Winchester took advantage of their masters' claims to disafforestment of lands in the New Forest in order to follow their lead in taking "venison throughout the year at will." Disorders occurred: in between the battles of Lewes and Eynsham (i.e. between May 1264 and August 1265) when Robert, count of Ferrières, broke into the park and warren of Northampton; at the death of Henry III in 1272, when many men invaded Alrewas Hay in the forest of Cannock "shooting almost all day with clamor and tumult"; and during the "disorders between King Edward II and his magnates" (i.e., 1321–26), when the rolls of expeditation of dogs were taken away and lost. See *Calendar of New Forest Documents, 1244–1334*, pp. 94–95, 193; *Select Pleas of the Forest*, p. 40; *Collections for a History of Staffordshire*, p. 133.

190 *Sherwood Forest Book*, pp. 125, 132, 134–36. On the Inquisition's use of the written word, particularly in its ability to search and retrieve names of past heretical offenders, see James B. Given, *Inquisition and Medieval Society: Power, Discipline, and Resistance in Languedoc* (Ithaca, N.Y.: Cornell University Press, 1997), pp. 25–51.

191 Rackham, *History of the Countryside*, p. 181. Rackham envisions an overall cull rate of one animal per every 25 acres, based on yield of fallow deer from Hatfield Forest in Essex, which was intensively hunted by the crown for its venison needs.

192 Birrell, "Deer and Deer Farming," pp. 123–24; Birrell, "Procuring, Preparing, and Serving Venison," pp. 186–87.

193 *Sherwood Forest Book*, p. 140; Birrell, "Deer and Deer Farming," p. 124; Young, *Royal Forests of Medieval England*, p. 101; *Honor and Forest of Pickering*, 2:133, 138, 141.
194 Birrell, "Hunting and the Royal Forest," pp. 456–57; idem, "Deer and Deer Farming," p. 125; Rackham, *Ancient Woodland*, p. 181.
195 *Statutes of the Realm, (1225–1713)*, 9 vols. (London: G. Eyre and A. Strahan, 1810–22), 2:65.
196 *Statutes of the Realm*, 2:505–6; Stamper, "Woods and Parks," p. 147.
197 *Calendar of New Forest Documents: The Fifteenth to the Seventeenth Centuries*, pp. ix, 24.
198 *Ordonnances des Roys de France de la troisième race*, 21 vols. (Paris: Imprimerie royale, 1723–1849), 8:117–18.
199 Sudhoff, "Pestschriften," AGM 16 (1924–25):8; *Dictionary of the Middle Ages*, 5:361.
200 *Calendar of New Forest Documents: The Fifteenth to the Seventeenth Centuries*, pp. 8, 30.
201 Nicholas Orme, "Medieval Hunting: Fact and Fancy," in *Chaucer's England: Literature in Historical Context*, ed. Barbara A. Hanawalt (Medieval Studies at Minnesota, 4, 1992); Anne Rooney, *Hunting in Middle English Literature* (Woodbridge, Suffolk: D.S. Brewer and Boydell Press, 1993); Almond, *Medieval Hunting*; William Perry Marvin, *Hunting Law and Ritual in Medieval English Literature* (Woodbridge, Suffolk: D.S. Brewer, 2006); Lucien-Jean Bord and Jean-Pierre Mugg, *La Chasse au Moyen Âge: Occident latin, VIe–XVe siècle* (Aix-en-Provence: Editions du Gerfaut, 2008). For printed medieval English treatises on hunting, see: Edward, Duke of York, *Master of Game*; Rachel Hands, *English Hawking and Hunting in* The Boke of St. Albans (Oxford: Oxford University Press, 1975), pp. 57–81; *The Tretyse off Huntyng*, ed. Anne Rooney (Scripta: Mediaeval and Renaissance Texts and Studies, 19, 1987); William Twiti, *The Art of Hunting*, ed. David Scott-Macnab (Heidelberg: Universitätsverlag, 2009).
202 Susan Crane, "Ritual Aspects of the Hunt à Force," in *Engaging with Nature: Essays on the Natural World in Medieval and Early Modern Europe*, eds. Barbara A. Hanawatt and Lisa J. Kiser (Notre Dame, IN: University of Notre Dame Press, 2008), p. 79; Sykes, "Impact of the Normans on Hunting," pp. 174–75; Cummins, *Hound and the Hawk*, pp. 32–46. According to Cummins, the eight stages of hunting a hart included the quest, assembly, relays, moving or unharboring, chase, death, unmaking, and *curée*, or rewarding of the hounds.
203 Anne Rooney characterizes the late medieval English hunting manuals (one of which she edited herself) as "virtually useless as practical texts," being more concerned with esoteric details such as the proper sequence of horn signals and hunting cries, although the French manuals on which the English texts are based tend to be more utilitarian. See Rooney, *Hunting in Middle English Literature*, p. 11.
204 Birrell, "Who Poached the King's Deer?" p. 18.
205 *Select Pleas of the Forest*, pp. 44, 74.
206 *Select Pleas of the Forest*, pp. 95–96.
207 Birrell, "Who Poached the King's Deer?" p. 21; idem, "Peasant Deer Poachers," pp. 80–81. For examples of demands for shares of venison as bribes (sometimes refused), see: *Select Pleas of the Forest*, pp. 8, 106, 112; *Collections for a History of Staffordshire*, pp. 49, 135–36. On November 20, 1275, two men disputed a stag that was chased out of Cannock Forest and brought down in a fishpond owned by the nuns of Brewood: Even though John of la Wytemore had shot the beast and dragged it out of the pond, he in fact ended up with nothing except perhaps the hide, while John Child of Chillington, who claimed to have stalked the stag "a long way," got one half of the skinned carcass and the nuns got the other half. See *Collections for a History of Staffordshire*, p. 126.
208 *Collections for a History of Staffordshire*, pp. 124–25.
209 *Select Pleas of the Forest*, pp. 92, 102, 113.
210 *Honor and Forest of Pickering*, 3:224–26.
211 *Calendar of New Forest Documents: The Fifteenth to the Seventeenth Centuries*, p. 24; *Victoria History of the Counties of England: Yorkshire*, 1:506.
212 Birrell, "Hunting and the Royal Forest," pp. 453–54; Birrell, "Procuring, Preparing, and Serving Venison," p. 183.
213 Edward, duke of York, *Master of Game*, p. 30; Birrell, "Hunting and the Royal Forest," p. 452. *The Master of Game*, p. 22, also makes disparaging remarks on nets and snares in the hunting of hares.

Notes    273

214 *Collections for a History of Staffordshire*, p. 48.
215 Richard Thomas, "Chasing the Ideal? Ritualism, Pragmatism and the Later Medieval Hunt in England," in *Breaking and Shaping Beastly Bodies: Animals as Material Culture in the Middle Ages*, ed. Aleksander Pluskowski (Oxford: Oxbow Books, 2007), pp. 131–42; Naomi Sykes, "Taking Sides: The Social Life of Vension in Medieval England," in *Breaking and Shaping Beastly Bodies*, pp. 150–55; Sykes, "Impact of the Normans on Hunting," pp. 171–74.
216 *Calendar of New Forest Documents, 1244–1334*, pp. 107–8, 169, 171; *Select Pleas of the Forest*, pp. 3–4, 6–7, 21, 23, 30, 37, 74, 86, 96, 115; *Records of Feckenham Forest*, pp. 27–28; *Collections for a History of Staffordshire*, p. 118, 127. *Tretyse off Huntyng*, pp. 55–56; Hands, *English Hawking and Hunting*, pp. 76–79; Birrell, "Peasant Deer Poachers," p. 81.
217 Sykes, "Impact of the Normans on Hunting," pp. 166–68.
218 Birrell, "Who Poached the King's Deer," pp. 21–22.
219 Cummins, *Hound and the Hawk*, pp. 47–67.
220 Birrell, "Who Poached the King's Deer?" p. 15.
221 *Select Pleas of the Forest*, pp. 28, 79–81.
222 *Select Pleas of the Forest*, p. 77, 143.
223 *Select Pleas of the Forest*, pp. 38–40. These poachers also shot at the foresters as they attempted to arrest them.
224 *Honor and Forest of Pickering*, 2:60; 4:xxxviii.
225 Gilbert, *Hunting and Hunting Reserves in Medieval Scotland*, pp. 91–182; Cummins, pp. 267–70; Roland Bechmann, *Trees and Man: The Forest in the Middle Ages* (New York: Paragon House, 1990), pp. 241–56; Bord and Mugg, *La Chasse au Moyen Âge*, pp. 25–30.
226 Birrell, "Peasant Deer Poachers," p. 88; Almond, *Medieval Hunting*, p. 111.
227 Sykes, "Taking Sides," pp. 125–59; Sykes, "Impact of the Normans on Hunting," p. 172.
228 *Collections for a History of Staffordshire*, p. 47; Birrell, "Who Poached the King's Deer," p. 19. For other examples, see: *Calendar of New Forest Documents, 1244–1334*, p. 73; *Records of Feckenham Forest*, pp. 60–61, 64, 66, 68, 73, 91; *Collections for a History of Staffordshire*, p. 132.
229 *Honor and Forest of Pickering*, 2:63–64. Around the same time and place, Peter de Marle IV was fined over £2 for a poaching offense that was specified to have been committed by his wife, Eleanor. See *Honor and Forest of Pickering*, 4:58.
230 *Calendar of New Forest Documents, 1244–1334*, p. 71.
231 *Victoria History of the Counties of England: A History of the County of Essex*, volume 2, eds. William Page and J. Horace Round (London: Constable, 1907), p. 118; Stamper, "Woods and Parks," p. 144; *Honor and Forest of Pickering*, 4:10.
232 Rooney, *Hunting in Middle English Literature*, p. 8; *Dictionary of the Middle Ages*, 5:359–61.
233 Schlag, *Hunting Book of Gaston Phébus*, pp. 27,42; Edward, duke of York, *Master of Game*, pp. 51, 133–34, 181; *Livres du Roy Modus et de la Royne Ratio*, 1:19–20.
234 Jill Mann, *From Aesop to Reynard: Beast Literature in Medieval Britain* (Oxford: Oxford University Press, 2009), p. 160; Nona C. Flores, "The Mirror of Nature Distorted: The Medieval Artist's Dilemma in Depicting Animals," in *The Medieval World of Nature: A Book of Essays*, ed. Joyce E. Salisbury (New York and London: Garland Publishing, 1993), pp. 3–6; Ron Baxter, *Bestiaries and their Users in the Middle Ages* (Stroud, Gloucestershire: Sutton Publishing, 1998), pp. 183–209; Hassig, *Medieval Bestiaries*, pp. 167–81.
235 Albarella and Thomas, "They Dined on Crane," pp. 23–38; Serjeantson, "Birds," pp. 129–34, 145–47; Stone, "Consumption and Supply of Birds," pp. 155–61; Thomas, "Food and the Maintenance of Social Boundaries," pp. 138–41, 144–45; D. Serjeantson and C.M. Woolgar, "Fish Consumption in Medieval England," in *Food in Medieval England: Diet and Nutrician*, eds. C.M. Woolgar, D. Serjeantson, and T. Waldron (Oxford: Oxford University Press, 2006), pp. 118–26; Dyer, *Everyday Life*, pp. 100–111; Dyer, *Standards of Living*, pp. 58–62; Christopher Dyer, "The Consumption of Fresh-Water Fish in Medieval England," Food in *Medieval Fish, Fishers and Fishponds in England*, 2 parts, eds. Michael Aston (BAR British Series, 182(i), 1988), 1:27–38; Fabienne Pigière et al., "Status as Reflected in Food Refuse of the Late Medieval Noble and Urban

Households at Namur (Belgium)," in *Behaviour Behind Bones: The Zooarchoelogy of Ritual, Religion, Status and Identity*, eds. Anton Ervynck, Sharyn Jones O'Day, and Wim Van Neer (Oxford: Oxbow Books, 2003), pp. 238–41.
236 Robin S. Oggins, *The Kings and their Hawks* (New Haven, CT: Yale University Press, 2004), pp. 1–6; idem, "Falconry and Medieval Views of Nature," in *The Medieval World of Nature: A Book of Essays*, ed. Joyce E. Salisbury (New York and London: Garland Publishing, 1993), pp. 48–56.
237 Frederick II states openly at the beginning of his work that it was based on "knowledge acquired from our own experience or gleaned from others," and he was not afraid to deviate from authority, such as Aristotle, especially when it conflicted with his "hard-won experience" of some 30 years in falconry. See Frederick II, *The Art of Falconry, being the De Arte Venandi cum Avibus of Frederick II of Hohenstaufen*, trans. and eds. Casey A. Wood and F. Marjorie Fyfe (Stanford, CA: Stanford University Press, 1943), pp. 3–4.
238 Frederick II, *Art of Falconry*, p. 129.
239 Frederick II, *Art of Falconry*, p. 6; Wietske Prummel, "Evidence of Hawking (Falconry) from Bird and Mammal Bones," *International Journal of Osteoarchaeology* 7 (1997):333.
240 Albertus Magnus, *On Animals*, 2:1581; Oggins, *Kings and their Hawks*, p. 6.
241 Albertus Magnus, *On Animals*, 2:1615; Cummins, *Hound and the Hawk*, pp. 208–9; Oggins, "Falconry and Medieval Views of Nature," p. 49; idem, *Kings and their Hawks*, p. 2. Medical treatment of falcons also forms the main focus of earlier treatises on falconry, such as Adelard of Bath's *De Avibus*. See Adelard of Bath, *Conversations with his Nephew*, ed. and trans. Charles Burnett (Cambridge: Cambridge University Press, 1998), pp. 248–67.
242 Albertus Magnus, *On Animals*, 2:1582, 1630, 1645; Cummins, *Hound and the Hawk*, pp. 195–96, 242–47; Almond, *Medieval Hunting*, pp. 103–5; *Dictionary of the Middle Ages*, 5:361. For a contemporary description of a variety of nets, snares, bird-lime, and cross-bows used in fowling, readers should also consult Crescenzi, *Liber ruralium commodorum*, pp. 370–79.
243 James H. Barrett, Alison M. Locker, and Callum M. Roberts, "The Origins of Intensive Marine Fishing in Medieval Europe: The English Evidence," *Proceedings of the Royal Society of London* (2004): 2417–20; Maryanne Kowaleski, "The Expansion of the South-Western Fisheries in Late Medieval England," *Economic History Review*, new series, 53 (2000):430–31, 439–40; Richard C. Hoffmann, "Medieval Fishing," in *Working with Water in Medieval Europe: Technology and Resource-Use*, ed. Paolo Squatriti (Leiden: Brill, 2000), pp. 336–40; idem, "Frontier Foods," pp. 140–54; Paolo Squatriti, *Water and Society in Early Medieval Italy, A.D. 400–1000* (Cambridge: Cambridge University Press, 1998), pp. 103–9; Inge Bødker Enghoff, "A Medieval Herring Industry in Denmark and the Importance of Herring in Eastern Denmark," *Archaeofauna* 5 (1996):43–47; Dirk Heinrich, "Temporal Changes in Fishery and Fish Consumption between Early Medieval Haithabu and its Successor, Schleswig," in *Animals and Archaeology: Volume 2. Shell Middens, Fishes and Birds*, eds. Caroline Grigson and Juliet Clutton-Brock (BAR International Series, 183, 1983), p. 156.
244 Hoffmann, "Medieval Fishing," pp. 365–72; idem, "Economic Development and Aquatic Ecosystems in Medieval Europe," *American Historical Review* 101 (1996):638–65; Angelika Lampen, "Medieval Fish Weirs: The Archaeological and Historical Evidence," *Archaeofauna* 5 (1996):129–34; *Medieval Fish, Fisheries and Fishponds in England*, 2 parts, ed. Michael Aston (BAR British Series, 182(i), 1988); An Lentacker, Wim Van Neer, and Jean Plumier, "Historical and Archaeozoological Data on Water Management and Fishing during Medieval and Post-Medieval Times at Namur (Belgium)," in *Environment and Subsistence in Medieval Europe*, eds. Guy De Boe and Frans Verhaeghe (I.A.P. Rapporten, 9, 1997), pp. 84–85.
245 Printed editions of these works are available in: Richard C. Hoffman, *Fishers' Craft and Lettered Art: Tracts on Fishing from the End of the Middle Ages* (Toronto: University of Toronto Press, 1997), pp. 76–98, 218–306; Willy L. Braekman, *The Treatise on Angling in the Boke of St. Albans (1496): Background, Context and Text of "The treatyse of fysshynge wyth an Angle"* (Scripta: Mediaeval and Renaissance Texts and Studies, 1, 1980); Hands, *English Hawking and Hunting*. These late medieval works on sport fishing mask its earlier existence going back to the twelfth century (see Richard C. Hoffman, "Fishing for Sport in Medieval Europe: New Evidence," *Speculum* 60 (1985):877–902).

246 Hands, *English Hawking and Hunting*, pp. 53–55; Hoffman, *Fishers' Craft and Lettered Art*, p. 97. The fact that the *Boke of St. Albans* assigns hawks and falcons even to the lower classes of society, such as a goshawk for a yeoman, a hobby for a "young man," and a tiercel, or a male hawk (said to be a "third" of the weight of the female, which was usually used in falconry), for a "powere man," indicates perhaps a greater penetration of the sport among the people. In the "burlesque comparison of fish" from *How to Catch Fish and Birds by Hand*, the pike is equated with "a robber," which accords well with its reputation among encyclopedists as a "water wolf" among fishes, who with its deadly jaws would "besiege" other fishes (including its own kind) and exercise "a tyranny in the waters." See: Richard C. Hoffmann, "The Protohistory of Pike in Western Culture," in *The Medieval World of Nature: A Book of Essays*, ed. Joyce E. Salisbury (New York and London: Garland Publishing, 1993), pp. 61–76; Neckham, *De Naturis Rerum*, pp. 147–48; Albertus Magnus, *On Animals*, 2:1688–89.

247 Reginald of Durham, *Libellus de Vita et Miraculis S. Godrici, Heremitae de Finchale* (London: Surtees Society, 1847), pp. 44, 67–69, 98–99, 365–66; Dominic Alexander, *Saints and Animals in the Middle Ages* (Woodbridge, Suffolk: Boydell Press, 2008), pp. 160–65; *Beasts and Saints*, trans. Helen Waddell (London: Constable, 1934), pp. 90–91. For a re-interpretation of St. Francis of Assisi's relationship with animals, see: David Salter, *Holy and Noble Beasts: Encounters with Animals in Medieval Literature* (Woodbridge, Suffolk: D.S. Brewer, 2001), pp. 39–52; Roger D. Sorrell, *St. Francis of Assisi and Nature: Tradition and Innovation in Western Christian Attitudes toward the Environment* (Oxford: Oxford University Press, 1988), pp. 75–79.

248 Usama ibn Munqidh, *The Book of Contemplation: Islam and the Crusades*, trans. Paul M. Cobb (London: Penguin Books, 2008), pp. 224–31.

249 Jan M. Ziolkowski, *Talking Animals: Medieval Latin Beast Poetry, 750–1150* (Philadelphia: University of Pennsylvania Press, 1993), p. 135.

250 Randall, *Images in the Margins*, plate LXXIV.

251 E. Fuller Torrey and Robert H. Yolken, *Beasts of the Earth: Animals, Humans, and Disease* (New Brunswick, N.J.: Rutgers University Press, 2005), pp. 23–67; Swabe, *Animals, Disease and Human Society*, pp. 44–49.

252 Alfred Jay Bollet, *Plagues and Poxes: The Impact of Human History on Epidemic Disease* (New York: Demos Medical Publishing, 2004), pp. 141–96.

253 Clive A. Spinage, *Cattle Plague: A History* (New York: Kluwer Academic/Plenum Publishers, 2003), pp. 333–34.

254 Ian Kershaw, "The Great Famine and the Agrarian Crisis in England, 1315–22," *Past and Present* 59 (1973):26–28.

255 Kershaw, "Great Famine," p. 29.

256 Kershaw, "Great Famine," pp. 20–26; William Chester Jordan, *The Great Famine: Northern Europe in the Early Fourteenth Century* (Princeton: Princeton University Press, 1996), pp. 36–39.

257 *The Black Death*, trans. and ed. Rosemary Horrox (Manchester: Manchester University Press, 1994), pp. 77–78.

258 Gentile da Foligno, *Consilium contra Pestilentiam* (Colle di Valdelsa, c. 1479), p. 2; MS 1227, Universitätsbibliothek, Leipzig, fol. 146v.

259 Foligno, *Consilium*, pp. 32, 37–38; Pietro di Tussignano, *Consilium pro Peste Evitanda*, reprinted in *The Fasciculus Medicinae of Johannes de Ketham, Alemanus: Facsimile of First Edition of 1491*, trans. Charles Singer (Birmingham, AL: Classics of Medicine Library, 1988), cols. 4–5; Codex Latin 363, Bayerische Staatsbibliothek, Munich, fols. 123r., 124v.; MS 1227, Universitätsbibliothek, Leipzig, fols. 147v., 150v.-151r.; Sudhoff, "Pestschriften," AGM 5 (1912):394; 9 (1916):128–30; 11 (1919):57–58; 16 (1924–25):172; 17 (1925):40–41. Several other plague doctors mention these points but without giving much discussion of the reasoning behind them. See: Jacme d'Agramont, "Regiment de Preservacio a Epidimia o Pestilencia e Mortaldats," trans. M.L. Duran-Reynals and C.-E.A. Winslow, *Bulletin of the History of Medicine*, 23 (1949):73; Sudhoff, "Pestschriften," AGM 5 (1912):368–71; 7 (1913):82, 85; 8 (1915):188–89; 9 (1916):70; 16 (1924–25):64, 108; 17 (1925):78–79.

260 Sudhoff, "Pestschriften," AGM 16 (1924–25):25.

261 Gentile da Foligno asserts that the pestilence "lasts for a longer time in the finer air before it is extinguished" (*sed in aere subtiliore durat longiore tempore antequam extinguatur*), but this seems to be a copyist's error. See Foligno, *Consilium*, p. 38.
262 Sudhoff, "Pestschriften," AGM 11 (1919):58. For other treatises besides the ones discussed in the text that mention fishes washing up on the shore as a sign of plague or their corruption by bad air infecting the water, see: H. Émile Rébouis, *Étude Historique et Critique sur la Peste* (Paris: Alphonse Picard, 1888), p. 88; Sudhoff, "Pestschriften," AGM 5 (1912):84, 377; 11 (1919):167; 16 (1924–25):82, 108.
263 Taha Dinanah, "Die Schrift von Abi G'far Ahmed ibn 'Ali ibn Mohammed ibn 'Ali ibn Hatimah aus Almeriah über die Pest," *Archiv für Geschichte der Medizin* 19 (1927):35–36, 38. Many other doctors cited in n. 262 above also gave dust on trees as a sign of plague.
264 Agramont, "Regiment de Preservacio a Epidimia," p. 61.
265 Sudhoff, "Pestschriften," AGM 11 (1919):123–24; 16 (1924–25):61; 17 (1925):24.
266 Albertus Magnus, *On Animals*, 2: 1712–13.
267 Sudhoff, "Pestschriften," AGM 17 (1925):79, 128.
268 Frederick W. Gibbs, "Medical Understandings of Poison circa 1250–1600," (Ph.D. dissertation, University of Wisconsin—Madison, 2009), p. 120.
269 Marsilio Ficino, *Contro alla Peste* (Florence: Appresso i Giunti, 1576), pp. 4–5. A fifteenth-century German treatise also compared plague contagion passing among men to what happens with "diseased sheep." See Sudhoff, "Pestschriften," AGM 4 (1911):423.
270 Sudhoff, "Pestschriften," AGM 9 (1916):128; Tussignano, *Consilium*, col. 4–5.
271 Foligno, *Consilium*, p. 37; Sudhoff, "Pestschriften," AGM 9 (1916):129–30; Tussignano, *Consilium*, col. 5.
272 Sudhoff, "Pestschriften," AGM 17 (1925):78–79.
273 L.-A. Joseph Michon, ed., *Documents Inédits sur la Grande Peste de 1348* (Paris, 1860), p. 73; Sudhoff, "Pestschriften," AGM 11 (1919):146.
274 Codex Latin 363, Bayerische Staatsbibliothek, Munich, fol. 123r.; MS 1227, Universitätsbibliothek, Leipzig, fol. 147r.
275 Agramont, "Regiment de Preservacio a Epidimia," p. 62. The three degrees of life are based on Aristotle's *De Anima* and *Parva Naturalia*: See Sobol, "Shadow of Reason," pp. 109–10.
276 Plague treatises that advise flight from the plague include: R. Simonini, "Il Codice di Mariano di Ser Jacopo spora 'Rimedi Abili nel Tempo di Pestilenza'," *Bollettino dell'Istituto Storico Italiano dell'Arte Sanitaria* 9 (1929):164–65; *I Trattati in Volgare della Peste e dell'acqua Ardente di Michele Savonarola*, ed. Luigi Belloni (Rome: Congresso Nazionale della Società Italiana de Medicina Interna, 1953), p. 6; Ficino, *Contro alla Peste*, p. 14; Sudhoff, "Pestschriften," AGM 4 (1911):403, 420; 5 (1912):38, 51, 57, 349, 353, 374; 6 (1913):315–16, 335, 361; 7 (1913):100; 8 (1915):182, 207, 237, 258, 264; 11 (1919):53, 75; 14 (1922–23):2, 85, 92, 130, 139–40, 159–60; 16 (1924–25):15, 29, 35, 41, 56, 59, 61, 65–66, 84, 110, 135, 158; 17 (1925):25, 34, 81–82, 96, 107, 112, 261. The Florentine doctor, Tommaso del Garbo, in his plague *Consiglio* observed that a question could be asked that if everyone took his advice to flee the pestilence, "what will the sick do without somebody to do all those things necessary to their spirit and their body?" His answer was that certain people—doctors, priests, notaries of wills, cleaning ladies, relatives—"can and should visit" the sick, and he prescribed precautions to take especially for them, as did other doctors like Primus of Görlitz and John of Speyer. Nonetheless, John of Speyer, John of Saxony, and an anonymous Bohemian treatise from c. 1450 cited abandonment of plague patients by servants and even family members as one of the contributory causes of plague, echoing the observations made by many chroniclers of the plague. See *Consiglio contro a Pistolenza per Maestro Tommaso del Garbo*, ed. Pietro Ferrato (Bologna: Presso Gaetano Romagnoli, 1866), pp. 13–15, 22–25; Sudhoff, "Pestschriften," AGM 14 (1922–23):159; 16 (1924–25):26; 17 (1925):83, 126–28.
277 This view in a shorter casebook of Foligno's represents an about face from his earlier position in favor of flight in the *Long Consilium*. See: Foligno, *Consilium*, p. 5; Sudhoff, "Pestschriften," AGM 5 (1912):333.

278 Sudhoff, "Pestschriften," AGM 7 (1913):84–85. See also the closely related treatise of Johannes von Glogau, in Sudhoff, "Pestschriften," AGM 9 (1916):68.
279 Sudhoff, "Pestschriften," AGM 8 (1915):207.
280 *Urkunden und Akten der Stadt Strassburg*, ed. Wilhelm Wiegand et al., 15 vols. (Strasbourg: K.J. Trübner, 1879–1933), 5:166. A French physician at the University of Montpellier, Bernard of Gordon, had already suggested in the early fourteenth century that compound drugs be tested "on birds, then on dumb animals" before trying them out on humans, "lest they should prove to be poisonous and so fatal." See M.R. McVaugh, "Quantified Medical Theory and Practice at Fourteenth-Century Montpellier," *Bulletin of the History of Medicine* 43 (1969):403.
281 John Aberth, *The Black Death: The Great Mortality of 1348–1350. A Brief History with Documents* (Boston and New York: Bedford/St. Martin's, 2005), pp. 156–57.
282 Gibbs, "Medical Understandings of Poison," p. 90.
283 Sudhoff, "Pestschriften," AGM 17 (1925):50.
284 *I Trattati in Volgare della Peste*, p. 8.
285 Sudhoff, "Pestschriften," AGM 5 (1912):395; 8 (1915):178; 11 (1919):85–86, 162.
286 Sudhoff, "Pestschriften," AGM 16 (1924–25):27.
287 For a more expansive discussion of the issue of flight from plague, see Aberth, *From the Brink of the Apocalypse*, pp. 191–98.
288 Agramont, "Regiment de Preservacio a Epidimia," pp. 60, 66, 73; H. Pinkhof, *Abraham Kashlari, over Pestachtige Koorsten* (Amsterdam, 1891), p. 22; Sudhoff, "Pestschriften," AGM 5 (1912):84, 347, 393; 6 (1913):354; 7 (1913):82–83; 8 (1915):188–89; 9 (1916):142; 11 (1919):126, 133, 151, 167; 14 (1922–23):130; 16 (1924–25):7, 25, 39, 42–43, 82, 108–9, 175, 178; 17 (1925):50.
289 Sudhoff, "Pestschriften," AGM 14 (1922–23):159–60.
290 MS Vatic. Lat. 4589, fols. 140r.-v., 142v.-143r.
291 John Aberth, *Plagues in World History* (Lanham, MD: Rowman and Littlefield, 2011), pp. 44–50.
292 Cohen, "Law, Folklore and Animal Lore," pp. 16–17.
293 Cohen, "Law, Folklore and Animal Lore," p. 18.
294 J.J. Finkelstein, "The Ox that Gored," *Transactions of the American Philosophical Society* 71 (1981):20–32.
295 Finkelstein, "The Ox that Gored," pp. 7–14, 52–55; Cohen, "Law, Folklore and Animal Lore," pp. 15–16.
296 Finkelstein, "The Ox that Gored," pp. 58–60.
297 Finkelstein, "The Ox that Gored," pp. 73–81.
298 E.P. Evans, *The Criminal Prosecution and Capital Punishment of Animals* (London: William Heinemann, 1906), pp. 313–34.
299 Cohen, "Law, Folklore and Animal Lore," p. 17; Salisbury, *The Beast Within*, p. 39.
300 Finkelstein, "The Ox that Gored," pp. 64–73; Darren Oldridge, *Strange Histories: The Trial of the Pig, the Walking Dead, and Other Matters of Fact from the Medieval and Renaissance Worlds* (London and New York: Routledge, 2007), pp. 40–55.
301 Evans, *Criminal Prosecution and Capital Punishment of Animals*, pp. 335–43, 346–51, 354–55, 358–59; Cohen, "Law, Folklore and Animal Lore," pp. 11–12.
302 Cohen, "Law, Folklore, and Animal Lore," p. 17.
303 Jean Boutillier, *Somme Rural, ou le Grand Coustumier* (Paris: Barthelemy Mace, 1621), p. 267.
304 Cohen, "Law, Folklore and Animal Lore," p. 21; Keith Thomas, *Man and the Natural World: Changing Attitudes in England, 150–1800* (Oxford: Oxford University Press, 1983), pp. 75–79.
305 Salisbury, *The Beast Within*, pp. 105–9, 122–26; Mann, *From Aesop to Reynard*, pp. 220–61.
306 For interpretations of the judicial aspects of the Reynard romance, see: E. Rombauts, "Grimbeert's Defense of Reinaert in *Van den Vos Reynaerde*, an Example of *Oratio Iudicialis?*" in *Aspects of the Medieval Animal Epic: Proceedings of the International Conference, Louvain, May 15–17, 1972*, eds. E. Rombauts and A. Welkenhuysen (Louvain and the Hague: Leuven University Press and Martinus Nijhoff, 1975), pp. 129–41; Jean Subrenat, "Rape and

Adultery: Reflected Facets of Feudal Justice in the Roman de Renart," in *Reynard the Fox: Social Engagement and Cultural Metamorphoses in the Beast Epic from the Middle Ages to the Present*, ed. Kenneth Varty (New York and Oxford: Berghahn Books, 2000), pp. 17–35. The translation of the *Roman de Renart* that I have used is by D.D.R. Owen, *The Romance of Reynard the Fox* (Oxford: Oxford University Press, 1994).

307 Kenneth Varty, *Reynard the Fox: A Study of the Fox in Medieval English Art* (Leicester: Leicester University Press, 1967), pp. 43–50 and plate 124–44; Kenneth Varty, "Further Examples of the Fox in Medieval English Art," in *Aspects of the Medieval Animal Epic: Proceedings of the International Conference, Louvain, May 15–17, 1972*, eds. E. Rombauts and A. Welkenhuysen (Louvain and the Hague: Leuven University Press and Martinus Nijhoff, 1975), pp. 251–56.

308 Cohen, "Law, Folklore and Animal Lore," pp. 15–37.

309 Léon Ménabréa, *De l'Origine de la Forme et de l'Esprit des Jugements Rendus au Moyen-Age contre les Animaux* (Chambéry: Puthod, 1846), pp. 153–57. A very similar argument was made by the thirteenth-century French jurist, Philippe de Beaumanoir, in his condemnation of animal trials in the *Coutumes de Beauvaisis* (The Legal Customs of the Beauvaisis): See Philippe de Beaumanoir, *Coutumes de Beauvaisis*, ed. A. Salmon, 3 vols. (Paris: Alphonse Picard, 1899–1900), 2:481.

310 Ménabréa, *De l'Origine de la Forme*, pp. 157–60.

311 For example, I have personally visited L'Insectarium at Montreal, Quebec, and the Butterfly Conservatory in Niagra Falls, Ontario, while nearby where I live is the "Spider Web Farm" in Williamstown, Vermont.

312 One exception was the case of Protestant heretics tried in 1540 at Arles before the Parlement of Provence, presided over by Bartholomé Chassenée, who was reminded by a member of the tribunal of his earlier defense of the rats of Autun.

313 Some of the methods of execution of animals, however, such as by hanging them upside down by their hindquarters, were also used against Jews, indicating that in this case, animals and Jews were viewed as virtually one and the same.

314 Evans, *Criminal Prosecution and Capital Punishment of Animals*, pp. 287–306; Cohen, "Law, Folklore and Animal Lore," pp. 28–31.

315 Finkelstein, "The Ox that Gored," pp. 70–71.

316 Salisbury, *The Beast Within*, pp. 85–86; Joyce E. Salisbury, "Bestiality in the Middle Ages," in *Sex in the Middle Ages: A Book of Essays*, ed. Joyce E. Salisbury (New York and London: Garland, 1991), p. 173–74; Maureen A. Tilley, "Martyrs, Monks, Insects, and Animals," in *The Medieval World of Nature: A Book of Essays*, ed. Joyce E. Salisbury (New York and London: Garland Publishing, 1993), pp. 96–97.

317 Medieval romances on Alexander's life explain that Olympias was seduced by the king of Egypt, Neptenabus, who cast a magic spell on her that caused her to dream of being impregnated by a "dragon" and appeared to her in this guise. See David Salter, *Holy and Noble Beasts: Encounters with Animals in Medieval Literature* (Woodbridge, Suffolk: D.S. Brewer, 2001), pp. 123–33.

318 Salisbury, *The Beast Within*, p. 85.

319 Salisbury, *The Beast Within*, pp. 86–94; Salisbury, "Bestiality in the Middle Ages," pp. 177–79.

320 St. Thomas Aquinas, *Summa Theologiae*, II.II, question 154, article 12.

321 *Earliest Norwegian Laws*, p. 252.

322 Salisbury, *The Beast Within*, pp. 93–100; Salisbury, "Bestiality in the Middle Ages," pp. 179–80, 182–83; Evans, *Criminal Prosecution and Capital Punishment of Animals*, p. 147.

323 Benedict Carpov, *Practica Nova Rerum Criminalium*, (Lipsia, 1669), p. 394; Evans, *Criminal Prosecution and Capital Punishment of Animals*, p. 151.

324 Salisbury, "Bestiality in the Middle Ages," p. 179; Salisbury, *The Beast Within*, pp. 100–101.

325 Salisbury, *The Beast Within*, pp. 95–96; Salisbury, "Bestiality in the Middle Ages," pp. 180–81; Gerald of Wales, *The first version of the Topography of Ireland*, trans. John J. O'Meara (Dundalk: Dundalgan Press, 1951), pp. 56–59.

326 *Witchcraft in Europe, 400–1700: A Documentary History*, eds. Alan Charles Kors and Edward Peters, 2nd edn. (Philadelphia: University of Pennsylvania Press, 2001), pp. 193–99.
327 Salisbury, *The Beast Within*, pp. 96–99; Salisbury, "Bestiality in the Middle Ages," pp. 181–82.
328 For examples of such trials, see: Richard Kieckhefer, *European Witch Trials: Their Foundations in Popular and Learned Culture, 1300–1500* (Berkeley and Los Angeles: University of California Press, 1976), pp. 25, 32–33; *Witchcraft in Europe*, pp. 115–16, 160, 165.
329 Suetonius, *The Twelve Caesars*, rev. edn., trans. Robert Graves and Michael Grant (Harmondsworth, Middlesex: Penguin Books, 1979), pp. 49–50.
330 *Witchcraft in Europe*, pp. 53, 65–66, 69.
331 Dominic Alexander, *Saints and Animals in the Middle Ages* (Woodbridge, Suffolk: Baydell Press, 2008), Salter, *Holy and Noble Beasts*; Helen Waddell, *Beasts and Saints* (London: Constable, 1934).
332 Albert the Great believed a wolf's penis could both incite and inhibit intercourse: See Albertus Magnus, *On Animals*; 1519–20; Pluskowski, *Wolves and the Wilderness*, p. 116. According to unmaking rituals in medieval hunting manuals, a stag's testicles and penis were among the first organs to be eviscerated after a hunt, but this would presumably be done only after the deer had already been killed.
333 Richard Kieckhefer, "Erotic Magic in Medieval Europe," in *Sex in the Middle Ages: A Book of Essays*, ed. Joyce E. Salisbury (New York and London: Garland Publishing, 1991), pp. 32–37, 43; Stephen Wilson, *The Magical Universe: Everyday Ritual and Magic in Pre-Modern Europe* (London and New York: Hambledon and London, 2000), p. 416; Richard Kieckhefer, *Forbidden Rites: A Necromancer's Manual of the Fifteenth Century* (University Park, PA: Pennsylvania State University Press, 1997), pp. 82, 85–86.
334 Kieckhefer, "Erotic Magic in Medieval Europe," p. 30, 34, 37–38, 41; Kieckhefer, *European Witch Trials*, p. 57–58; Kieckhefer, *Forbidden Rites*, pp. 48–49, 54, 74; *Witchcraft in Europe*, pp. 123, 132, 161; *The Sorcery Trial of Alice Kyteler*, eds. L.S. Davidson and J.O. Ward (Binghamton, N.Y.: Medieval and Renaissance Texts and Studies, 1993), pp. 62–63.
335 *Witchcraft in Europe*, pp. 157–58, 190.
336 Kieckhefer, *European Witch Trials*, pp. 50, 55, 63.
337 Michael D. Bailey, *Magic and Superstition in Europe: A Concise History from Antiquity to the Present* (Lanham, MD: Rowman and Littlefield, 2007), p. 166; Wilson, *Magical Universe*, p. 417.
338 Brian P. Levack, *The Witch-Hunt in Early Modern Europe*, 3rd edn. (Harlow, U.K.: Pearson/Longman, 2006), p. 93.
339 *Witchcraft in Europe*, p. 71.
340 Kieckhefer, *European Witch Trials*, pp. 71–72.
341 *Witchcraft in Europe*, pp. 102, 109, 136.
342 *Witchcraft in Europe*, pp. 158, 168–69, 191.
343 Hyland, *The Medieval Warhorse*; Andrew Ayton, *Knights and Warhorses: Military Service and the English Aristocracy under Edward III* (Woodbridge, Suffolk: Boydell Press, 1994), p. 194–96; Davis, *The Medieval Warhorse*.

### Afterword

1 For a facsimile reproduction of the *Roman de Fauvel*, see *Le Roman de Fauvel: In the Edition of Mesire Chaillou de Pesstain*, eds. Edward H. Roesner, François Avril, and Nancy Freeman Regalado (New York: Broude Brothers, 1990).

# Bibliography

## General Works

Aberth, John. *From the Brink of the Apocalypse:* Confronting Famine, War, Plague, and Death in the *Later Middle Ages*, 2nd edn. London and New York: Routledge, 2010.

Baillie, M.G.L. "Putting Abrupt Environmental Change back into Human History." In *Environments and Historical Change: The Linacre Lectures (1998)*, edited by Paul Slack, 46–75. Oxford: Oxford University Press, 1999.

Barbour, Ian G. *Western Man and Environmental Ethics: Attitudes Toward Nature and Technology*. Reading, MA: Addison-Wesley Publishing Co., 1973.

Bilsky, Lester J., ed. *Historical Ecology: Essays on Environment and Social Change*. Port Washington, N.Y.: Kennikat Press, 1980.

Bjork, Robert E., ed. *The Oxford Dictionary of the Middle Ages*. 4 vols. Oxford: Oxford University Press, 2010.

Botkin, Daniel. *Discordant Harmonies: A New Ecology for the Twenty-First Century*. New York: Oxford University Press, 1990.

Bowlus, Charles R. "Ecological Crises in Fourteenth Century Europe." In *Historical Ecology: Essays on Environment and Social Change*, edited by Lester J. Bilsky, 86–99. Port Washington, N.Y.: Kennikat Press, 1980.

Britnell, Richard. *Britain and Ireland, 1050–1530: Economy and Society*. Oxford: Oxford University Press, 2004.

Bynum, Caroline Walker, and Paul Freedman, eds. *Last Things: Death and the Apocalypse in the Middle Ages*. Philadelphia: University of Pennsylvania Press, 2000.

Chenu, M.-D. *Nature, Man, and Society in the Twelfth Century*: Essay on New Theological Perspectives in the *Latin West*, edited and translated by Jerome Taylor and Lester K. Little. Chicago: University of Chicago Press, 1968.

Collingwood, R.G. *The Idea of Nature*. Oxford: Clarendon Press, 1945.

De Boe, Guy, and Frans Verhaeghe, eds. *Environment and Subsistence in Medieval Europe*. Zellik: I.A.P. Rapporten, 9, 1997.

Dubos, René. *A God Within*. New York: Charles Scribner's Sons, 1972.

——"A Theology of Earth." In *Western Man and* Environmental Ethics: Attitudes Toward Nature and *Technology*, edited by Ian G. Barbour, 43–54. Reading, MA: Addison-Wesley Publishing Co., 1973.

——"Franciscan Conservation versus Benedictine Stewardship." In *Ecology and Religion in History*, edited by David and Eileen Spring, 114–36. New York: Harper and Row, 1974.

Duby, Georges. *Rural Economy and Country Life in the Medieval West*, translated by Cynthia Postan. London: Edward Arnold, 1968.

Dyer, Christopher. *An Age of Transitions? Economy and Society in England in the Later Middle Ages*. Oxford: Clarendon Press, 2005.
——*Everyday Life in Medieval England*. London: Hambledon Press, 1994.
——*Making a Living in the Middle Ages: The People of Britain, 850–1520*. New Haven, CT: Yale University Press, 2002.
——*Standards of Living in the Middle Ages: Social Change in England, c. 1200–1520*. Cambridge: Cambridge University Press, 1989.
Fraser, Veronica. "The Goddess Natura in the Occitan Lyric." In *The Medieval World of Nature: A Book of Essays*, edited by Joyce E. Salisbury, 129–44. New York and London: Garland Publishing, 1993.
Glotfelty, Cheryll, and Harold Fromm, eds. *The Ecocriticism Reader: Landmarks in Literary Ecology*. Athens, GA: The University of Georgia Press, 1996.
Green, Miranda J. *Dictionary of Celtic Myth and Legend*. London: Thames and Hudson, 1992.
Gunn, J.D., ed. *The Years without Summer: Tracing A.D. 536 and Its Aftermath*. Oxford: Archaeopress, 2000.
Hallam, H.E., ed. *The Agrarian History of England and Wales, volume 2: 1042–1350*. Cambridge: Cambridge University Press, 1988.
Hanawalt, Barbara A., and Lisa J. Kiser, eds. *Engaging* with Nature: Essays on the Natural World in Medieval *and Early Modern Europe*. Notre Dame, IN: University of Notre Dame Press, 2008.
Henschel, A.W. "Document zur Geschichte des schwarzen Todes". In *Archiv für die gesammte Medicin*, edited by Heinrich Haeser. 10 vols., 2:26–59. Jena, 1841–49.
Herlihy, David. "Attitudes Toward the Environment in Medieval Society." In *Historical Ecology: Essays on Environment and Social Change*, edited by Lester J. Bilsky, 100–116. Port Washington, N.Y.: Kennikat Press, 1980.
Hoffman, Richard C. "Homo et Natura, Homo in Natura: Ecological Perspectives on the European Middle Ages." In *Engaging with Nature: Essays on the Natural World in Medieval and Early Modern Europe*, edited by Barbara A. Hanawalt and Lisa J. Kiser, 11–38. Notre Dame, IN: University of Notre Dame Press, 2008.
Hughes, J. Donald. "Early Greek and Roman Environmentalists." In *Historical Ecology: Essays on Environment and Social Change*, edited by Lester J. Bilsky, 45–59. Port Washington, N.Y.: Kennikat Press, 1980.
Jolly, Karen. "Father God, Mother Earth: Nature- Mysticism in the Anglo-Saxon World." In *The Medieval World of Nature: A Book of Essays*, edited by Joyce E. Salisbury, 211–52. New York and London: Garland Publishing, 1993.
Klyza, Christopher McGrory, and Stephen C. Trombulak. *The Story of Vermont: A Natural and Cultural History*. Hanover, N.H.: University Press of New England, 1999.
Miller, Edward, ed. *Agrarian History of England and Wales. Volume 3: 1348–1500*. Cambridge: Cambridge University Press, 1991.
Niavis, Paulus. *Judicum Jovis*. Leipzig: Martin Landsberg, c. 1492–95.
Roesner, Edward H., and François Avril, and Nancy Freeman Regalado, eds. *Le Roman de Fauvel: In the Edition of Mesire Chaillou de Pesstain*. New York: Broude Brothers, 1990.
Salisbury, Joyce E., ed. *The Medieval World of Nature: A Book of Essays*. New York and London: Garland Publishing, 1993.
Schofield, P.R. "Medieval Diet and Demography." In *Food in Medieval England: Diet and Nutrition*, edited by C.M. Woolgar, D. Serjeantson, and T. Waldron, 239–53. Oxford: Oxford University Press, 2006.
Slack, Paul, ed. *Environments and Historical Change: The Linacre Lectures (1998)*. Oxford: Oxford University Press, 1999.

Smoller, Laura A. "Of Earthquakes, Hail, Frogs, and Geography: Plague and the Investigation of the Apocalypse." In *Last Things: Death and the Apocalypse in the Middle Ages*, edited by Caroline Walker Bynum and Paul Freedman, 156–87. Philadelphia: University of Pennsylvania Press, 2000.
Spring, David and Eileen, eds. *Ecology and Religion in History*. New York: Harper and Row, 1974.
Strayer, Joseph R., ed. *Dictionary of the Middle Ages*. 13 vols. New York: Scribner, 1982–89.
Thomas, Keith. *Man and the Natural World: Changing Attitudes in England, 1500–1800*. Oxford: Oxford University Press, 1983.
White, Jr., Lynn. "Continuing the Conversation." In
Western Man and Environmental Ethics: *Attitudes Toward Nature and Technology*, edited by Ian G. Barbour, 55-64. Reading, MA: Addison-Wesley Publishing Co., 1973.
———"The Historical Roots of our Ecologic Crisis." *Science*, 155 (1967): 1203–7. Also reprinted in *The Ecocriticism Reader: Landmarks in Literary Ecology*, edited by Cheryll Glotfelty and Harold Fromm, 3–14. Athens, GA: The University of Georgia Press, 1996.
Woolgar, C.M., D. Serjeantson, and T. Waldron, eds. *Food in Medieval England: Diet and Nutrition*. Oxford: Oxford University Press, 2006.

## Part I

Aberth, John. *The Black Death: The Great Mortality of 1348–1350. A Brief History with Documents*. Boston and New York: Bedford/St. Martin's Press, 2005.
Agramont, Jacme d'. "Regiment de Preservacio a Epidimia o Pestilencia e Mortaldats," translated by M.L. Duran-Reynals and C.-E.A. Winslow. *Bulletin of the History of Medicine* 23 (1949): 57–89.
Albala, Ken. *Eating Right in the Renaissance*. Berkeley and Los Angeles: University of California Press, 2002.
Albert the Great. *On Animals: A Medieval* Summa Zoologica, translated by Kenneth F. Kitchell, Jr. and Irven Michael Resnick. 2 vols. Baltimore: Johns Hopkins University Press, 1999.
———*On the Causes of the Properties of the Elements*, translated by Irven M. Resnick. Milwaukee, WI: Marquette University Press, 2010.
Alexandre, Pierre. *Le Climat en Europe au Moyen Age*. L'Ecole des Hautes Etudes en Sciences Sociales, *Recherches d'Histoire et de Sciences Sociales*, 24, 1987.
Anglicus, Bartholomaeus. *On the Properties of Things: John Trevisa's Translation: A Critical Text*. 2 vols. Oxford: Clarendon Press, 1975.
Armstrong, Edward A. *Saint Francis: Nature Mystic: The* Derivation and Significance of the Nature Stories in *the Franciscan Legend*. Berkeley and Los Angeles: University of California Press, 1973.
Arrizabalaga, Jon. "Facing the Black Death: Perceptions and Reactions of University Medical Practitioners." In *Practical Medicine from Salerno to the Black Death*, edited by L. García-Ballester, R. French, J. Arrizabalaga, and A. Cunningham, 237–88. Cambridge:
Cambridge University Press, 1994.
Assman, Erwin, ed. *Godeschalcus und Visio Godeschalci*. Neumünster: Karl Wachholtz, 1979.
Astill, Grenville, and Annie Grant, eds. *The Countryside of Medieval England*. Oxford: Basil Blackwell, 1988.
———and Annie Grant. "Efficiency, Progress and Change." In *The Countryside of Medieval England*, edited by Grenville Astill and Annie Grant, 213–33. Oxford: Basil Blackwell, 1988.
———and John Langdon, eds. *Medieval Farming and* Technology: The Impact of Agricultural Change in *Northwest Europe*. Leiden: Brill, 1997.

Attfield, R. "Christian Attitudes to Nature." *Journal of the History of Ideas* 44 (1983): 369–86.

Avicenna. *The Canon of Medicine: First Book*, translated by O. Cameron Gruner. New York: Augustus M. Kelley Publishers, 1970.

——*Liber Canonis*. Hildesheim: Georg Olms, 1964.

Bachrach, Bernard S., and David Nicholas, eds. *Law, Custom, and the Social Fabric in Medieval Europe: Essays in Honor of Bryce Lyon*. Studies in Medieval Culture, 28, 1990.

Barney, A., W.J. Lewis, J.A. Beach, and Oliver Berghof, eds. *The Etymologies of Isidore of Seville*. Cambridge: Cambridge University Press, 2006.

Bassett, Steven, ed. *Death in Towns: Urban Responses to the Dying and the Dead, 100–1600*. London and New York: Leicester University Press, 1992.

Beauvais, Vincent of. *Speculum Quadruplex, sive Speculum Maius*. 4 vols. Douai: Balthazar Belleri, 1624.

Beck, Patrice, Philippe Braunstein, and Michel Philippe. "Wood, Iron, and Water in the Othe Forest in the Late Middle Ages: New Findings and Perspectives." In *Technology and Resource Use in Medieval Europe:
Cathedrals, Mills, and Mines*, edited by Elizabeth
Bradford Smith and Michael Wolfe, 173–84. Aldershot, UK and Brookfield, VT: Ashgate, 1997.

Bede, *A History of the English Church and People*, edited by Leo Sherley-Price. Harmondsworth, Middlesex: Penguin Books, 1955.

Behringer, Wolfgang. "Weather, Hunger and Fear." *German History* 13 (1995): 1–27.

Belloni, Luigi, ed. *I Trattati in Volgare della Peste e dell'acqua Ardente di Michele Savonarola*. Rome: Congresso Nazionale della Società Italiana de Medicina Interna, 1953.

Benedictow, Ole. *The Black Death, 1346–1353: The Complete History*. Woodbridge, Suffolk: Boydell Press, 2004.

Benoit, Paul, and Josephine Rouillard. "Medieval Hydraulics in France." In *Working with Water in Medieval Europe: Technology and Resource-Use*, edited by Paolo Squatriti, 161–216. Leiden: Brill, 2000.

Beresford, G. "Climatic Change and its Effect upon the Settlement and Desertion of Medieval Villages in Britain." In *Consequences of Climatic Change*, edited by C. Delano Smith and M. Parry, 30–39. Nottingham: University of Nottingham Press, 1981.

Brown, Neville. "Approaching the Medieval Optimum, 212 to 1000 A.D." In *Water, Environment and Society in Times of Climatic Change*, edited by Arie S. Issar and Neville Brown, 69–95. Dordrecht: Kluwer Academic Publishers, 1998.

——*History and Climate Change: A Eurocentric Perspective*. London and New York: Routledge, 2001.

Brunner, Karl. "Continuity and Discontinuity of Roman Agricultural Knowledge in the Early Middle Ages." In Agriculture in the Middle Ages: Technology, Practice, *and Representation*, edited by Del Sweeney, 21–40. Philadelphia: University of Pennsylvania Press, 1995.

Buckland, Paul C. "The North Atlantic Environment." In *Vikings: The North Atlantic Saga*, edited by William W. Fitzhugh and Elisabeth I. Ward, 146–53. Washington D.C.: Smithsonian Institution Press, 2000.

Campbell, Bruce M.S. "Economic Rent and the Intensification of English Agriculture, 1086–1350." In *Medieval Farming and Technology: The Impact of Agricultural Change in Northwest Europe*, edited by Grenville Astill and John Langdon, 225–50. Leiden: Brill, 1997.

Cantimpré, Thomas of. *Liber de Natura Rerum. Volume 1: Text*, edited by Helmut Boese. Berlin: Walter de Gruyter, 1973.

Carmichael, Ann G. *Plague and the Poor in Renaissance Florence*. Cambridge: Cambridge University Press, 1986.

Cipolla, Carlo M. *Public Health and the Medical Profession in the Renaissance*. Cambridge: Cambridge University Press, 1976.
Cohen, Jeremy. *"Be Fertile and Increase, Fill the Earth* and Master It": The Ancient and Medieval Career of a *Biblical Text*. Ithaca, N.Y.: Cornell University Press, 1989.
Cohn, Jr., Samuel K. *The Black Death Transformed: Disease and Culture in Early Renaissance Europe*. London and New York: Arnold and Oxford University Press, 2003.
Conches, William of. *Philosophia Mundi*, edited by Gregor Maurach. Pretoria: University of South Africa Press, 1974.
Cosman, Madeleine Pelner. *Fabulous Feasts: Medieval Cookery and Ceremony*. New York: George Braziller, 1976.
Crescenzi, Pietro de. *Liber ruralia commodorum*. Augsburg: Johann Schüssler, 1471.
Crossley, David. "The Archaeology of Water Power in Britain before the Industrial Revolution." In *Technology and Resources, and Mines*, edited by Elizabeth Bradford Smith and Michael Wolfe, 109–24. Aldershot, UK and Brookfield, VT: Ashgate, 1997.
De Backer, C. "Maatregelen Tegen de Pest te Diest in de Vijftiende en Zestiende Eeuw." *Verhandelingen-Koninklijke Academie voor Geneeskunde van Belgie* 61 (1999): 273–300.
Derr, Thomas Sieger. "Religion's Responsibility for the Ecological Crisis: An Argument Run Amok." *Worldview* 18 (1975): 39–45.
Dinānah, Taha. "Die Schrift von Abi G'far Ahmed ibn 'Ali ibn Mohammed ibn 'Ali ibn Hatimah aus Almeriah über die Pest." *Archiv für Geschichte der Medizin* 19 (1927): 27–81.
Dutton, Paul Edward. *Carolingian Civilization: A Reader*. Peterborough, Ont.: Broadview Press, 1993.
——*Charlemagne's Mustache and Other Cultural Clusters of a Dark Age*. New York and Basingstoke, Hampshire: Palgrave Macmillan, 2004.
——"Thunder and Hail over the Carolingian Countryside." In *Agriculture in the Middle Ages: Technology, Practice, and Representation*, edited by Del Sweeney. Philadelphia: University of Pennsylvania Press, 1995.
Dyer, Christopher. "Documentary Evidence: Problems and Enquiries." In *The Countryside of Medieval England*, edited by Grenville Astill and Annie Grant, 12–35. Oxford: Basil Blackwell, 1988.
——"Medieval Farming and Technology: Conclusion." In Medieval Farming and Technology: The Impact of *Agricultural Change in Northwest Europe*, edited by Grenville Astill and John Langdon, 293–312. Leiden: Brill, 1997.
Fagan, Brian. *The Great Warming: Climate Change and the Rise and Fall of Civilizations*. New York: Bloomsbury Press, 2008.
——*The Little Ice Age: How Climate Made History, 1300-1850*. New York: Basic Books, 2000.
Ficino, Marsilio. *Contro alla Peste*. Florence: Appresso I Giunti, 1576.
Filotas, Bernadette. *Pagan Survivals, Superstitions and* Popular Cultures in Early Medieval Pastoral *Literature*. Toronto: Pontifical Institute of Mediaeval Studies, 2005.
Fitzhugh, William W., and Elisabeth I. Ward, eds. *Vikings: The North Atlantic Saga*. Washington D.C.: Smithsonian Institution Press, 2000.
Flohn, Hermann, and Roberto Fantechi, eds. *The Climate of* Europe: *Past, Present and Future. Natural and Man-Induced Climatic Changes: A European Perspective*. Dordrecht, Holland: D. Reidel Publishing, 1984.
Foligno, Gentile da. *Consilium contra Pestilentiam*. Colle di Valdelsa, c. 1479.
French, Roger, *Canonical Medicine: Gentile da Foligno and Scholasticism*. Leiden: Brill, 2001.

Frenzel, Burkhard, Christian Pfister, and Birgit Gläser, eds. *European Climate Reconstructed from Documentary Data: Methods and Results*. Akademie der Wissenschaften und der Literatur, Paleoclimate Research, 7, 1992.

García-Ballester, L., R. French, J. Arrizabalaga, and A. Cunningham, eds. *Practical Medicine from Salerno to the Black Death*. Cambridge: Cambridge University Press, 1994.

Gari, Lutfallah, "Arabic Treatises on Environmental Pollution up to the End of the Thirteenth Century." *Environment and History* 8 (2002): 475–88.

Gibbs, Frederick W. "Medical Understandings of Poison circa 1250–1600." Ph.D. dissertation, University of Wisconsin-Madison, 2009.

Gimpel, Jean. *The Medieval Machine: The Industrial Revolution of the Middle Ages*. New York: Holt, Rinehart and Winston, 1976.

Glick, Thomas F., and Helena Kirchner. "Hydraulic Systems and Technologies of Islamic Spain: History and Archaeology." In *Working with Water in Medieval Europe: Technology and Resource-Use*, ed. Paolo Squatriti, 267–330. Leiden: Brill, 2000.

Goldstein, B.R., and D. Pingree. *Levi ben Gerson's Prognostication for the Conjunction of 1345*. Transactions of the American Philosophical Society, 80, 1990.

Gottschall, Dagmar. "Conrad of Megenberg and the Causes of the Plague: A Latin Treatise on the Black Death Composed ca. 1350 for the Papal Court in Avignon." In *La Vie Culturelle, Intellectuelle et Scientifique a la Cour des Papes d'Avignon*, edited by Jacqueline Hamesse, 319–32. Brepols, 2006.

Grant, Annie. "Animal Resources." In *The Countryside of Medieval England*, edited by Grenville Astill and Annie Grant, 149–85. Oxford: Basil Blackwell, 1988.

Grewe, Klaus. "Water Technology in Medieval Germany." In Working with Water in Medieval Europe: Technology and *Resource-Use*, edited by Paolo Squatriti, 129–60. Leiden: Brill, 2000.

Grove, A.T., and Oliver Rackham. *The Nature of Mediterranean Europe: An Ecological History*. New Haven: Yale University Press, 2001.

Grove, Jean M. *Little Ice Ages: Ancient and Modern, Volume 1*. 2nd edn. London and New York: Routledge, 2004.

Guilleré, Christian. "La Peste Noire a Gérone (1348)." *Annals de Institut de'Estudis Gironins* 27 (1984): 87-161.

Guillerme, André E. *The Age of Water: The Urban Environment in the North of France, A.D. 300–1800*. College Station, TX: Texas A&M University Press, 1988.

Hamesse, Jacqueline, ed. *La Vie Culturelle, Intellectuelle et Scientifique à la Cour des Papes d'Avignon*. Brepols, 2006.

Hansen, Joseph. *Quellen und Untersuchungen zur Geschichte der Hexenwahns und der Hexenverfolgung im Mittelalter*. Bonn, 1901.

Hatcher, John. *The Black Death: A Personal History*. Boston: Da Capo Press, 2008.

——— *English Tin Production and Trade before 1550*. Oxford: Clarendon Press, 1973.

Henderson, John. "The Black Death in Florence: Medical and Communal Responses." In *Death in Towns: Urban Responses to the Dying and the Dead, 100–1600*, edited by Steven Bassett, 136–50. London and New York: Leicester University Press, 1992.

Hill, Thomas D. "The Æcerbot Charm and its Christian User." *Anglo-Saxon England* 6 (1977): 213–21.

Hillenbrand, Carole. *The Crusades: Islamic Perspectives*. New York: Routledge, 2000.

Hoffman, Richard C. "Economic Development and Aquatic Ecosystems in Medieval Europe." *American Historical Review* 101 (1996): 630–69.

Holt, Richard. "Mechanization in Medieval England." In Technology and Resource Use in Medieval Europe: *Cathedrals, Mills, and Mines*, edited by Elizabeth Bradford Smith and Michael Wolfe, 139–57. Aldershot, UK and Brookfield, VT: Ashgate, 1997.

——"Medieval England's Water-Related Technologies." In *Working with Water in Medieval Europe: Technology and Resource-Use*, edited by Paolo Squatriti, 51–100. Leiden: Brill, 2000.

——*The Mills of Medieval England*. Oxford: Basil Blackwell, 1988.

Horrox, Rosemary, trans. and ed. *The Black Death*. Manchester: Manchester University Press, 1994.

Huntingdon, Henry of. *The History of the English*, edited by Thomas Arnold. London: Longman and Co., 1879.

Issar, Arie S., and Neville Brown, eds. *Water, Environment and Society in Times of Climatic Change*. Dordrecht: Kluwer Academic Publishers, 1998.

Jolly, Karen. "Father God, Mother Earth: Nature-Mysticism in the Anglo-Saxon World." In *The Medieval World of Nature: A Book of Essays*, edited by Joyce E. Salisbury, 211–52. New York and London: Garland Publishing, 1993.

Jordan, W.C. *The Great Famine: Northern Europe in the Early Fourteenth Century*. Princeton: Princeton University Press, 1996.

Kershaw, Ian. "The Great Famine and Agrarian Crisis in England, 1315–22." *Past and Present* 59 (1973): 3–50.

Kieckhefer, Richard. *European Witch Trials: Their Foundations in Popular and Learned Culture, 1300–1500*. Berkeley and Los Angeles: University of California Press, 1976.

——*Forbidden Rites: A Necromancer's Manual of the Fifteenth Century*. University Park, PA: Pennsylvania State University Press, 1997.

——*Magic in the Middle Ages*. Cambridge: Cambridge University Press, 1989.

Kors, Alan Charles, and Edward Peters, eds. *Witchcraft in Europe, 400–1700: A Documentary History*. 2nd edn. Philadelphia: University of Pennsylvania Press, 2001.

Krüger, Sabine. "Krise der Zeit als Ursache der Pest? Der Traktat *De mortalitate in Alamannia* des Konrad von Megenberg." In *Festschrift für Hermann Heimpel zum 70. Geburtstag am 19. September 1971*, 839–83. Göttingen: Vandenhoeck and Ruprecht, 1971.

Lamb, H.H. "An Approach to the Study of the Development of Climate and Its Impact in Human Affairs." In *Climate and History: Studies in Past Climates and Their Impact on Man*, edited by T.M.L. Wigley, M.J. Ingram, and G. Farmer, 291–309. Cambridge: Cambridge University Press, 1981.

——"Climate in the Last Thousand Years." In *The Climate of Europe: Past, Present and Future: Natural and Man-Induced Climatic Changes: A European Perspective*, edited by Hermann Flohn and Roberto Fantechi, 38–53. Dordrecht: D. Reidel Publishing, 1984.

——*Climate, History and the Modern World*. 2nd edn. London and New York: Routledge, 1995.

——*Climate: Present, Past and Future*. 2 vols. London: Methuen and Co., 1977.

Langdon, John. "Agricultural Equipment." In *The Countryside of Medieval England*, edited by Grenville Astill and Annie Grant, 86–107. Oxford: Basil Blackwell, 1988.

——*Horses, Oxen and Technological Innovation: The Use of Draught Animals in English Farming from 1066 to 1500*. Cambridge: Cambridge University Press, 1986.

Le Roy Ladurie, Emmanuel. *Times of Feast, Times of Famine: A History of Climate since the Year 1000*, translated by Barbara Bray. Garden City, N.Y.: Doubleday and Co., 1971.

Lentacker, An, Wim Van Neer, and Jean Plumier. "Historical and Archaeozoological Data on Water Management and Fishing during Medieval and Post-Medieval Times at Namur (Belgium)." In *Environment and Subsistence in Medieval Europe*, edited by Guy de Boe and Frans Verhaeghe, 83–94. Zellik: I.A.P. Rapporten, 9, (1997).

Lille, Alain de. *The Complaint of Nature*, translated by Douglas M. Moffat. New York: Henry Holt and Company, 1908.
McGovern, Thomas H. "The Demise of Norse Greenland." In *Vikings: The North Atlantic Saga*, edited by William W. Fitzhugh and Elisabeth I. Ward, 327–39. Washington D.C.: Smithsonian Institution Press, 2000.
MacMullen, Ramsey. *Christianity and Paganism in the Fourth to Eighth Centuries*. New Haven, CT: Yale University Press, 1997.
Magnusson, Roberta, and Paolo Squatriti. "The Technologies of Water in Medieval Italy." In *Working with Water in Medieval Europe: Technology and Resource-Use*, edited by Paolo Squatriti, 217–66. Leiden: Brill, 2000.
Maineri, Maino de. *Libellus de Preservation ab Epydimia*, edited by R. Simonini. Modena, 1923.
———*Regimen Sanitatis*. 3 vols., edited by Séamus Ó Ceithearnaigh. Baile Átha Cliath, 1942–44.
Mate, Mavis. "Agricultural Technology in Southeast England, 1348–1530." In *Medieval Farming and Technology: The Impact of Agricultural Change in Northwest Europe*, edited by Grenville Astill and John Langdon, 251–74. Leiden: Brill, 1997.
Maurus, Hrabanus. *De Universo: The Peculiar Properties of Words and the Mystical Significance. The Complete English Translation*, translated by Priscilla Throop. 2 vols. Charlotte, VT: MedievalMS, 2009.
Megenberg, Konrad von. *Buch der Natur, vol. 2: Critical Text*, edited by Robert Luff and George Steer. Tübingen: Max Niemeyer Verlag, 2003.
Mengus, Hieronymus. *Flagellum Daemonum*. Bologna: Johannes Rossium, 1578.
Merle, William. *Consideraciones Temperiei pro 7 Annis*, translated by G.J. Symons. London: Edward Stanford, 1891.
Michon, L.-A. Joseph, ed. *Documents Inédits sur la Grande Peste de 1348*. Paris, 1860.
Muendel, John. "Mills in the Florentine Countryside." In *Pathways to Medieval Peasants*, edited by J.A. Raftis. Toronto: Pontifical Institute of Mediaeval Studies, 1981.
Myrdal, Janken. "The Agricultural Transformation of Sweden, 1000–1300." In *Medieval Farming and Technology: The Impact of Agricultural Change in Northwest Europe*, edited by Grenville Astill and John Langdon, 147–71. Leiden: Brill, 1997.
Neckam, Alexander. *De Naturis Rerum*, edited by Thomas Wright. London: Rolls Series, 34, 1863.
Parry, Joseph Henry. *Registrum Roberti Mascall, episcopi Herefordensis, A.D. 1404–1416*. Canterbury and York Society, 21, 1917.
Perrato, Pietro, ed. *Consiglio contro a Pistolenza per Maestro Tommaso del Garbo*. Bologna: Presso Gaetano Romagnoli, 1866.
Peters, Edward. *Torture*. Exp. edn. Philadelphia: University of Pennsylvania Press, 1996.
Pinkhof, H. *Abraham Kashlari, over Pestachtige Koorsten*. Amsterdam, 1891.
Postan, M.M., and E.E. Rich, eds. *The Cambridge Economic History of Europe, volume II: Trade and Industry in the Middle Ages*. Cambridge: Cambridge University Press, 1952.
Poulsen, Bjørn. "Agricultural Technology in Medieval Denmark." In *Medieval Farming and Technology: The Impact of Agricultural Change in Northwest Europe*, edited by Grenville Astill and John Langdon, 115–45. Leiden: Brill, 1997.
Rackham, Oliver. *The History of the Countryside*. London: J.M. Dent and Sons, 1986.
Raftis, J.A., ed. *Pathways to Medieval Peasants*. Toronto: Pontifical Institute of Mediaeval Studies, 1981.
Rébouis, H. Émile. *Étude Historique et Critique sur la Peste*. Paris: Alphonse Picard, 1888.
Riché, Pierre. *Daily Life in the World of Charlemagne*, translated by Jo Ann McNamara. Philadelphia: University of Pennsylvania Press, 1978.

Salisbury, Joyce E. *The Medieval World of Nature: A Book of Essays*. New York and London: Garland Publishing, 1993.
Salvesen, Helge. "The Climate as a Factor of Historical Causation." In *European Climate Reconstructed from Documentary Data: Methods and Results*, edited by Burkhard Frenzel, Christian Pfister, and Birgit Gläser, 219–33. Akademie der Wissenschaften unde der Literatur, Paleoclimate Research, 7, 1992.
Sárraga, M.V. Amasuno. *La Peste en la Corona de Castilla Durante la Segunda Mitad del Siglo XIV*. Estudios de historia de la ciencia y de la técnica, no. 12, 1996.
Scott, S.P., ed. *The Visigothic Code*. Boston: Boston Book Co., 1910.
Silvester, Bernard. *Cosmographia*, edited by Peter Dronke. Leiden: E.J. Brill, 1978.
Simonini, R. "Il Codice di Mariano di Ser Jacopo sopra 'Rimedi Abili nel Tempo di Pestilenza'." *Bollettino dell'Istituto Storico Italiano dell'Arte Sanitaria* 9 (1929): 161–69.
Singer, Charles, trans. *The Fasciculus Medicinæ of Johannes de Ketham, Alemanus: Facsimile of First Edition of 1491*. Birmingham, Ala.: Classics of Medicine Library, 1988.
Slicher van Bath, B.H. *The Agrarian History of Western Europe, A.D. 500–1850*. London: Edward Arnold, 1963.
Smith, C. Delano, and M. Parry, eds. *Consequences of Climatic Change*. Nottingham: University of Nottingham Press, 1981.
Smith, Elizabeth Bradford, and Michael Wolfe, eds. Technology and Resource Use in Medieval Europe: *Cathedrals, Mills, and Mines*. Aldershot, UK and
Brookfield, VT: Ashgate, 1997.
Sorrell, Roger D. *St. Francis of Assisi and Nature: Tradition and Innovation in Western Christian Attitudes toward the Environment*. Oxford: Oxford University Press, 1988.
Squatriti, Paolo. "'Advent and Conquests' of the Water Mill in Italy." In *Technology and Resource Use in Medieval Europe: Cathedrals, Mills, and Mines*, edited by Elizabeth Bradford Smith and Michael Wolfe, 125–38. Aldershot, UK and Brookfield, VT: Ashgate, 1997.
———. "The Floods of 589 and Climate Change at the Beginning of the Middle Ages: An Italian Microhistory." *Speculum* 85 (2010): 799–826.
———. *Water and Society in Early Medieval Italy, A.D. 400-1000*. Cambridge: Cambridge University Press, 1998.
———, ed. *Working with Water in Medieval Europe: Technology and Resource-Use*. Leiden: Brill, 2000.
Stock, Brian. *Myth and Science in the Twelfth Century: A Study of Bernard Sylvester*. Princeton: Princeton University Press, 1972.
Sudhoff, Karl. "Ein anderer pestretetet." *Studien zur Geschichte der Medizin* 8 (1909): 193–99.
———. "Ein deutsches Pest-Regiment aus dem 14. Jahrhundert." *Archiv für Geschichte der Medizin* 2 (1908–9): 379–83.
———. "Ein Pestregimen aus dem Anfange des 15. Jahrhunderts." *AGM* 3 (1910): 407–8.
———. "Pestschriften aus den ersten 150 Jahren nach der Epidemie des 'schwarzen Todes' 1348." *Archiv für Geschichte der Medizin* 4 (1911):191–222, 389–424; 5 (1912):36–87, 332–96; 6 (1913):313–79; 7 (1913):57–114; 8 (1915):175–215, 236–89; 9 (1916):53–78, 117–67; 11 (1919):44–92, 121–76; 14 (1922–23):1–25, 79–105, 129–68; 16 (1924–25):1–69, 77–188; 17 (1925):12–139, 241–91.
———. "Ein weiteres deutsches Pest-Regiment aus dem 14. Jahrhundert und seine lateinische Vorlage, das Prager Sendschreiben 'Missum Imperatori' vom Jahre 1371." *AGM* 3 (1910): 144–53.
Sweeney, Del, ed. *Agriculture in the Middle Ages: Technology, Practice, and Representation*. Philadelphia: University of Pennsylvania Press, 1995.

Takács, Károly. "Medieval Hydraulic Systems in Hungary: Written Sources, Archaeology and Interpretation." In *People and Nature in Historical Perspective*, edited by József Laszlovszky and Péter Szabó, 289–311. Budapest: Central European University and Archaeolingua, 2003.

Tebrake, William H. "Ecology of Village Settlement in the Dutch Rijnland." In *Pathways to Medieval Peasants*, edited by J.A. Raftis, 2–26. Toronto: Pontifical Institute of Mediaeval Studies, 1981.

——— "Hydraulic Engineering in the Netherlands." In *Working with Water in Medieval Europe: Technology and Resource-Use*, edited by Paolo Squatriti, 101–27. Leiden: Brill, 2000.

——— *Medieval Frontier: Culture and Ecology in Rijnland*. College Station, TX: Texas A&M University Press, 1985.

Thoen, Erik. "The Birth of 'The Flemish Husbandry': Agricultural Technology in Medieval Flanders." In Medieval Farming and Technology: The Impact of *Agricultural Change in Northwest Europe*, edited by Grenville Astill and John Langdon, 68–88. Leiden: Brill, 1997.

Thorndike, Lynn. *A History of Magic and Experimental Science. Volumes 3 and 4: Fourteenth and Fifteenth Centuries*. New York: Columbia University Press, 1934.

Tilley, Maureen A. "Martyrs, Monks, Insects, and Animals." In *The Medieval World of Nature: A Book of Essays*, edited by Joyce E. Salisbury, 93–107. New York and London: Garland Publishing, 1993.

Toch, Michael. "Agricultural Progress and Agricultural Technology in Medieval Germany: An Alternative Model." In *Technology and Resource Use in Medieval Europe: Cathedrals, Mills, and Mines*, edited by Elizabeth Bradford Smith and Michael Wolfe, 158–69. Aldershot, UK and Brookfield, VT: Ashgate, 1997.

Toutain, J. *Les Cultes Païens dans l'Empire Romain*. 3 vols. Paris: Ernest Leroux, 1907–20.

Trexlar, Richard C. "Measures against Water Pollution in Fifteenth-Century Florence." *Viator* 5 (1974): 445–68.

Ullmann, Manfred. *Islamic Medicine*. Edinburgh:
University of Edinburgh Press, 1978.

Verhulst, Adriaan. "The 'Agricultural Revolution' of the Middle Ages Reconsidered." In *Law, Custom, and the* Social Fabric in Medieval Europe: Essays in Honor of *Bryce Lyon*, edited by Bernard S. Bachrach and David Nicholas, 17–28. Studies in Medieval Culture, 28, 1990.

Ward-Perkins, Bryan. *From Classical Antiquity to the Middle Ages: Urban Public Building in Northern and Central Italy, A.D. 300–850*. Oxford: Oxford
University Press, 1984.

Webb, J.F., trans. *Lives of the Saints*. Harmondsworth, Middlesex: Penguin Books, 1965.

White, Jr., Lynn. *Medieval Technology and Social Change*. Oxford: Clarendon Press, 1962.

Williamson, Tom. *Shaping Medieval Landscapes: Settlement, Society, Environment*. Macclesfield, Cheshire: Windgather Press, 2003.

Wilson, Stephen. *The Magical Universe: Everyday Ritual and Magic in Pre-Modern Europe*. London and New York: Hambledon and London, 2000.

Zunz, L. *Jubelschrift zum Neunzigsten Geburtstag*. Berlin,
1884.

## Part II

Albert the Great. *De Vegetabilibus*, edited by Karl Jessen. Berolini Reimer, 1867.

Anglicus, Bartholomaeus. *On the Properties of Things: John Trevisa's Translation: A Critical Text*. 2 vols. Oxford: Clarendon Press, 1975.

Astill, Grenville, and Annie Grant, eds. *The Countryside of Medieval England*. Oxford: Basil Blackwell, 1988.

Aston, Michael. *Interpreting the Landscape: Landscape Archaeology in Local Studies.* London: B.T. Batsford, 1985.

———, David Austin, and Christopher Dyer, eds. *The Rural Settlements of Medieval England: Studies Dedicated to Maurice Beresford and John Hurst.* Oxford: Basil Blackwell, 1989.

Attenborough, F.L., trans. and ed. *The Laws of the Earliest English Kings.* Cambridge: Cambridge University Press, 1992.

Aubin, Hermann. "Medieval Agrarian Society in its Prime The Lands East of the Elbe and German Colonization Eastwards." In *The Cambridge Economic History of Europe, volume I: The Agrarian Life of the Middle Ages,* edited by M.M. Postan, 449–86. Cambridge: Cambridge University Press, 1966.

Aubrun, Michel. "Droits d'usages forestiers et libertés paysannes (XIe-XIIIe siècle): Leur role dans la formation de la carte foncière." *Revue Historique* 280 (1988): 377–86.

Bannister, A.T., ed. *Registrum Ade de Orleton, episcopi Herefordensis, A.D. 1317–1327.* Canterbury and York Society, 1, 1907.

———, ed. *Registrum Johannis Stanbury, episcopi Herefordensis, A.D. 1453–1474.* Canterbury and York Society, 25, 1919.

Barney, Stephen A., W.J. Lewis, J.A. Beach, and Oliver Berghof, eds. *The Etymologies of Isidore of Seville.* Cambridge: Cambridge University Press, 2006.

Beauvais, Vincent of. *Speculum Quadruplex sive Speculum Maius.* 4 vols. Douai: Balthazar Belleri, 1624.

Bechmann, Roland. *Trees and Man: The Forest in the Middle Ages.* New York: Paragon House, 1990.

Bell, M. "Environmental Archaeology as an Index of Continuity and Change in the Medieval Landscape." In The Rural Settlements of Medieval England: Studies *Dedicated to Maurice Beresford and John Hurst,* edited by M.W. Beresford, John G. Hurst, Michael Aston, David Austin, and Christopher Dyer, 269–86. Oxford: Basil Blackwell, 1989.

Bellamy, John. *Robin Hood: An Historical Enquiry.* London: Croom Helm, 1985.

Beresford, Maurice, and John G. Hurst. *Deserted Medieval Villages.* London: Lutterworth Press, 1971.

Berman, Constance Hoffman. *Medieval Agriculture, the Southern French Countryside, and the Early Cistercians. A Study of Forty-three Monasteries.* Transactions of the American Philosophical Society, 76, 1986.

Birrell, Jean, ed. *Collections for a History of Staffordshire. The Forests of Cannock and Kinver: Select Documents, 1235–1372.* Staffordshire Record Society, 4th series, 18, 1999.

———"Common Rights in the Medieval Forest." *Past and Present* 117 (1987): 22–49.

———"Peasant Craftsmen in the Medieval Forest." *Agricultural History Review* 17 (1969): 91–107.

———, ed. *Records of Feckenham Forest, Worcestershire, c. 1236–1377.* Worcestershire Historical Society, new series, 21, 2006.

Bouchard, Constance Brittain. *Holy Entrepreneurs: Cistercians, Knights, and Economic Exchange in Twelfth-Century Burgundy.* Ithaca, N.Y.: Cornell University Press, 1991.

Boulton, Helen E., ed. *The Sherwood Forest Book.* Thoroton Society Record Series, 23, 1965.

Breeze, David J. "The Great Myth of Caledon." In *Scottish Woodland History,* edited by T.C. Smout, 47–51. Edinburgh: Scottish Cultural Press, 1997.

Brown, William, ed. *The Register of William Wickwane, Lord Archbishop of York, 1279–1285.* Surtees Society, 114, 1907.

Calendar of the Patent Rolls Preserved in the Public Record *Office: Edward III, 1327–1377.* 16 vols. London, 1891-1916.

Calendar of the Patent Rolls Preserved in the Public Record *Office: Henry III, A.D. 1232–1247*. London: HMSO, 1906.
Cantimpré, Thomas of. *Liber de Natura Rerum. Volume 1: Text*, edited by Helmut Boese. Berlin: Walter de Gruyter, 1973.
Cantor, Leonard, ed. *The English Medieval Landscape*. Philadelphia: University of Pennsylvania Press, 1982.
———. "Forests, Chases, Parks and Warrens." In *The English Medieval Landscape*, edited by Leonard Cantor, 56–85. Philadelphia: University of Pennsylvania Press, 1982.
Capes, William W., ed. *Registrum Ricardi de Swinfield, episcopi Herefordensis, A.D. 1283–1317*. Canterbury and York Society, 5, 1909.
Close Rolls of the Reign of Henry III Preserved in the *Public Record Office: A.D. 1234–1237*. London: HMSO, 1908.
Close Rolls of the Reign of Henry III Preserved in the *Public Record Office: A.D. 1237–1242*. London: HMSO, 1911.
Close Rolls of the Reign of Henry III Preserved in the *Public Record Office: A.D. 1256–1259*. London: HMSO, 1932.
Close Rolls of the Reign of Henry III Preserved in the *Public Record Office: A.D. 1261–1264*. London: HMSO, 1936.
Crescenzi, Pietro de. *Liber ruralia commodorum*. Augsburg: Johann Schüssler, 1471.
Crook, David. "The Records of Forest Eyres in the Public Record Office, 1179 to 1670." *Journal of the Society of Archivists* 17 (1996): 183–93.
Darby, H.C. *Domesday England*. Cambridge: Cambridge University Press, 1977.
de Maulde-la-Clavière, René. *Étude sur la condition forestière de l'Orleanais au moyen age et la renaissance*. Orléans: Herluison, 1871.
Devèze, Michel. "Forêts françaises et forêts allemandes: Étude historique comparée." *Revue Historique* 235 (1966): 362–80.
Dobson, R.B., and John Taylor. *Rymes of Robyn Hood: An Introduction to the English Outlaw*. Rev. edn. Stroud, Gloucestershire: Sutton, 1997.
Dougall, Martin, and Jim Dickson. "Old Managed Oaks in the Glasgow Area." In *Scottish Woodland History*, edited by T.C. Smout, 76–85. Edinburgh: Scottish Cultural Press, 1997.
Drew, Katherine Fischer, trans. *The Burgundian Code*. Philadelphia: University of Pennsylvania Press, 1972.
———, trans. *The Laws of the Salian Franks*. Philadelphia: University of Pennsylvania Press, 1991.
———, trans. *The Lombard Laws*. Philadelphia: University of Pennsylvania Press, 1973.
Dyer, Christopher. *Hanbury: Settlement and Society in a Woodland Landscape*. Leicester: Leicester University Press, 1991.
———. "'The Retreat from Marginal Land': The Growth and Decline of Medieval Rural Settlements." In *The Rural Settlements of Medieval England: Studies Dedicated to Maurice Beresford and John Hurst*, edited by Michael Aston, David Austin, and Christopher Dyer, 45–57. Oxford: Basil Blackwell, 1989.
Farrer, William, and J. Brownbill, eds. *Victoria History of the Counties of England: A History of the County of Lancaster*. 8 vols. London: A. Constable, 1906–14.
Fitzherbert, Anthony. *The Book of Husbandry*, edited by Walter W. Skeat. London: English Dialect Society, 1882.
Grant, Raymond. *The Royal Forests of England*. Stroud, Gloucestershire: Alan Sutton, 1991.
Griffiths, R.G., ed. *Registrum Thome de Cantilupo, episcopi Herefordensis, A.D. 1275–1282*. Canterbury and York Society, 2, 1907.

Hansen, Inge Lyse, and Chris Wickham, eds. *The Long Eighth Century: Production, Distribution and Demand*. Leiden: Brill, 2000.

Hansman, J. "Gilgamesh, Humbaba and the Land of the Erin-Trees." *Iraq* 38 (1976): 23–35.

Hart, Cyril E. *Royal Forest: A History of Dean's Woods as Producers of Timber*. Oxford: Clarendon Press, 1966.

Higounet, Charles. "Les forêts de l'Europe occidentale du Ve au XIe siècle." *Settimane di Studio* 13 (1965): 343–98.

Holt, J.C. *Robin Hood*. Rev. edn. London: Thames and Hudson, 1989.

Hooke, Della. "Early Medieval Estate and Settlement Patterns: The Documentary Evidence." In *The Rural Settlements of Medieval England: Studies dedicated to Maurice Beresford and John Hurst*, edited by Michael Aston, David Austin, and Christopher Dyer, 9–30. Oxford: Basil Blackwell, 1989.

——"Pre-Conquest Woodland: its Distribution and Usage." *Agricultural History Review* 37 (1989): 113-29.

Horrox, Rosemary, trans. and ed. *The Black Death*. Manchester: Manchester University Press, 1992.

Hutton, Ronald. *The Pagan Religions of the Ancient British Isles: Their Nature and Legacy*. Oxford: Blackwell, 1991.

Keen, Maurice. *The Outlaws of Medieval Legend*. Rev. edn. London and New York: Routledge, 1987.

Kirby, D.P. "The Old English Forest: Its Natural Flora and Fauna." In *Anglo-Saxon Settlement and Landscape: Papers presented to a Symposium, Oxford 1973*, edited by Trevor Rowley, 120–30. British Archaeological Reports, 6, 1974.

Knight, Stephen, ed. *Robin Hood: An Anthology of Scholarship and Criticism*. Woodbridge, Suffolk: D.S. Brewer, 1999.

Larson, Laurence M., trans. *The Earliest Norwegian Laws, being the Gulathing Law and the Frostathing Law*. New York: Columbia University Press, 1935.

Laszlovszky, József, and Péter Szabó, eds. *People and Nature in Historical Perspective*. Budapest: Central European University and Archaeolingua, 2003.

Linnard, William. *Welsh Woods and Forests: History and Utilization*. Cardiff: National Museum of Wales, 1982.

MacMullen, Ramsay. *Christianity and Paganism in the Fourth to Eighth Centuries*. New Haven, CT: Yale University Press, 1997.

Maurus, Hrabanus. *De Universo: The Peculiar Properties of Words and their Mystical Significance. The Complete English Translation*, translated by Priscilla Throop. 2 vols. Charlotte, VT: MedievalMS, 2009.

Megenberg, Konrad von. *Buch der Natur*, edited by Robert Luff and George Steer Tübingen: Niemeyer Verlag, 2003.

Meiggs, Russell. *Trees and Timber in the Ancient Mediterranean World*. Oxford: Clarendon Press, 1982.

Montanari, Massimo. *L'alimentazione contadina nell'alto Medioevo*. Naples: Liguori, 1979.

Neckam, Alexander. *De Naturis Rerum*, edited by Thomas Wright. London: Rolls Series, 34, 1863.

Noble, Thomas F.X., and Thomas Head, eds. *Soldiers of Christ: Saints and Saints' Lives from Late Antiquity and the Early Middle Ages*. University Park PA: Pennsylvania State University Press, 1995.

*Ordonnances des Roys de France de la troisième race*. 21 vols. Paris: Imprimerie royale, 1723–1849.

O'Sullivan, Aidan. "Woodland Management and the Supply of Timber and Underwood to Anglo-Norman Dublin." In *Environment and Subsistence in Medieval Europe*, edited by Guy de Boe and Frans Verhaeghe, 135–41. Zellik: I.A.P. Rapporten, 9, 1997.

Page, William, ed. *Victoria History of the Counties of England: A History of the County of Buckingham*. 4 vols. London: A. Constable, 1905–27.

———, ed. *Victoria History of the Counties of England: A History of Shropshire. Volume I*. London: A. Constable, 1908.

Parry, Joseph Henry, ed. *Registrum Johannis Gilbert, episcopi Herefordensis, A.D. 1375–1389*. Canterbury and York Society 18, 1915.

———, ed. *Registrum Johannis de Trillek, episcopi Herefordensis, A.D. 1344–1361*. Canterbury and York Society, 8, 1911.

Pilcher, Jon R., and Seán Mac an tSaoir, eds. *Wood, Trees and Forests in Ireland*. Dublin: Royal Irish Academy, 1995.

Postan, M.M., ed. *The Cambridge Economic History of Europe, volume I: The Agrarian Life of the Middle Ages*. Cambridge: Cambridge University Press, 1966.

Quelch, Peter R. "Ancient Trees in Scotland." In *Scottish Woodland History*, edited by T.C. Smout, 24–39. Edinburgh: Scottish Cultural Press, 1997.

Rackham, Oliver. *Ancient Woodland: Its History, Vegetation and Uses in England*. London: Edward Arnold, 1980.

———"Ecology and Pseudo-Ecology: The Example of Ancient Greece." In *Human Landscapes in Classical Antiquity: Environment and Culture*, edited by Graham Shipley and John Salmon, 16–43. London and New York: Routledge Press, 1996.

———*The History of the Countryside*. London: J.M. Dent and Sons, 1986.

———"Looking for Ancient Woodland in Ireland." In *Wood Trees and Forests in Ireland* edited by Jon R. Pilcher and Seán Mac an tSaoir, 1–12. Dublin: Royal Irish Academy, 1995.

———"The Medieval Countryside of England: Botany and Archaeology." In *Inventing Medieval Landscapes: Senses of Place in Western Europe*, edited by John Howe and Michael Wolfe, 13–32. Gainesville, FL: University Press of Florida, 2002.

———*Trees and Woodland in the British Landscape*. London: J.M. Dent and Sons, 1976.

*Register of Edward the Black Prince*. 4 vols. London: HMSO, 1930–33.

Rivers, Theodore John, trans. *Laws of the Alamans and Bavarians*. Philadelphia: University of Pennsylvania Press, 1977.

———*Laws of the Salian and Ripuarian Franks*. New York: AMS Press, 1986.

Rowley, Trevor, ed. *Anglo-Saxon Settlement and Landscape: Papers presented to a Symposium, Oxford 1973*. British Archaeological Reports, 6, 1974.

Saunders, Corinne J. *The Forest of Medieval Romance: Avernus, Broceliande, Arden*. Rochester, N.Y.: D.S. Brewer, 1993.

Sawyer, Peter. "Anglo-Saxon Settlement: The Documentary Evidence." In *Anglo-Saxon Settlement and Landscape: Papers presented to a Symposium, Oxford 1973*, edited by Trevor Rowley, 108–19. British Archaeological Reports, 6, 1974.

Scott, S.P., ed. *The Visigothic Code*. Boston: Boston Book Co. 1910.

Shaw, R. Cunliffe. *The Royal Forest of Lancaster*. Preston, 1956.

Shipley, Graham, and John Salmon, eds. *Human Landscapes in Classical Antiquity: Environment and Culture*. London and New York: Routledge, 1996.

Sillasoo, Ülle. "Plant Depictions in Late Medieval Religious Art." In *People and Nature in Historical Perspective*, edited by József Laszlovszky and Péter Szabó, 377–93. Budapest: Central European University and Archaeolingua, 2003.

Simmons, I.G., and M.J. Tooley, eds. *The Environment in British Prehistory*. Ithaca, N.Y.: Cornell University Press, 1981.

Smith, Catherine Delano. "Where was the 'wilderness' in Roman times?" In *Human Landscapes in Classical Antiquity: Environment and Culture*, edited by Graham Shipley and John Salmon. London and New York: Routledge, 1996.

Smout, T.C., "Highland Land-Use before 1800: Misconceptions, Evidence and Realities." In *Scottish Woodland History*, edited by T.C. Smout. Edinburgh: Scottish Cultural Press, 1977.

——, ed. *People and Woods in Scotland: A History*. Edinburgh: Edinburgh University Press, 2003.

——, ed. *Scottish Woodland History*. Edinburgh: Scottish Cultural Press, 1997.

Stagg, D.J., ed. *A Calendar of New Forest Documents, 1244-1334*. Hampshire Record Series, 3, 1979.

Stamper, Paul. "Woods and Parks." In *The Countryside of Medieval England*, edited by Grenville Astill and Annie Grant, 128–48. Oxford: Basil Blackwell, 1988.

Sudhoff, Karl. "Pestschriften aus den ersten 150 Jahren nach der Epidemie des 'schwarzen Todes' von 1348." *Archiv für Geschichte der Medizin* 11 (1919): 44–92, 121–76.

Szabó, Péter. *Woodland and Forests in Medieval Hungary*. Oxford: Archaeopress, 2005.

Taylor, Christopher. "The Anglo-Saxon Countryside." In *Anglo-Saxon Settlement and Landscape: Papers presented to a Symposium, Oxford 1973*, edited by Trevor Rowley, 5–15. British Archaeological Reports, 6, 1974.

Thompson, Alexander Hamilton. *Visitations of Religious Houses in the Diocese of Lincoln*. 3 vols. Publications of the Lincoln Record Society, 7, 14, 21, 1914–20.

Toutain, J. *Les Cultes Païens dans l'Empire Romain*. 3 vols. Paris: Ernest Leroux, 1907–20.

Turner, G.J., ed. *Select Pleas of the Forest*. London: Selden Society, 1901.

Turton, Robert Bell, ed. *The Honor and Forest of Pickering*. 4 vols. North Riding Record Society, new series, 1894–97.

Unwin, P.T.H. "The Changing Identity of the Frontier in Medieval Nottinghamshire and Derbyshire." In *Villages, Fields and Frontiers: Studies in European Rural Settlement in the Medieval and Early Modern Periods*, 339–51. Oxford: B.A.R. international series, 185, 1983.

Vera, F.W.M. *Grazing Ecology and Forest History*. Wallingford, UK and New York: CABI Publishing, 2000.

Villages, Fields and Frontiers: *Studies in European Rural Settlement in the Medieval and Early Modern Periods*. Oxford: B.A.R. international series, 185, 1983.

Wales, Gerald of. *The Journey through Wales*, translated by Lewis Thorpe. London: Penguin Books, 1978.

Wickham, Chris. *Framing the Early Middle Ages: Europe and the Mediterranean, 400–800*. Oxford: Oxford University Press, 2005.

——*The Inheritance of Rome: Illuminating the Dark Ages, 400–1000*. New York: Viking Penguin, 2009.

——*Land and Power: Studies in Italian and European Social History, 400–1200*. London: British School at Rome, 1994.

Williams, George H. *Wilderness and Paradise in Christian Thought: The Biblical Experience of the Desert in the History of Christianity and the Paradise Theme in the Theological Idea of the University*. New York: Harper and Brothers, 1962.

Williams, Michael. *Deforesting the Earth: From Prehistory to Global Crisis*. Chicago: University of Chicago Press, 2003.

Wilson, Dolores. "Multi-Use Management of the Medieval Anglo-Norman Forest." *Journal of the Oxford University History Society* 1 (2004): 1–16.

Winters, Robert K. *The Forest and Man*. New York: Vantage Press, 1974.

Wood, Ian. "Before or After Mission: Social Relations Across the Middle and Lower Rhine in the Seventh and Eighth Centuries." In *The Long Eighth Century: Production, Distribution and Demand*, edited by Inge Lyse Hansen and Chris Wickham, 149–66. Leiden: Brill, 2000.

Young, Charles. "Conservation Policies in the Royal Forests of Medieval England." *Albion* 10 (1978): 95-103.

——*The Royal Forests of Medieval England*. Philadelphia: University of Pennsylvania Press, 1979.

## Part III

Aberth, John. *The Black Death: The Great Mortality of 1348–1350. A Brief History with Documents*. Boston and New York: Bedford/St. Martin's Press, 2005.

——*Plagues in World History*. Lanham, MD: Rowman and Littlefield, 2011.

Agramont, Jacme d'. "Regiment de Preservacio a Epidimia o Pestilencia e Mortaldats," translated by M.L. Duran-Reynals and C.-E.A. Winslow. *Bulletin of the History of Medicine* 23 (1949): 57–89.

Albarella, Umberto, Pig Husbandry and Pork Consumption in Medieval England." In *Food in Medieval England: Diet and Nutrition*, edited by C.M. Woolgar, D. Serjeantson, and T. Waldron, 72–87. Oxford: Oxford University Press, 2006.

——and Simon J.M. Davis. "Mammals and Birds from Launceston Castle, Cornwall: Decline in Status and the Rise of Agriculture." *Circaea: The* Journal of the Associations for Environmental *Archaeology* 12 (1996): 1–26.

——, and Richard Thomas. "They Dined on Crane: Bird Consumption, Wild Fowling and Status in Medieval England." *Acta Zoologica Cracoviensa* 45 (2002): 23-38.

Albert the Great. *On Animals: A Medieval* Summa Zoologica, translated by Kenneth F. Kitchell, Jr. and Irven Michael Resnick. 2 vols. Baltimore: Johns Hopkins University Press, 1999.

Alexander, Dominic. *Saints and Animals in the Middle Ages*. Woodbridge, Suffolk: Boydell Press, 2008.

Almond, Richard. *Medieval Hunting*. Stroud, Gloucestershire: Sutton Publishing, 2003.

Anderson, J.K. *Hunting in the Ancient World*. Berkeley: University of California Press, 1985.

Anglicus, Bartholomaeus. *On the Properties of Things: John Trevisa's Translations: A Critical Text*. 2 vols. Oxford: Clarendon Press, 1975.

Archer, Rowena E., and Simon Walker, eds. *Rulers and Ruled in Late Medieval England: Essays Presented to Gerald Harriss*. London: Hambledon Press, 1995.

Astill, Grenville, and Annie Grant, eds. *The Countryside of Medieval England*. Oxford: Basil Blackwell, 1988.

——, and Annie Grant. "Efficiency, Progress and Change." In *The Countryside of Medieval England,* edited by Grenville Astill and Annie Grant, 213–33. Oxford: Basil Blackwell, 1988.

——, and John Langdon, eds. *Medieval Farming and Technology: The Impact of Agricultural Change in Northwest Europe*. Leiden: Brill, 1997.

Aston, T.H., ed. *Medieval Fish, Fisheries and Fishponds in England*. 2 parts. BAR British Series, 182 (i), 1988.

——, P.R. Coss, Christopher Dyer, and Joan Thirsk, eds. *Social Relations and Ideas: Essays in Honour of R.H. Hilton*. Cambridge: Cambridge University Press, 1983.

Attenborough, F.L., trans. and ed. *The Laws of the Earliest English Kings*. Cambridge: Cambridge University Press, 1922.

Ayton, Andrew. *Knights and Warhorses: Military Service and the English Aristocracy under Edward III*. Woodbridge, Suffolk: Boydell Press, 1994.

Bailey, Mark. "The Rabbit and the Medieval East Anglian Economy." *Agriculture History Review* 36 (1988): 1–20.
Bailey, Michael D. *Magic and Superstition in Europe: A Concise History from Antiquity to the Present.* Lanham, MD: Rowman and Littlefield, 2007.
Barney, Stephen A., W.J. Lewis, J.A. Beach, and Oliver Berghof, eds. *The* Etymologies *of Isidore of Seville.* Cambridge: Cambridge University Press, 2006.
Barrett, James H., Alison M. Locker, and Callum M. Roberts. "The Origins of Intensive Marine Fishing in Medieval Europe: The Engish Evidence." *Proceedings of the Royal Society of London* (2004): 2417–21.
Barthélemy, Dominique, and Philippe Contamine. "The Use of Private Space." In *A History of Private Life. Volume II: Revelations of the Medieval World*, edited by Georges Duby and translated by Arthur Goldhammer, 444-60. Cambridge, MA: Harvard University Press, 1988.
Bath, Adelard of. *Conversations with his Nephew*, edited and translated by Charles Burnett. Cambridge: Cambridge University Press, 1998.
Bäuerliche Sachkultur des Spätmittelalters: Internationaler Kongress, Krems an der Donau 21 bis 24 *September 1982*, 307–20. Vienna: Verlag der Österreichischen Akademie der Wissenschaften, 1984.
Baxter, Ron. *Bestiaries and their Users in the Middle Ages.* Stroud, Gloucestershire: Sutton Publishing, 1998.
Beaumanoir, Philippe de. *Coutumes de Beauvaisis*, edited by A. Salmon. 3 vols. Paris: Alphonse Picard, 1899-1900.
Beauvais, Vincent of. *Speculum Quadruplex sive SpeculumMaius.* 4 vols. Douai: Balthazar Belleri, 1624.
Belloni, Luigi, ed. *I Trattati in Volgare della Peste edell'acqua Ardente di Michele Savonarola.* Rome: Congressio Nazionale della Società Italiana de Medicina Interna, 1953.
Benecke, Norbert. "On the Utilization of the Domestic Fowl in Central Europe from the Iron Age up to the Middle Ages." *Archaeofauna* 2 (1993): 21–31.
Birrell, Jean. "Aristocratic Poachers in the Forest of Dean: Their Methods, their Quarry, and their Companions." *Transactions of the Bristol and Gloucestershire Archaeological Society* 119 (2001): 147–54.
———, ed. *Collections for a History of Staffordshire.* The Forests of Cannock and Kinver: Select Documents, *1235–1372.* Staffordshire Record Society, 4th series, 18, 1999.
———"Deer and Deer Farming in Medieval England." *Agriculture History Review* 40 (1992): 112-26.
———"Hunting and the Royal Forest." In *L'Uomo e la Foresta, secc. XIII-XVIII*, edited by Simonetta Cavaciocchi. Prato: Istituto Internazionale di Storia Economica "F. Datini," 1996.
———"Peasant Deer Poachers in the Medieval Forest." In Progress and Problems in Medieval England: Essays in *Honour of Edward Miller*, edited by Richard Britnell and John Hatcher, 68–88. Cambridge: Cambridge University Press, 1996.
———"Procuring, Preparing, and Serving Venison in Late Medieval England." In *Food in Medieval England: Diet and Nutrition*, edited by C.M. Woolgar, D. Serjeantson, and T. Waldron, 176–90. Oxford: Oxford University Press, 2006.
———, ed. *Records of Feckenham Forest, Worcestershire, c. 1236–1377.* Worcestershire Historical Society, new series, 21, 2006.
———"Who Poached the King's Deer? A Study in Thirteenth Century Crime." *Midland History* 7 (1982): 9–25.
Bollet, Alfred Jay. *Plagues and Poxes: The Impact of Human History on Epidemic Disease.* New York: Demos Medical Publishing, 2004.

Bord, Lucien-Jean, and Jean-Pierre Mugg. *La Chasse au Moyen Âge: Occident latin, VIe-XVe siècle.* Aix-en-Provence: Editions du Gerfaut, 2008.

Boulton, Helen E., ed. *The Sherwood Forest Book.* Thoroton Society Record Series, 23, 1965.

Bourbon, Etienne de. *Anecdotes historiques*, edited by A. Lecoy de Marche. Paris: Librairie Renouard, 1877.

Boutillier, Jean. *Somme Rural, ou le Grand Coustumier.* Paris: Barthelemy Mace, 1621.

Braekman, Willy L. *The Treatise on Angling in the Boke of St. Albans (1496): Background, Context and Text of "The treatyse of fysshynge wyth an Angle".* Scripta: Mediaeval and Renaissance Texts and Studies, 1, 1980.

Britnell, Richard, and John Hatcher, eds. *Progress and Problems in Medieval England: Essays in Honour of Edward Miller.* Cambridge: Cambridge University Press, 1996.

Budiansky, Stephen. *The Character of Cats: The Origins, Intelligence, Behavior, and Stratagems of Felis silvestris catus.* New York: Viking Penguin, 2002.

——— *The Covenant of the Wild: Why Animals Chose Domestication.* New York: William Morrow and Company, 1992.

Buglione, Antonietta. "People and Animals in Northern Apulia from Late Antiquity to the Early Middle Ages: Some Considerations." In *Breaking and Shaping Beastly Bodies: Animals as Material Culture in the Middle Ages*, edited by Aleksander Pluskowski, 189–216. Oxford: Oxbow Books, 2007.

Campbell, Bruce M.S. "Economic Rent and the Intensification of English Agriculture, 1086–1350." In *Medieval Farming and Technology: The Impact of Agricultural Change in Northwest Europe*, edited by Grenville Astill and John Langdon, 225–50. Leiden: Brill, 1997.

——— *English Seigniorial Agriculture, 1250–1450.* Cambridge: Cambridge University Press, 2000.

Cantimpré, Thomas of. *Liber de Natura Rerum. Volume 1: Text*, edited by Helmut Boese. Berlin: Walter de Gruyter, 1973.

Cantor, Leonard, M. ed. *The English Medieval Landscape.* Philadelphia: University of Pennsylvania Press, 1982.

———"Forests, Chases, Parks and Warrens." In *The English Medieval Landscape*, edited by Leonard Cantor, 56–85. Philadelphia: University of Pennsylvania Press, 1982.

———, and J. Hatherly. "The Medieval Parks of England." *Geography* 64 (1979): 71–85.

Capes, W.W., ed. *Registrum Thome de Cantilupo, Episcopi Herefordensis, A.D. 1275–1282.* Hereford: Cantilupe Society, 1906.

Carpov, Benedict. *Practica Nova Rerum Criminalium.* Lipsia, 1669.

Cavaciocchi, Simonetta, ed. *L'Uomo e la Foresta, secc. XIII-XVIII.* Prato: Istituto Internazionale di Storia Economica "F. Datini," 1996.

Chauliac, Guy de. *Inventarium sive Chirurgia Magna. Volume One: Text*, edited by Michael R. McVaugh. Leiden: E.J. Brill, 1997.

Clutton-Brock, Juliet, *Domesticated Animals from Early Times.* London and Austin, TX: British Museum and University of Texas Press, 1981.

——— *Horse Power: A History of the Horse and the Donkey in Human Societies.* Cambridge, MA: Harvard University Press, 1992.

———, and Caroline Grigson, eds. *Animals and Archaeology: Volume 2. Shell Middens, Fishes and Birds.* BAR International Series, 183, 1983.

———, and Caroline Grigson, eds. *Animals and Archaeology: Volume 1. Hunters and their Prey.* BAR International Series, 163, 1983.

Cohen, Esther. "Law, Folklore and Animal Lore." *Past and Present* 110 (1986): 6–37.

Cohen, Jeremy. *"Be Fertile and Increase, Fill the Earth and Master It"*: The Ancient and Medieval Career of a *Biblical Text*. Ithaca, N.Y.: Cornell University Press, 1989.

Comet, Georges. "Technology and Agricultural Expansion in the Middle Ages: The Example of France north of the Loire." In *Medieval Farming and Technology: The Impact of Agricultural Changes in Northwest Europe*, edited by Grenville Astill and John Langdon, 11–39. Leiden: Brill, 1997.

Cox, John Charles. *The Royal Forests of England*. London: Methuen and Co., 1905.

Coy, Jennie. "The Provision of Fowls and Fish for Towns." In *Diet and Crafts in Towns: The Evidence of Animal Remains from the Roman to the Post-Medieval Periods*, edited by D. Serjeantson and T. Waldron, 25–40. BAR British Series, 199, 1989.

Crabtree, Pam Jean. "Animals as Material Culture in Middle Saxon England: The Zooarchaeological Evidence for Wool Production at Brandon." In *Breaking and Shaping* Beastly Bodies: Animals as Material Culture in the *Middle Ages*, edited by Aleksander Pluskowski, 161–69. Oxford: Oxbow Books, 2007.

———"Zooarchaeology at Early Anglo-Saxon West Stow." In Medieval Archaeology: Papers of the Seventeenth Annual Conference of the Center for Medieval and Early *Renaissance Studies*, edited by Charles L. Redman, 203-15. SUNY Binghamton, Medieval and Renaissance Texts and Studies, 60, 1989.

Crane, Susan. "Ritual Aspects of the Hunt à Force." In Engaging with Nature: Essays on the Natural World in *Medieval and Early Modern Europe*, edited by Barbara A. Hanawalt and Lisa J. Kiser, 63–84. Notre Dame, IN: University of Notre Dame Press, 2008.

Crescenzi, Pietro de. *Liber ruralia commodorum*. Augsburg: Johann Schlösser, 1471.

Cummins, John. *The Hound and the Hawk: The Art of Medieval Hunting*. New York: St. Martin's Press, 1988.

———"Veneurs s'en vont en Paradis: Medieval Hunting and the 'Natural' Landscape." In *Inventing Medieval Landscapes: Senses of Place in Western Europe*, edited by John Howe and Michael Wolfe, 33–56. Gainesville, FL: University Press of Florida, 2002.

Darby, H.C. *Domesday England*. Cambridge: Cambridge University Press, 1977.

Davidson, L.S., and J.O. Ward, eds. *The Sorcery Trial of Alice Kteler*. Binghamton, N.Y.: Medieval and Renaissance Texts and Studies, 1993.

Davis, R.H.C. *The Medieval Warhorse: Origin, Development and Redevelopment*. London: Thames and Hudson, 1989.

De Venuto, Giovanni. "Animals and Economic Patterns in Medieval Apulia (South Italy): Preliminary Findings." In *Breaking and Shaping Beastly Bodies: Animals as Material Culture in the Middle Ages*, edited by Aleksander Pluskowski, 217–34. Oxford: Oxbow Books, 2007.

Dinānah, Taha. "Die Schrift von Abi G'far Ahmed ibn 'Ali ibn Mohammed ibn 'Ali ibn Hatimah aus Almeriah über die Pest." *Archiv für Geschichte der Medizin* 19 (1927): 27–81.

Ditchfield, P.H., and William Page, eds. *Victoria History of the Counties of England: A History of Berkshire*. 4 vols. London: A. Constable, 1906–24.

Drew, Katherine Fischer, trans. *The Burgundian Code*. Philadelphia: University of Pennsylvania Press, 1972.

———, trans. *The Laws of the Salian Franks*. Philadelphia: University of Pennsylvania Press, 1991.

———, trans. *The Lombard Laws*. Philadelphia: University of Pennsylvania Press, 1973.

Duby, Georges, ed., and Arthur Goldhammer, trans. A History of Private Life. Volume II: Revelations of *the Medieval World*. Cambridge, MA: Harvard University Press, 1988.

Durham, Reginald of. *Libellus de Vita et Miraculis S. Godrici, Heremitae de Finchale*. London: Surtees Society, 1847.

Dyer, Christopher. "Consumption of Fresh-Water Fish in Medieval England." In *Medieval Fish, Fisheries and Fishponds in England*. 2 parts, edited by Michael Aston, 27–38. BAR British Series, 182 (i), 1988.

——"Documentary Evidence: Problems and Enquiries." In *The Countryside of Medieval England*, edited by Grenville Astill and Annie Grant, 12–35. Oxford: Basil Blackwell, 1988.

——"English Diet in the Later Middle Ages." In *Social Relations and Ideas: Essays in Honour of R.H. Hilton*, edited by T.H. Aston, P.R. Coss, Christopher Dyer, and Joan Thirsk, 191–216. Cambridge: Cambridge University Press, 1983.

——"English Peasant Buildings in the Later Middle Ages (1200–1500)," in Christopher Dyer, *Everyday Life in Medieval England*. London: Hambledon Press, 1994.

——"Seasonal Patterns in Food Consumption in the Later Middle Ages." In *Food in Medieval England: Diet and Nutrition*, edited by C.M. Woolgar, D. Serjeantson, and T. Waldron, 201–14. Oxford: Oxford University Press, 2006.

——"Sheepcotes: Evidence for Medieval Sheepfarming." *Medieval Archaeology* 39 (1995): 136–64.

Edward, Duke of York. *The Master of Game: The Oldest English Book on Hunting*, edited by Wm. A. and F. Baillie-Grohman. London: Chatto and Windus, 1909.

Enghoff, Inge Bødker. "A Medieval Herring Industry in Denmark and the Importance of Herring in Eastern Denmark." *Archaeofauna* 5 (1996): 43–47.

Ervynck, Anton, Sharyn Jones O'Day, and Wim Van Neer, eds. Behavior Behind Bones: The Zooarchaeology of Ritual, *Religion, Status and Identity*. Oxford: Oxbow Books, 2003.

Evans, Edward Payson. *Animal Symbolism in Ecclesiastical Architecture*. New York: Henry Holt and Co., 1896.

——*The Criminal Prosecution and Capital Punishment of Animals*. London: William Heinemann, 1906.

Ferrato, Pietro, ed. *Consiglio contro a Pistolenza per Maestro Tommaso del Garbo*. Bologna: Presso Gaetano Romagnoli, 1866.

Ficino, Marsilio. *Contro alla Peste*. Florence: Appresso I Giunti, 1576.

Filotas, Bernadette. *Pagan Survivals, Superstitions and Popular Culture in Early Medieval Pastoral Literature*. Toronto: Pontifical Institute of Mediaeval Studies, 2005.

Finkelstein, J.J. "The Ox that Gored." *Transactions of the American Philosophical Society* 71 (1981): 1–89.

Fitzherbert, Anthony. *The Book of Husbandry*, edited by Walter W. Skeat. London: English Dialect Society, 1882.

Flores, Nona C., ed. *Animals in the Middle Ages*. New York and London: Routledge, 1996.

——, "The Mirror of Nature Distorted: The Medieval Artist's Dilemma in Depicting Animals." In *The Medieval World of Nature: A Book of Essays*, edited by Joyce E. Salisbury, 3–45. New York and London: Garland Publishing, 1993.

Foard, Glenn, David Hall, and Tracey Britnell. *The Historic Landscape of Rockingham Forest: Its Character and Evolution from the 10th to the 20th Centuries*. Rockingham Forest Trust and Northamptonshire County Council, 2003.

Foligno, Gentile da. *Consilium contra Pestilentiam*. Colle di Valdelsa, c. 1479.

Foreman, Martin, ed. *Further Excavations at the Dominican Priory, Beverly, 1986–89*. Sheffield Excavation Reports, 4, 1996.

Fornasari, M, ed. *Collection Canonum in V Libris*. Continuatio Mediaevalis, VI, 1976.

Fox, Robin Lane. "Ancient Hunting: From Homer to Polybios." In *Human Landscapes in Classical Antiquity: Environment and Culture*, edited by Graham Shipley and John Salmon, 119–53. London and New York: Routledge, 1996.

Gibbs, Frederick W. "Medical Understandings of Poison circa 1250–1600." Ph.D. dissertation, University of Wisconsin-Madison, 2009.

Gilbert, John M. *Hunting and Hunting Reserves in Medieval Scotland.* Edinburgh: John Donald Publishers, 1979.

Gilchrist, Roberta. "The Animal Bones." In *Further Excavations at the Dominican Priory, Beverly, 1986–89,* edited by Martin Foreman, 228–31. Sheffield Excavation Reports, 4, 1996.

Given, James B. *Inquisition and Medieval Society: Power, Discipline, and Resistance in Languedoc.* Ithaca, N.Y.: Cornell University Press, 1997.

Glaber, Rodulfus. *The Five Books of the Histories,* edited and translated by John France. Oxford: Clarendon Press, 1989.

Glosecki, Stephen O. "Movable Beasts: The Manifold Implications of Early Germanic Animal Imagery." In *Animals in the Middle Ages,* edited by Nona C. Flores, 3–23. New York and London: Routledge, 1996.

Grant, Annie. "Animal Resources." In *The Countryside of Medieval England,* edited by Grenville Astill and Annie Grant, 149–85. Oxford: Basil Blackwell, 1988.

Grant, Raymond. *The Royal Forests of England.* Stroud, Gloucestershire: Alan Sutton, 1991.

Groenman-van Waateringe, W., and L.H. van Wijngaarden-Bakker, eds. *Farm Life in a Carolingian Village: A Model Based on Botanical and Zoological Data from an Excavated Site.* Assen/Maastricht: Van Gorcum, 1987.

Hall, A.R., and H.K. Kenward, eds. *Urban-Rural Connexions: Pespectives from Environmental Archaeology.* Oxford: Oxbow Books, 1994.

Hamerow, Helena. *Early Medieval Settlements: The Archaeology of Rural Communities in Northwest Europe, 400–900.* Oxford: Oxford University Press, 2002.

Hanawalt, Barbara A., ed. *Chaucer's England: Literature in Historical Context.* Medieval Studies at Minnesota, 4, 1992.

Hands, Rachel, ed. *English Hawking and Hunting in* The Boke of St. Albans. Oxford: Oxford University Press, 1975.

Hare, J.N., ed. *Battle Abbey: The Eastern Range and the Excavations of 1978–80.* London: Historic Buildings and Monuments Commission for England, 1985.

Harris, B.E., ed. *Victoria History of the Counties of England: A History of the County of Chester.* 3 vols. Oxford: Oxford University Press, 1979.

Hart, Cyril E. *Royal Forest: A History of Dean's Woods as Producers of Timber.* Oxford: Clarendon Press, 1966.

Hassig, Debra. *Medieval Bestiaries: Text, Image, Ideology,* Cambridge: Cambridge University Press, 1995.

Hatting, Tove. "Cats from Viking Age Odense." *Journal of Danish Archaeology* 9 (1990):179–93.

Heinrich, Dirk. "Temporal Changes in Fishery and Fish Consumption between Early Medieval Haithabu and its Successor, Schleswig." In *Animals and Archaeology: Volume 2. Shell Middens, Fishes and Birds,* edited by Juliet Clutton-Brock and Caroline Grigson, 151–56. BAR International Series, 183, 1983.

Hicks, Michael, ed. *Revolution and Consumption in Late Medieval England.* Woodbridge, Suffolk: Boydell Press, 2001.

Hoffmann, Richard C. "Economic Development and Aquatic Ecosystems in Medieval Europe." *American Historical Review* 101 (1996): 630–69.

——. *Fishers' Craft and Lettered Art: Tracts on Fishing from the End of the Middle Ages.* Toronto: University of Toronto Press, 1997.

——. "Fishing for Sport in Medieval Europe: New Evidence." *Speculum* 60 (1985): 877–902.

——. "Frontier Foods for Late Medieval Consumers: Culture, Economy, Ecology." *Environment and History* 7 (2001): 131–67.

———"Medieval Fishing." In *Working with Water in Medieval Europe: Technology and Resource-Use*, edited by Paolo Squatriti, 331–93. Leiden: Brill, 2000.

———"The Protohistory of Pike in Western Culture." In *The Medieval World of Nature: A Book of Essays*, edited by Joyce E. Salisbury, 61–76. New York and London: Garland Publishing, 1993.

Horrox, Rosemary, trans. and ed. *The Black Death*. Manchester: Manchester University Press, 1994.

Howe, John, and Michael Wolfe, eds. *Inventing Medieval Landscapes: Senses of Place in Western Europe*. Gainesville, FL: University Press of Florida, 2002.

Hyland, Ann. *The Horse in the Middle Ages*. Stroud, Gloucestershire: Sutton Publishing, 1999.

———*The Medieval Warhorse: From Byzantium to the Crusades*. London: Grange Books, 1994.

Ibn Munqidh, Usama. *The Book of Contemplation: Islam and the Crusades*, translated by Paul M. Cobb. London: Penguin Books, 2008.

IJzereef, G.F. "The Animal Remains." In *Farm Life in a Carolingian Village: A Model Based on Botanical and Zoological Data from an Excavated Site*, edited by W. Groenman-van Waatering and L.H. van Wijngaarden-Bakker, 39–51. Assen/Maastricht: Van Gorcum, 1987.

Jordan, William Chester. *The Great Famine: Northern Europe in the Early Fourteenth Century*. Princeton: Princeton University Press, 1996.

Kershaw, Ian. "The Great Famine and the Agrarian Crisis in England, 1315–22." *Past and Present* 59 (1973): 3–50.

Kieckhefer, Richard. "Erotic Magic in Medieval Europe." In *Sex in the Middle Ages: A Book of Essays*, edited by

Joyce E. Salisbury, 30–55. New York and London: Garland Publishing, 1991.

———*European Witch Trials: Their Foundations in Popular and Learned Culture, 1300–1500*. Berkeley and Los Angeles: University of California Press, 1976.

———*Forbidden Rites: A Necromancer's Manual of the Fifteenth Century*. University Park, PA: Pennsylvania State University Press, 1997.

Kors, Alan Charles, and Edward Peters, eds. *Witchcraft in Europe, 400–1700: A Documentary History*. 2nd edn. Philadelphia: University of Pennsylvania Press, 2001.

Kowaleski, Maryanne. "The Expansion of the South-Western Fisheries in Late Medieval England." *Economic History Review*. New series 53 (2000): 429–54.

Labarge, Margaret Wade. *A Small Sound of the Trumpet: Women in Medieval Life*. Boston, MA: Beacon Press, 1986.

Lamond, Elizabeth, trans. *Walter of Henley's Husbandry*. London: Longmans, Green, and Co., 1890.

Lampen, Angelika. "Medieval Fish Weirs: The Archaeological and Historical Evidence." *Archaeofauna* 5 (1996): 129-34.

Langdon, John. "Agricultural Equipment." In *The Countryside of Medieval England*, edited by Grenville Astill and Annie Grant, 86–107. Oxford: Basil Blackwell, 1988.

———*Horses, Oxen and Technological Innovation: The Use of Draught Animals in English Farming from 1066 to 1500*. Cambridge: Cambridge University Press, 1986.

———"Was England a Technological Backwater in the Middle Ages?" In *Medieval Farming and Technology: The Impact of Agricultural Change in Northwest Europe*, edited by Grenville Astill and John Langdon, 275–92. Leiden: Brill, 1997.

Larson, Laurence M., trans. *The Earliest Norwegian Laws, being the Gulathing Law and the Frostathing Law*. New York: Columbia University Press, 1935.

Lentacker, An, Wim Van Neer, and Jean Plumier. "Historical and Archaeozoological Data on Water Management and Fishing during Medieval and Post-Medieval Times at Namur

(Belgium)." In *Environment and Subsistence in Medieval Europe*, edited by Guy De Boe and Frans Verhaeghe, 83–94. I.A.P. Rapporten, 9, 1997.
Levack, Brian P. *The Witch-Hunt in Early Modern Europe*. 3rd edn. Harlow, UK: Pearson/Longman, 2006.
Locker, A. "Animal and Plant Remains." In *Battle Abbey: The Eastern Range and the Excavations of 1978–80*, edited by J.N. Hare, 183–87. London: Historic Buildings and Monuments Commission for England, 1985.
Longchamp, Nigel. *A Mirror for Fools: The Book of Burnel the Ass*, translated by J.H. Mozley. Oxford: Blackwell, 1961.
Luff, Rosemary M., and Marta Moreno García. "Killing Cats in the Medieval Period. An Unusual Episode in the History of Cambridge, England." *Archaeofauna* 4 (1995): 93–114.
MacCracken, Henry Noble, ed. *The Minor Poems of John Lydgate. Part II: Secular Poems*. London: Early English Text Society, 192, 1934.
McVaugh, Michael R. "Quantified Medical Theory and Practice at Fourteenth-Century Montpellier." *Bulletin of the History of Medicine* 43 (1969): 397–413.
Mann, Jill. *From Aesop to Reynard: Beast Literature in Medieval Britain*. Oxford: Oxford University Press, 2009.
Marvin, William Perry. *Hunting Law and Ritual in Medieval English Literature*. Woodbridge, Suffolk: D.S. Brewer, 2006.
Maurus, Hrabanus. *De Universo: The Peculiar Properties of Words and the Mystical Significance. The Complete English Translation*, translated by Priscilla Throop. 2 vols. Charlotte, VT: MedievalMS, 2009.
Megenberg, Konrad von. *Buch der Natur, vol. 2: critical text*, edited by Robert Luff and George Steer. Tübingen: Max Niemeyer Verlag, 2003.
Ménabréa, Léon. *De l'Origine de la Forme et de l'Esprit des Jugements Rendus au Moyen-Age contre les Animaux*. Chambéry: Puthod, 1846.
Menjot, D., ed. *Manger et Boire au Moyen Age*. Nice: Les Belles Lettres, 1984.
Michon, L.-A. Joseph, ed. *Documents Inédits sur la Grande Peste de 1348*. Paris, 1860.
Montanari, Massimo. "Rural Food in Late Medieval Italy." In *Bäuerliche Sachkultur des Spätmittelalters:* Internationaler Kongress, Krems an der Donau 21 bis 24 *September 1982*, 307–20. Vienna: Verlag der Österreichischen Akademie der Wissenschaften, 1984.
Neckam, Alexander. *De Naturis Rerum*, edited by Thomas Wright. London: Rolls Series, 34, 1863.
Noodle, Barbara. "The Animal Bones." In *Excavations in Medieval Southampton, 1953–69. Volume 1: The Excavation Reports*, edited by Colin Platt and Richard Coleman-Smith, 332–40. Leicester: Leicester University Press, 1975.
Oggins, Robin S. "Falconry and Medieval Views of Nature." In *The Medieval World of Nature: A Book of Essays*, edited by Joyce E. Salisbury, 47–60. New York and London: Garland Publishing, 1993.
———. *The Kings and their Hawks*. New Haven, CT: Yale University Press, 2004.
Oldridge, Darren. *Strange Histories: The Trial of the Pig, the Walking Dead, and Other Matters of Fact from the Medieval and Renaissance Worlds*. London and New York: Routledge, 2007.
*Ordonnances des Roys de France de la troisième race*. 21 vols. Paris: Imprimerie royale, 1723–1849.
Origen. *Contra Celsum*, translated by Henry Chadwick. Cambridge: Cambridge University Press, 1953.
Orme, Nicholas. "Medieval Hunting: Fact and Fancy." In *Chaucer's England: Literature in Historical Context*, edited by Barbara A. Hanawalt. Medieval Studies at Minnesota, 4, 1992.
Owen, Aneurin, ed. *Ancient Laws and Institutes of Wales*. London, 1841.

Owen, D.D.R. *The Romance of Reynard the Fox*. Oxford: Oxford University Press, 1994.
Page, William, ed. *Victoria History of the Counties of England: A History of the County of Derby*. 2 vols. London: A. Constable, 1905.
———, and J. Horace Round, eds. *Victoria History of the Counties of England: A History of the County of Essex, volume 2*. London: A. Constable, 1907.
Parry, Joseph Henry, ed. *Registrum Johannis de Trillek, Episcopi Herefordensis, A.D. 1344–1361*. Canterbury and York Society, 8, 1912.
Pellegrin, Pierre. *Aristotle's Classification of Animals:* Biology and the Conceptual Unity of the Aristotelian *Corpus*, translated by Anthony Preus. Berkeley and Los Angeles: University of California Press, 1982.
Persson, G. "Consumption, Labour and Leisure in the Late Middle Ages." In *Manger et Boire au Moyen Age*, edited by D. Menjot, 211–23. Nice: Les Belles Lettres, 1984.
Pigière, Fabienne, et al. "Status as Reflected in Food Refuse of Late Medieval Noble and Urban Households at Namur (Belgium)." In *Behavior Behind Bones: The Zooarchaeology of Ritual, Religion, Status and Identity*, edited by Anton Ervynck, Sharyn Jones O'Day, and Wim Van Neer, 233–43. Oxford: Oxbow Books, 2003.
Pinkhof, H. *Abraham Kashlari, over Pestachtige Koorsten*. Amsterdam, 1891.
Platt, Colin, and Richard Coleman-Smith, eds. *Excavation in Medieval Southampton, 1953–69. Volume 1: The Excavation Reports*. Leicester: Leicester University Press, 1975.
Pluskowski, Aleksander, ed. *Breaking and Shaping Beastly Bodies: Animals as Material Culture in the Middle Ages*. Oxford: Oxbow Books, 2007.
———*Wolves and the Wilderness in the Middle Ages*. Woodbridge, Suffolk: Boydell Press, 2006.
Popescu, Elizabeth Shepherd, et al. *Norwich Castle: Excavations and Historical Survey, 1987–98. Part I: Anglo-Saxon to c. 1345*. East Anglian Archaeology, Report no. 132, 2009.
Postan, M.M. "Village Livestock in the Thirteenth Century." *Economic History Review*. New series 15 (1962): 228–35.
Power, Eileen. *Medieval English Nunneries, c. 1275 to 1535*. Cambridge: Cambridge University Press, 1922.
Preus, Anthony. *Science and Philosophy in Aristotle's Biological Works*. Hildesheim, Germany, and New York: Georg Olms Verlag, 1975.
Prummel, Wietske. "Evidence of Hawking (Falconry) from Bird and Mammal Bones." *International Journal of Osteoarchaeology* 7 (1997): 333–38.
Rackham, Oliver. *Ancient Woodland: Its History, Vegetation and Uses in England*. London: Edward Arnold, 1980.
———*The History of the Countryside*. London: J.M. Dent and Sons, 1986.
———*Trees and Woodland in the British Landscape*. London: J.M. Dent and Sons, 1976.
Randall, Lilian M.C. *Images in the Margins of Gothic Manuscripts*. Berkeley and Los Angeles: University of California Press, 1966.
Rangarajan, Mahesh. *India's Wildlife History: An Introduction*. Delhi: Ranthambhore Foundation, 2001.
Rawlinson, Henry. *Babylonian and Assyrian Literature*. New York: P.F. Collier and Son, 1901.
Rébouis, H. Émile. *Étude Historique et Critique sur la Peste*. Paris: Alphonse Picard, 1888.
Redman, Charles L., ed. *Medieval Archaeology: Papers of the Seventeenth Annual Conference of the Center for Medieval and Early Renaissance Studies*. SUNY Binghamton, Medieval and Renaissance Texts and Studies, 60, 1989.
Reeves, Compton. *Pleasures and Pastimes in Medieval England*. Oxford: Oxford University Press, 1998.
*Register of Edward the Black Prince*. 4 vols. London: HMSO, 1930–33.

Rivers, Theodore John, trans. *Laws of the Alamans and Bavarians*. Philadelphia: University of Pennsylvania Press, 1977.

———, trans. *Laws of the Salian and Ripuarian Franks*. New York: AMS Press, 1986.

Rombauts, E., "Grimbeert's Defense of Reinaert in *Van den Vos Reynaerde*, an Example of *Oratio Iudicialis*?" In *Aspects of the Medieval Animal Epic: Proceedings of the International Conference, Louvain, May 15–17, 1972*, edited by E. Rombauts and A. Welkenhuysen, 129-41. Louvain and the Hague: Leuven University Press and Martinus Nijhoff, 1975.

——— and A. Welkenhuysen, eds. *Aspects of the Medieval Animal Epic: Proceedings of the International Conference, Louvain, May 15–17, 1972*. Louvain and the Hague: Leuven University Press and Martinus Nijhoff, 1975.

Rooney, Anne. *Hunting in Middle English Literature*. Woodbridge, Suffolk: D.S. Brewer and Boydell Press, 1993.

———, ed. *The Tretyse off Huntyng*. Scripta: Mediaeval and Renaissance Texts and Studies, 19, 1987.

Salisbury, Joyce E. *The Beast Within: Animals in the Middle Ages*. New York and London: Routledge, 1994.

———"Bestiality in the Middle Ages." In *Sex in the Middle Ages: A Book of Essays*, edited by Joyce E. Salisbury, 173–86. New York and London: Garland Publishing, 1991.

———"Human Animals of Medieval Fables." In *Animals in the Middle Ages*, edited by Nona C. Flores. New York and London: Routledge, 1996.

———, ed. *Sex in the Middle Ages: A Book of Essays*. New York and London: Garland Publishing, 1991.

Salter, David. *Holy and Noble Beasts: Encounters with Animals in Medieval Literature*. Woodbridge, Suffolk: D.S. Brewer, 2001.

Schlag, Wilhelm. *The Hunting Book of Gaston Phébus: Manuscrit français 616, Paris, Bibliothèque nationale*. London: Harvey Miller Publishers, 1998.

Schmitt, Jean-Claude. *The Holy Greyhound: Guinefort, Healer of Children since the Thirteenth Century*. Cambridge and Paris: Cambridge University Press and La Maison des Sciences de l'Homme, 1983.

Scott, S.P., ed. *The Visigothic Code*. Boston: Boston Book Co., 1910.

Serjeantson, D. "Birds: Food and a Mark of Status." In *Food in Medieval England: Diet and Nutrition*, edited by C.M. Woolgar, D. Serjeantson, and T. Waldron, 131-47. Oxford: Oxford University Press, 2006.

——— and T. Waldron, eds. *Diet and Crafts in Towns: The Evidence of Animal Remains from the Roman to the Post-Medieval Periods*. BAR British Series, 199, 1989.

———, and C.M. Woolgar. "Fish Consumption in Medieval England." In *Food in Medieval England: Diet and Nutrition*, edited by C.M. Woolgar, D. Serjeantson, and T. Waldron, 102–30. Oxford: Oxford University Press, 2006.

Shipley, Graham, and John Salmon, eds. *Human Landscapes in Classical Antiquity: Environment and Culture*. London and New York: Routledge, 1996.

Simonini, R. "Il Codice di Mariano di Ser Jacopo sopra 'Rimedi Abili nel Tempo di Pestilenza'." *Bollettino dell'Istituto Storico Italiano dell'Arte Sanitaria* 9 (1929): 161–69.

Singer, Charles, trans. *The Fasciculus Medicinae of Johannes de Ketham, Alemanus: Facsimile of First Edition of 1491*. Birmingham, Ala.: Classics of Medicine Library, 1988.

Slicher Van Bath, B.H. *The Agrarian History of Western Europe, A.D. 500–1850*, translated by Olive Ordish. New York: St. Martin's Press, 1963.

Sobol, Peter G. "The Shadow of Reason: Explanations of Intelligent Animal Behavior in the Thirteenth Century." In *The Medieval World of Nature: A Book of Essays*, edited by Joyce E. Salisbury, 109–28. New York and London: Garland Publishing, 1993.

Sorrell, Roger D. *St. Francis of Assisi and Nature: Tradition and Innovation in Western Christian Attitudes toward the Environment.* Oxford: Oxford University Press, 1988.

Spinage, Clive A. *Cattle Plague: A History.* New York: Kluwer Academic/Plenum Publishers, 2003.

Squatriti, Paolo, *Water and Society in Early Medieval Italy, A.D. 400–1000.* Cambridge: Cambridge University Press, 1998.

———, ed. *Working with Water in Medieval Europe: Technology and Resource-Use.* Leiden: Brill, 2000.

Stagg, D.J., ed. *A Calendar of New Forest Documents, 1244-1334.* Hampshire Record Series, 3, 1979.

———, ed. *A Calendar of New Forest Documents: The Fifteenth to the Seventeenth Centuries.* Hampshire Record Series, 5, 1983.

Stamper, Paul. "Woods and Parks." In *The Countryside of Medieval England*, edited by Grenville Astill and Annie Grant, 128–48. Oxford: Basil Blackwell, 1988.

*Statutes of the Realm (1225–1713).* 9 vols. London: G. Eyre and A. Strahan, 1810–22.

Stone, D.J. "Consumption and Supply of Birds in Late Medieval England." In *Food in Medieval England: Diet and Nutrition*, edited by C.M. Woolgar, D. Serjeantson, and T. Waldron, 148–61. Oxford: Oxford University Press, 2006.

Subrenat, Jean. "Rape and Adultery: Reflected Facets of Feudal Justice in the Roman de Renart." In *Reynard the Fox: Social Engagement and Cultural Metamorphoses in the Beast Epic from the Middle Ages to the Present*, edited by Kenneth Varty, 17–35. New York and Oxford: Berghahn Books, 2000.

Sudhoff, Karl. "Ein chirurgisches Manual des Jean Pitard, Wundarztes König Philipps des Schönen von Frankreich." *Archiv für Geschichte der Medizin* 2 (1908–9): 189-278.

———"Pestschriften aus den ersten 150 Jahren nach der Epidemie des 'schwarzen Todes' von 1348," *Archiv für Geschichte der Medizin* 4 (1911):191–222, 389–424; 5 (1912):36–87, 332–96; 6 (1913):313–79; 7 (1913):57–114; 8 (1915):175–215, 236–89; 9 (1916):53-78, 117–67; 11 (1919):44–92, 121–76; 14 (1922–23):1-25, 79–105, 129–68; 16 (1924–25):1–69, 77–188; 17 (1925):12–139, 241–91.

Suetonius. *The Twelve Caesars.* Rev. edn., translated by Robert Graves and Michael Grant. Harmondsworth, Middlesex: Penguin Books, 1979.

Swabe, Joanna. *Animals, Disease and Human Society: Human-Animal Relations and the Rise of Veterinary Medicine.* London and New York: Routledge, 1999.

Sykes, Naomi J. "From *Cu* and *Sceap* to *Beffe* and *Motton*: The Management, Distribution, and Consumption of Cattle and Sheep in Medieval England." In *Food in Medieval England: Diet and Nutrition*, edited by C.M. Woolgar, D. Serjeantson, and T. Waldron, 56–71. Oxford: Oxford University Press, 2006.

———"The Impact of the Normans on Hunting Practices in England." In *Food in Medieval England: Diet and Nutrition*, edited by C.M. Woolgar, D. Serjeantson, and T. Waldron, 162–75. Oxford: Oxford University Press, 2006.

———"Taking Sides: The Social Life of Venison in Medieval England." In *Breaking and Shaping Beastly Bodies: Animals as Material Culture in the Middle Ages*, edited by Aleksander Pluskowski, 149–60. Oxford: Oxbow Books, 2007.

te Velde, H. "A Few Remarks upon the Religious Significance of Animals in Ancient Egypt." *Numen* 27 (1980): 76–82.

Thirsk, Joan. *Alternative Agriculture: A History from the Black Death to the Present Day.* Oxford: Oxford University Press, 1997.

Thomas, Richard M. "Chasing the Ideal? Ritualism, Pragmatism and the Later Medieval Hunt in England." In *Breaking and Shaping Beastly Bodies: Animals as Material Culture in the Middle Ages*, edited by Aleksander Pluskowski. Oxford: Oxbow Books, 2007.

———"Food and the Maintenance of Social Boundaries in Medieval England." In *The Archaeology of Food and Identity*, edited by Katheryn C. Twiss. Southern Illinois University: Center for Archaeological Investigations, Occasional Paper No. 34, 2007.

Tilander, Gunnar, ed. *Les Livres du Roy Modus et de la Royne Ratio.* 2 vols. Paris: Société des Anciens Textes Français, 1932.

Tilley, Maureen A. "Martyrs, Monks, Insects, and Animals." In *The Medieval World of Nature: A Book of Essays*, edited by Joyce E. Salisbury, 93–107. New York and London: Garland Publishing, 1993.

Tillott, P.M., ed. *Victoria History of the Counties of England: A History of Yorkshire.* Oxford: Oxford University Press, 1961.

Torrey, E. Fuller, and Robert H. Yolken. *Beasts of the Earth: Animals, Humans, and Disease.* New Brunswick, N.J.: Rutgers University Press, 2005.

Toynbee, J.M.C. *Animals in Roman Life and Art.* Ithaca, N.Y.: Cornell University Press, 1973.

Trow-Smith, Robert. *A History of British Livestock Husbandry to 1700.* London: Routledge and Kegan Paul, 1957.

Turner, G.J., ed. *Select Pleas of the Forest.* London: Selden Society, 1901.

Turton, Robert Bell, ed. *The Honor and Forest of Pickering.* 4 vols. North Riding Record Society, 1894–97.

Twiss, Katheryn C., ed. *The Archaeology of Food and Identity.* Southern Illinois University: Center for Archaeological Investigations, Occasional Paper No. 34, 2007.

Twiti, William. *The Art of Hunting*, edited by David Scott-Macnab. Heidelberg: Universitätsverlag, 2009.

Valenti, Marco, and Frank Salvadori. "Animal Bones: Synchronous and Diachronic Distribution as Patterns of Socially Determined Meat Consumption in the Early and High Middle Ages in Central and Northern Italy." In *Breaking and Shaping Beastly Bodies: Animals as Material Culture in the Middle Ages*, edited by Aleksander Pluskowski, 170–88. Oxford: Oxbow Books, 2007.

van Dam, Petra J.E.M. "New Habitats for the Rabbit in Northern Europe, 1300–1600." In *Inventing Medieval Landscapes: Senses of Place in Western Europe*, edited by John Howe and Michael Wolfe, 57–69. Gainesville, FL: University Press of Florida, 2002.

Varty, Kenneth. "Further Examples of the Fox in Medieval English Art." In *Aspects of the Medieval Animal Epic: Proceedings of the International Conference, Louvain, May 15–17, 1972*, edited by E. Rombauts and A. Welkenhuysen, 251–56. Louvain and the Hague: Leuven University Press and Martinus Nijhoff, 1975.

———*Reynard the Fox: A Study of the Fox in Medieval English Art*. Leicester: Leicester University Press, 1967.

———, ed. *Reynard the Fox: Social Engagement and Cultural Metamorphoses in the Beast Epic from the Middle Ages to the Present.* New York and Oxford: Berghahn Books, 2000.

Un Village au Temps de Charlemagne: Moines et Paysans de *l'Abbaye de Saint-Denis du VIIe siècle à l'An Mil*. Paris: Ministère de la Culture. 1988.

Waddell, Helen, trans. *Beasts and Saints.* London: Constable, 1934.

Wales, Gerald of. *The first version of the Topography of Ireland*, trans. John J. O'Meara. Dundalk: Dundalgan Press, 1951.

White, T.H., ed. *The Book of Beasts, being a Translation from a Latin Bestiary of the Twelfth Century.* New York: G.P. Putnam's Sons, 1954.

Wiegand, Wilhelm, et al., eds. *Urkunden und Akten der Stadt Strassburg*. 15 vols. Strasbourg: K. J. Trübner, 1879–1933.

Wilson, Bob. "Mortality Patterns, Animal Husbandry and Marketing in and around Medieval and Post-Medieval Oxford." In *Urban-Rural Connexions: Perspectives from Environmental Archaeology*, edited by A.R. Hall and H.K. Kenward, 103–15. Oxford: Oxbow Books, 1994.

Wilson, Stephen. *The Magical Universe: Everyday Ritual and Magic in Pre-Modern Europe*. London and New York: Hambledon and London, 2000.

Wood, Casey A., and Florence Marjorie Fyfe, trans. and eds. The Art of Falconry, being the De Arte Venandi cum *Avibus of Frederick II of Hohenstaufen*. Stanford, CA: Stanford University Press, 1943.

Woolgar, C.M., "Diet and Consumption in Gentry and Noble Households: A Case Study from around the Wash." In Rulers and Ruled in Late Medieval England: Essays *Presented to Gerald Harriss*, edited by Rowena E. Archer and Simon Walker, 17–31. London: Hambledon Press, 1995.

——"Fast and Feast: Conspicuous Consumption and the Diet of the Nobility in the Fifteenth Century." In *Revolution and Consumption in Late Medieval England*, edited by Michael Hicks, 7–25. Woodbridge, Suffolk: Boydell Press, 2001.

——*The Great Household in Late Medieval England*. New Haven, CT: Yale University Press, 1999.

——"Meat and Dairy Products in Late Medieval England." In *Food in Medieval England: Diet and Nutrition*, edited by C.M. Woolgar, D. Serjeantson, and T. Waldron, 88–101. Oxford: Oxford University Press, 2006.

——, D. Serjeantson, and T. Waldron, eds. *Food in Medieval England: Diet and Nutrition*. Oxford: Oxford University Press, 2006.

Young, Charles R. *The Royal Forests of Medieval England*. Philadelphia: University of Pennsylvania Press, 1979.

Ziolkowski, Jan M. *Talking Animals: Medieval Latin Beast Poetry, 750–1150*. Philadelphia: University of Pennsylvania Press, 1993.

# Index

Abd-el-Latif 14
Abnil-Hercules 177
Abū Sahl al-Masīhī 17
Adam (Old Testament figure) 5, 24, 155
Adam de la Bould, steward of Feckenham Forest 106
Adam of Bremen 25
*Admonitio generalis* (General admonition), of Charlemagne 21
*The Advocate* 218
Aelfgifu of Northampton 24
Aelian 225–26
Aereda 20
Aesop 221
Æthelstan, king of England 153
afforestation 98–99, 102, 121
Africa 18, 149–50, 169
agisters 104–5
Agnes of Bednall, Lady 199
Agobard, archbishop of Lyons 20–22, 52, 76
Agotheol, archangel 23
agriculture 3, 23, 27–30, 34, 46, 50–52, 84–86, 91–92, 94–98, 101, 110, 137–38, 150–52, 155–57, 160–63, 182, 232
Agricultural Revolution 29, 138, 150, 163
air xv, 8–9, 11–15, 17–18, 20, 22, 25, 28, 37, 41–47, 50, 52–55, 57–64, 66–67, 70–76, 132, 144, 209–14, 216–17, 224, 229, 231, 233
*On Airs, Waters, and Places,* by Hippocrates 10, 12, 16–17
Aix-en-Provence, France 80
Alain of Lille 6, 41–42, 48
Alamannic laws 152, 170–71, 180
Alamanns 152, 180
Alan de Neville, chief forester 101
albedo effect 50
Albert the Great 18, 42, 46, 53, 57–58, 133–34, 146, 158–60,

166, 172–73, 201, 211, 230
Alberti, Leon Battista 8
Albich, Sigmund 59, 61, 71, 212
Albiorix 19
Albumasar 43, 56
alder trees 85, 96, 107–9, 114, 116, 124–25
Alderotti, Taddeo 212
Aldwinkle, England 118
*Alexander* 172
Alexander of Hales 226
Alexander the Great 79, 178, 225
Alfonso V, king of Portugal 185
Alfonso de Córdoba 66–68, 73
Alfred the Great, king of Wessex 90
Algazel 55
Allexton, England 109
*Almagest,* by Ptolemy 59
Almaric le Despenser, lord of Oldberrow 186
Almería, Spain 53, 55, 210
aloe trees 133
Alps 201
Alrewas Hay, forest of Cannock 107–8, 189
Alta Foresta, bailiwick of, England 106
Altamira, Spain 141
Amand, Saint 81
Amanti, Dominico, bishop of Brescia 216–17
Amanus Mountains 77
Amator, Saint 83
Ambrose, Saint 132–33, 146, 155, 172, 225
Amice de Clare, countess of Wight 199
Amon 142
Anatolia, *see* Turkey
Anaximander 11
Anaxogoras 133–34
*Ancrene Riwle* 175
al-Andalus, *see* Spain
Andeis 20
Andraste 79
Andrew of Calabria, Duke 58

Index 309

Anglesey, island of, Wales 79
Anglicus, Bartholomaeus 18, 42, 44–45, 57, 132, 158, 173
*Anglo-Saxon Chronicle* 181
Anglo-Saxon laws 83, 87, 152, 219
Anglo-Saxons 5–6, 23, 88, 90–93, 99, 151–52, 155, 170, 180, 184
animal trials 10
animals xv-vi, 5–6, 8–13, 15–16, 19–20, 28, 45, 47, 52, 54, 61, 63–65, 70–71, 73, 78, 83, 85, 87, 90–91, 98–99, 102, 104–5, 108, 110, 115, 117–18, 120, 123–25, 133–38, 233; as beasts of burden (farm) 148–69; as food 142, 148–49, 151, 153, 157, 160, 162–65, 168–69, 180, 182–83, 185–87, 191–92, 195, 200–202, 206, 216; as pets 142, 145, 169–76, 231–32; diseases of 138, 157–62, 206–17, 223, 232; human attitudes towards and comparison with 141–48, 164, 166–69, 171, 179, 204, 208–16, 220–21, 223–27, 229, 232–33; domestication of 141–42, 148–50, 169, 177, 206–7; hunting of, *see* hunting; magical use of 179, 207, 228–32; on trial 153, 217–24, 229, 232; protection of 178, 189–91, 205–6; worship of 141–42, 144, 150, 179
animism 4, 141
antelope 142, 148
ants, *see* insects
Anubis 142
Apennine Mountains, Italy 36, 86
Apollinaris, Sidonius 86
Apollo 1
Appenzell, Switzerland 230
apple trees, *see* fruit trees
Apulia 151, 174
aqueducts 31–32
Aquinas, Thomas, Saint 42, 48, 146, 153, 216, 221–22, 226-27, 231
Arabia 27
Aragon 68
Ardennes, forest of, France 79, 88, 179
Arduinna 79, 179
Aristotle 8, 11, 15, 42, 44–46, 53, 69, 79, 132–34, 142, 146-7, 158, 172, 211–13, 222, 225–26
Arixus 20
Arles, Council of 80
Arno River 66
Arnold of Villa Nova 210, 212
Arpeninus 20
Artemis 143, 178

Artio 179
Arthur, King 128–29
Asclepios 179
ash trees 83–85, 114, 124–25, 133, 135
Ashoka, emperor of India 178
Ashurbanipal, king of Assyria 177
Asia 12–13, 18, 149–50, 169, 207
Asia Minor 79
assarts 28, 92–96, 99–102, 104–6, 110, 114–15, 122, 124, 137, 192
Assize of Clarendon 100
Assize of the Forest, *see* Assize of Woodstock
Assize of Woodstock 99, 101
Assyrian Empire 177
Assyrians 78, 150, 169
astrology 55–56, 59
astronomy 55, 59
Astwood, England 106
Athens, Greece 86, 219
Atlantic Ocean 26, 34, 49
attachment courts, of the forest 105, 115–16
Audley, James, Lord 197
augury, *see* divination
Augustine of Hippo, Saint 5, 7, 22–23, 48, 145–46, 221, 225
aurochsen 85, 141, 149–50, 155, 176
Austrasia, kingdom of 91
Austria 20, 51
*Autobiography,* by Guibert of Nogent 128
Avenzoar 56
Averroes 69–70, 147, 213
Avicenna 14–17, 42–43, 45, 53–56, 58–59, 61, 63, 69–70, 132, 146, 158, 172–73, 207, 209, 211, 217
Avignon, France 55–56, 60, 68, 72

baboons 225
Bacchus 7
Bacon, Roger 146
badgers 166, 185, 231
Baeserta 20
Baghdad, Iraq 14
Bahrain 77
Bailly, Gaspar, 224
Bakony Forest, Hungary 127
Baltic Sea 51
Baltic States 52, 127
Barcelona, Spain 66
Bardney, England: abbey of 120
Barking, England: abbess of 199
barley 27, 84, 160
Basil of Caesarea 45, 172
Basilisk 15
Bastet 170

Bath, England 19, 79
bathing 31, 43, 85
bats 229
Bavaria 171
Bavarian laws 90, 152, 170
Bavarians 90, 152
bears xvi, 171, 178–80, 184–85, 199, 206, 213, 225
beasts, *see* animals
Beaulieu, England 195; abbot of 118
beavers 184–85
Bede the Venerable 24
Bedyng, Elizabeth 195
beech trees 47, 80, 90, 107, 124, 149
bees 124, 142, 144, 161
beetles, *see* insects
Belgium 51–52, 65
Belorussia 127
Beni Hasan, Egypt 170
bercelets 186
Berchtold, Master 67
Berkeley, England: sorceress of 231
Berkshire 93, 132, 182
Bernard of Clairvaux, Saint 93, 128
Bernard of Frankfurt 58, 71
Bernardino of Siena 231
Bernesmor, England: wood of 189
Berneval-le-Grand, France 181
*Beowulf* 5
Besançon, France 165
Beskwood Hay, Sherwood Forest 117
bestiality 6, 218, 224–29, 232
bestiaries 10, 148, 172, 176, 205
Białowieża Forest 127
birch trees 84, 107–8, 114, 125, 133
birds 15, 61, 63, 76, 87, 142, 146, 151, 165, 169, 171, 175, 177–80, 182, 185, 193, 201, 204; eating of 185, 200–201, 206–7, 209–14, 216–17, 228–29, 231
bison 141, 171, 176, 180
Black Death xiv, 1–4, 7, 9–10, 15–18, 30, 43–44, 46, 49, 51-2, 58, 64, 66–67, 69–70, 73, 92, 103, 111, 123–24, 127, 137–39, 162–63, 182, 192–93, 208, 210–11, 213, 215, 223, 232–33
black deer 193
blackthorn trees 125
Blake, William 76
Blakey Moor, England 198
Blanche, Baroness Wake of Liddell 199
Blasius of Barcelona 59
boars 142, 147, 149–50, 155, 158, 176–77, 179–80, 182–85, 206, 213
Bohemia 7, 216

*Boke of St. Albans* 204
Bologna, Italy 161
Boltingen, Switzerland 230
Bolton, England: priory of 207
Boniface, Saint 83, 90–91
*Book of Burnel the Ass, The*, *see Speculum Stultorum*
*Book of Husbandry, The*, by Sir Anthony Fitzherbert 134, 161
Book of Kells 176
*Book of St. Albans, The*, 166–67
botany, origins of 79
Botte, Henry, of Pillaton Hall 107
Boutillier, Jean 220
Bradfield Woods, England 114
Brambelwode, England 111
bream 204
Bremous, Edmund 195
Brescia, Italy 216
brewing industry 32, 65
Brian de Stapilton, Sir 175
Brigit, Saint 19
Brigstock, England 117; park of 198
Bristol, England 65, 187
Britain 20, 84–85, 92–93, 151, 170, 184–85
Bronze Age 84–85
Bubastis, Egypt 170
*Buch der Natur* (Book of Nature), by Konrad of Megenberg 10, 56–58, 68, 166
Buckinghamshire 182
bucks, *see* fallow deer
buffaloes 155, 171, 177, 180
bulls, *see* cattle
Burchard of Worms 19, 229, 231
Burgundian laws 90, 153, 170, 180, 219
Burgundians 90
Burgundy 19; duke of 168
Burmer, John 186
Bury St. Edmunds, abbey of 114
Burynges, England 107
butchering, of animals for meat 32, 64–65, 164–65, 176, 179, 196
Buxton, England 79
Byzantine Empire 83

Cabochien Revolt, of 1413 127
Caesar, Julius 228
Caesarius of Arles, Bishop 80
Cairo, Egypt 14
Calais, France 168
Caledonian Forest, Scotland 123
Callirius 79
Cambridge, England 34, 169, 174

Cambridgeshire 111
camels 207
Campania 44
canals 32, 34
Canne, Italy 174
cannibalism 52
Cannock, England: forest of 96, 107, 116, 131, 189, 196–97, 199
*Canon of Medicine*, by Avicenna 14, 16–17, 59, 70, 172, 209, 211
Canterbury, England 52
*Canticle of the Sun*, by St. Francis of Assisi 8, 47
*Capitulare Aquisgranense* (Capitularly of Aachen), issued by Charlemagne 90
*Capitulare de Villis* (Capitulary concerning the Royal Domains), issued by Charlemagne 90
Car Dyke, England 34
Carcassonne, France 67–68
Cardon, William, forester of Alrewas Hay 107
Carinthia 57
Carloman 91
Carolina code 227
Carolingian Renaissance 24
Carolingians 83, 88, 90–91, 99, 125, 170, 180
carp 204
Carpathian Basin, Hungary 34, 127
Carpov, Benedict 227
Carthage, Council of 80
Cassius Dio 79
Cassy, Alice 175
caterpillars, *see* insects
Cathwulf 21
cats 141–42, 169–71, 173–76, 211, 216, 231
cattle 29, 52, 105, 115, 122, 124–25, 135, 138, 141–42, 147-53, 155–58, 160–64, 166–67, 169, 171, 177, 182, 187, 215-16, 225–27, 230; diseases of 157, 160–62, 206–8, 212, 215; on trial 218–20
*Cattle Raid of Cooley* 150
Caxton, William 221
cedar trees 77, 83
Celsus 143–45, 147
Celtic culture 5, 19–20, 25, 79–80, 82–83, 150, 155, 179, 207, 225, 228
*Centilogium*, by Ptolemy 59
Ceres 7
Cernunnos 179
Chalin de Vinario, Raymond 55–56, 59, 63

charcoal, manufacturing of 104–5, 107–8, 111, 117, 119, 122–24, 128
Charlemagne 21, 83, 90, 96, 99, 174
Charles V, king of France 175
Charles VI, king of France 193
Charon 7
Chartres, France 128
chases, hunting 130, 181
Chaucer, Geoffrey 129
cheetahs 169
Chelmsford, England 231
cherry trees, *see* fruit trees
Cheshire 184, 189
Chester, England 189
chestnut trees 79
Chichele, Henry, archbishop of Canterbury 52
chickens 20, 148–53, 156, 158, 161–62, 165, 171–72, 214–15, 218
Cholmeley, Richard, Sir, keeper of Pickering Forest 189
Chrétien de Troyes 128
Christianity xvi, 4–5, 19–23, 25, 31, 47–49, 80–82, 143, 145, 217
Cicero 86, 178
cinnamon trees 133
Cistercian order of monks 6, 34, 48, 93–94, 124, 128
cisterns 31, 34
Cîteaux, monastery of 94
Clement VI, Pope 56
climate 12–13, 26–28, 49, 51, 61, 63, 85
climate change xiv, 50, 233
Clipston, England 109
Clipstone, England 208
clouds 42, 44–47, 61, 73–76
clove trees 133
Clovis, king of the Franks 152
Cnut the Great 24–25
coal 31, 42, 66, 119
cod 27, 181, 201–2
Coelius, adile of Rome 178
Colchester, England 65, 79, 174
*Colliget*, by Averroes 213
Colliure, France 68
Columbanus, Saint 22
Columella, Lucius 79, 150
Columella, Marcus 150
comets 59–60
coneys, *see* rabbits
Connacht 150, 227
*Consiglio contro la pestilenza*, by Marsilio Ficino 211
Constance, Lake 22

312  Index

Constantine Porphyrogenitus, emperor of Byzantium 160
Constantine the African 44–45, 55
Constantinople 89
*Contra Celsum,* by Origen 143
Copho 55
coppicing, of trees 79, 85, 90, 94–96, 111, 114–15, 120, 122–25, 127, 135
Corbeil, France 227
Corby, England 116
Cornwall 31; earl of 191
*Corrector,* by Burchard of Worms 229, 231
cosmic ray fluxes, from outer space 4
*Cosmographia* (Cosmography), by Bernard Silvester 6, 41
Coterel gang 131
Cotton, Matilda, wife of Sir Anthony Fitzherbert 167
Cowick, England 175
cows, *see* cattle
Coydrath, Wales 124
crabs 165, 172
cranes 179, 185, 217
Crauncester, Edmund 117
Cremona, Italy 167
Cristoforo de Honestis 215
*Critias,* by Plato 86
crocodiles 142, 177
Crouchback, Edmund, earl of Lancaster 119
Crowland, England: abbey of 208
crows 179, 207, 228
Crusades 27–28
cuckoos 172
cypress trees 77, 83

dace 204
Dalton, John, keeper of forest of Pickering 116–17, 189
Damascus, Syria 14
dams 33–34, 102, 109
Danes 5
Danube River 57
Dark Age Cold Period 27
Davis, Dean, governor of Vermont xiv
Dda, Hwel, king of Wales 171
Dean, England: forest of 96, 105, 111, 115, 119, 121–22, 132, 184, 187
*De Animalibus,* by Albert the Great 159, 201
*De Animalibus,* by Aristotle 146, 211
*De Arboribus* (On Trees), by Columella 79
*De Arte Venandi cum Avibus* (On the Art of Hunting with Birds), by Frederick II 201

*De Causis Proprietatum Elementorum, see Liber de Causis* Proprietatum Elementorum
*De Judicio Solis in Conviviis Saturni* (On the Judgment of the Sun at the Feasts of Saturn), by Simon of Corvino 7
*De Medicina Equorum* (On the Medicine of Horses), by Jordanus Ruffus 160
*De Mirabilibus Mundi* (On the Wonders of the World), by Solinus 172
*De Mortalitate in Alamannia* (On the Mortality in Germany), by Konrad of Megenberg 68
*De Mundi Universitate* (Concerning the World Universe), *see* Cosmographia
*De Re Rustica,*(On Rustic Matters), by Columella 79, 150
*De Re Rustica,*(On Rustic Matters), by Varro 150
*De Rerum Natura,*(On the Nature of Things), by Lucretius 17-18
*De Rerum Natura,*(On the Nature of Things), by Rabanus Maurus, *see De Universo*
*De Universo,* (On the Universe), by Rabanus Maurus 18, 132, 171
*De Venenis,*(On Poisons), by Pietro d'Abano 69
*Debate of the Horse, Goose, and Sheep, The* 168
*Decretum,* by Burchard of Worms 229
*Decretum,* by Gratian 222
Deepdale Springs, England 110
deer 87, 105, 108, 122, 135, 141, 148, 155, 161, 171, 176-7, 179–83, 185–99, 204, 230; farming of 182, 190; hunting of 183, 186–99, 205
Deerhurst, England 175
*Defender of Ladies,* by Martin Le Franc 231
deforestation, *see* forest, destruction and clearance of
Delamere, England: forest of 189
Delos, Greece 178
Democritus 134
demons 43, 73–76, 82, 128, 144, 221, 227–28, 231
dendrochronology 4, 26, 50, 125
Denmark 24–25, 163, 170, 174, 201
Derby, England: earl of 189
Derbyshire 93, 189
Descartes, René 147
Despenser, Elizabeth, Lady 175
devil 43, 68, 73–76, 91, 223, 228, 231–32
Devon 31
Devonshire Tapestries 195
*Dialogue Concerning the Exchequer,* by Richard FitzNigel 101
Diana, *see* Artemis

*Didascalicon,* by Hugh of St. Victor 229
Diemer River 64
Diest, Belgium 64
diet 62, 138, 180
Dijon, France 19
dikes 33
Diodorus Siculus 170, 179
Dioscorides 55, 132
disafforestment 98, 102–4, 119–23, 130, 182, 192
disease 10, 12–14, 16–17, 28, 44, 52, 55–56, 58, 60, 64–68, 70–73, 86, 133, 137, 141, 157–62, 164–65, 168, 174, 182, 192, 195, 206–17, 223–24
*Le Dite de Hosbondrie* (Sayings of Husbandry) 160
divination 228–29
Divine Mind, *see* Noys
does, *see* fallow deer
dogs xvi, 141–42, 144, 149–50, 158, 161–62, 164, 169–77, 180, 186–89, 192–94, 197–99, 201, 206, 211, 214–16, 228–31, 233; diseases of 172; expeditation of 187; on trial 218
Doliche, Mount, Syria 20
Dolichenus 20
dolphins 179, 217
Domesday Book 28, 36, 91–92, 97–99, 111, 121, 155–56
donkeys 148–50, 155, 161, 166–68, 218, 225, 231
dovecotes 159, 161, 163
doves 179, 230
Downham, England 182
dragons 137
drainage technology 32–34, 38, 94–96; pollution from 63, 66
Driby, England 50
Droitwich, England 122, 187
druidism 5, 80, 82
Dublin, Ireland 125
Duby, Georges 93, 96
ducks 142, 150–51, 158, 161, 172, 177
Durham, England, bishop of 156, 205
dust veils 4, 50, 60, 233

eagles 144, 168, 178–79, 225, 230
earth xv, 1, 4–5, 7, 9, 11, 13–15, 17–18, 20, 22–23, 25, 27-8, 41–42, 44–46, 48, 52–55, 57–63, 71, 76, 80, 91, 132, 136–37, 144, 207, 209, 211–12, 214, 216, 224, 229, 233

earthquakes 2, 42, 44–45, 50, 54, 57–58, 66, 71
Easingwold, England 98
East Anglia 33, 79, 84, 92, 97, 138, 151, 156
Eastnor, England 183
Eburones tribe 80
ecology 32, 192; *see also* new ecology Medieval West), by Georges Duby 93
Eden 5, 24, 136
Edmund, earl of Cornwall 191
Edward, duke of York 199
Edward, the Black Prince 111, 189–90
Edward I, king of England 98, 102–3, 109, 115, 190
Edward II, king of England 102–3, 120, 196
Edward III, king of England 121, 190
eels 204, 218–19
*Egil's Saga* 25
Egypt 27, 141–42, 144, 148, 150, 169–70, 176
Eigil of Fulda 90
Einhard 24
elder trees 85, 125
Eleanor de Montfort, countess of Leicester 175
Eleanor of Castile, queen of England 190
Eleanor of Provence, queen of England 196, 199
elements, natural 9, 21–25, 41–50, 52–53, 61, 76; *see also* air, earth, water
elephants 142, 144, 178
Elizabeth de Burgh, Lady Clare 123
Elizabeth of York, queen of England 175
elks 183
elm trees 80, 84–85, 114
Ely, England: bishop of 182
Emma of Normandy 25
Emmeldburg, England 199
Empedocles 8, 143
enclosures 138; *see also* purprestures
Endelechia 6, 41
England 8, 19, 24, 26–28, 31, 33–34, 36–38, 49, 51–52, 79, 84-5, 87–88, 91–93, 96–98, 100, 103, 114, 122–27, 130, 137–38, 155–58, 160, 162–63, 174, 180–87, 190, 195–97, 207–8
Enlightenment 220
environment: attitudes towards xiv-vi, 2, 4–10, 30, 72–76, 233; collaboration with 6, 9, 24, 28, 41–49; conservation of 8; exploitation of 3–4, 6–8, 24, 27–28, 30, 37, 41–49; human impacts upon 26, 28, 76, 96; impact upon humans 12–13, 16,

18, 76; *see also* air, animals, earth, forest, water
environmental history xiii, 8
environmental movement xv
*Epic of Gilgamesh* 77–78, 176
epidemics, *see* Black Death, plague
*Epistola et Regimen* (Letter and Regimen), by Alfonso de Córdoba 66
epizootic 15
Epona 150
*Eric et Enide,* by Chrétien de Troyes 128
erosion, of soil 8, 28, 51, 86
*Errores Gazariorum* (Errors of the Cathars) 73
Erzebirge, *see* Ore Mountains
Essex 132, 156, 199; forest of 196
Estonia 52
estovers 98, 124
ether 11
Etienne de Bourgon 175
*Etymologies,* by Isidore of Seville 18, 42–43, 132, 155, 166, 171, 229
Euphrates River 78
Europe 1–2, 4, 12–13, 18, 26–33, 49, 51, 64, 69, 84, 86–87, 89, 93, 96–97, 103, 125, 127, 130, 137–38, 149–51, 156, 163, 169–70, 179, 181, 207, 220, 223, 225, 233
Eve (Old Testament figure) 5, 24
Evrard de Breteuil, viscount of Chartres 128
Exeter, England 65, 164; cathedral of 176
Exodus, Book of 1, 11, 218, 224
eyres, forest 94–96, 99–109, 113–18, 120–21

Fagus 80
Falaise, France 220
falconry 166, 199–201, 204–6
falcons 142, 180, 201, 233; diseases of 201
fallow deer 183, 186, 190–91, 196–97, 199
famine 2–3, 5, 9, 49, 51–52
farming, *see* agriculture
Farne Islands, England 24
Fauns 7
Feckenham, England: forest of, 85, 94–96, 105–6, 110, 122, 138, 186–87, 189–90
Fens, England 33
ferrets 142, 169, 193
Ficino, Marsilio 9, 211
Filliol, Anthony 221–22
fire 11, 18, 41, 48, 58, 62, 84, 86, 90, 132
fishes 44–45, 49, 58, 61–63, 142, 146, 161, 172, 178–80, 182, 185, 201–4, 210, 213, 216; eating of 162, 200, 211, 232

fishing 10, 22, 27, 66, 177–78, 180, 200–204
Fitzherbert, Anthony, Sir 134–35, 161, 166–67
FitzNigel, Richard 101
FitzWarin, William, of Hadzor 106
Flagellants 69
Flambard, Ranulf, bishop of Durham 205
Flanders 29, 37, 51, 81, 157
flax 60, 63
fleas, *see* insects
Fleet River 65
flies, *see* insects
floods 18–19, 23, 33, 45–46, 49–51, 57, 86, 138, 233
Florence, Italy 36, 64, 66, 211
fog 50, 53, 60
Folville gang 131
Fontenay-aux-Roses, France 219
forest xv, 62, 233; clearance and destruction of xiii, 6, 8, 10, 28, 77–78, 83–97, 99–110, 114–17, 119–24, 131–32; definition of 87–88, 98, 127; extent of 84–92; management of 79, 84–85, 87–88, 93, 97, 111–19, 124–27, 130, 134–35, 138–39, 233; preservation of 78–79, 90, 94, 97–98, 114, 116–17, 119, 122–25, 138; regrowth of xiii-iv, 10, 79, 84, 89, 93, 108, 110, 115–17, 119, 122–23, 132, 135, 137–39; romantic depiction of 127–31, 136, 139; veneration of 79–81
Forest Assize, of 1198 101
Forest Charter, of 1217 102–3, 109, 119–20
forest law, *see* royal forest, of England
foresters 90, 99–100, 103–10, 113, 115–20, 126–27, 182, 186, 190, 192, 196–98
Forkbeard, Svein, king of Denmark 25
*Formicarius* (Ant-Colony), by Johannes Nider 73, 230–31
Fortune, Lady 7
fowling 119, 200–201, 205
foxes 144, 166, 177–78, 185, 199, 231
Fracastoro, Girolamo 73
France 8, 18, 20, 24, 31, 34, 36–37, 51–53, 66, 68, 73–76, 79–80, 82–83, 87–88, 94, 97, 125–27, 130, 137, 151, 158, 170, 175, 179–80, 183, 193, 199, 208, 219–20, 221
Francis of Assisi, Saint xv, 5, 8, 47–49, 205
Frankish laws 87, 90, 152–53, 170–71, 180
Franks 87, 90, 152, 180
Frederick II, Holy Roman emperor 160, 201
Fredsam, John 195

Freiberg, Germany 31
Frigid River, battle of 52
frogs 15–16, 44, 165, 207, 209, 214, 216, 229
fruit trees 90, 124–25, 132–33
fruits 13, 16, 43, 48, 58, 61, 78, 124, 134, 210, 214–15
Frytham, bailiwick of, England 107
Fulda, monastery of 90–91

Gabriele de Mussis 1–2, 9
Gailey Hay, forest of Cannock 107, 117
Galatia 79
Galen 13, 53–55, 59, 69, 132, 212, 214
Gallus, Irish monk 22
Galtres, England: forest of 190
Garscadden Wood, Scotland 124
Gascony 49
Gaul, *see* France
gazelles 142, 148
Geber 55
Geddington, England 117
geese 142, 150–51, 158, 161, 168, 177
Geismar, Germany 83
*Generation of Animals,* by Aristotle 142
Genesis, Book of 11, 23–24, 42, 58, 143, 145, 155, 218
Geneva, Switzerland 211
Gentile da Foligno 14, 16, 53–55, 59, 61, 63, 70–71, 209, 212–14, 216
Geoffrey of Sudborough, chaplain of Aldwinkle 118
Gerald of Wales 128, 227
Gerard de Sabloneta 14
Germain d'Auxerre, Saint 83
*Germania,* by Tacitus 180
Germanic culture 6, 22–23, 79, 151–52, 180, 225
Germanic kingdoms 90
Germanic law 21, 83, 170–71, 180, 219, 225
Germanic migrations 27
Germany 8, 20, 22, 32, 38, 47, 49, 51–53, 79, 83, 86–87, 91, 96, 125, 137, 151, 183, 208
Gérouville, Belgium 79
*Gest of Robyn Hode, A* 130
al-Ghazali, *see* Algazel
Giffard, Walter, archbishop of York 117
Gilbert de Clare, earl of Gloucester 181
Gilgamesh, king of Uruk 176
Giovanni della Penna 71
giraffes 178
Glaber, Rodulfus 181

glaciers 2, 49
Glasgow, Scotland 124
Glencree, Ireland 125
global warming xiv, 2–4, 26, 49, 97
Gloucester, England 32, 115, 187; earl of 181
Gloucestershire 96, 105, 111, 175
Glyndŵr, Owain 124
goats 22, 104–5, 141–42, 148–50, 152–53, 155, 158, 161–62, 172, 177, 187, 206, 227–28, 231
Godric of Finchale 205
*Golden Legend, The,* by Jacobus de Voragine 83
goshawks 206
Goslar, Germany 31
Göttingen, Germany 63
Grabfeld, Germany 91
grafting, of trees 79, 132–34
grain: milling of 34–38; yields of 29, 51–52
grapes: harvests of 26–27, 76, 223; vines of 134
Grasse, France 67
Gratian 222
Great Famine, of 1315–22 2, 7, 18, 51–52, 138, 157, 207–8, 233
Great Flood: in Vermont of 1927 xiv; in Vermont of 2011 xiv
Great Mortality, *see* Black Death, plague
Great Ocean Conveyer Belt, *see* thermohaline circulation
*Great Surgery,* by Gui de Chauliac 68
Great Warming, *see* Medieval Warm Period
Great Wind, of 1362 233
Greece 8, 17, 25, 142, 150, 170, 176–77, 219
Greeks, ancient 1, 11, 13, 16–18, 20, 25, 67, 78, 83, 169, 225
Greenland 26–27, 49
Gregory I, the Great, Pope 80, 145
Gregory III, Pope 151
Grenoble, France 79
grey partridges 185
greyhounds 175, 186–87, 198–99
groves, sacred 19, 25, 78–80, 83
Guaineri, Antonio 73
Guibert of Nogent 128
Guildford, park of, England 111
Guinefort, Saint 175
Guldenelak, England 118
Gundobad, king of Burgundians 90

Hadzor, England 106
hail 20–22, 45–46, 50, 76

Halatte, forest of, France 80
Haly Abbas 13–14, 16–17, 43
Hammurabi, king of Babylon 218; code of 218–19
Hampshire 93, 96, 98, 117, 175, 186, 190, 193
Hanbury, England 85, 96, 122, 138
Hardwick Wood, England 111
hares 155, 161, 166, 177, 183, 185, 191, 193, 199, 206, 231
Harrie, Germany 47
Hartmann, Johannes 147
harts, *see* red deer
Harz Mountains 31
Hatfield, England: chase of 196; forester of 196
Havering, England: park of 199
hawking, *see* falconry
hawks 124, 144, 171, 180, 211
hawthorn trees 114, 116
hays, deer 104, 106–8, 115–17, 122, 183, 189–90, 193
hazel trees 84–85, 108–9, 114, 116, 119, 124
Hebrews, *see* Jews
Heidelburg, Germany 204
Hellespont 25
hemp 63
Heneley, England 131
Henri de Ferrières 185, 199
Henry I, king of England 98, 100, 108, 120
Henry II, king of England 99–102
Henry III, king of England 93, 102, 108–9, 111, 116, 119-20, 132, 187, 189, 191, 196
Henry VII, king of England 121, 189, 196
Henry VIII, king of England 196
Henry of Grosmont, earl of Lancaster 198–99
Henry of Huntingdon 24
hens, *see* chickens
herbs 62, 68, 134
Hereford, England: bishops of 119–20, 181, 183; diocese of 19; sheriff of 131
Hermon, Mount 77
Herodotus 13, 25, 170
herons 185, 200
herring 181, 201–2
Hersfeld, Germany 90
*Hexameron,* by Ambrose of Milan 172
*Hexameron,* by Basil of Caesarea 45, 172
High Peak, England: forest of 189
Higounet, Charles 87–88
hinds, *see* red deer

*Hippiatrica* (Horse-medicine) 160
Hippocrates of Cos 8, 10, 12–13, 16–17, 43, 56, 61
hippopotami 142, 177–78
*Historia de Morbo* (History of the Disease), by Gabriele de Mussis 1
*Historia Naturalis* (Natural History), by Pliny the Elder 18, 78, 172
*History of Animals,* by Aristotle 142
Hittites 225
Hobbes, Thomas 49
Hofgeismar, Germany: forest of 137
holly trees 85, 107, 109, 116, 125
Holm Fen Post, England 33
Holland 32–34, 38, 52, 96, 151, 170
Holstein region, Germany 47
Holy Trinity, priory of, Dublin 125
Honorius of Autun 41
Hope Mansel Wood, England 115
Hopwas Hay, forest of Cannock 107
Hopwas Pass, forest of Cannock 131
Hopwas Wood, England 108
horses 29–30, 36, 64, 76, 104–5, 115, 119, 141–42, 148–53, 155–58, 160–61, 163, 166–68, 187, 201, 207, 215, 218, 225, 228, 230, 232–33; diseases of 159–61, 206
Horus 142
Houker, Stephen 195
*Hour of the Pig, The, see The Advocate*
*How to Catch Fish and Birds by Hand* 204
Huby, England 98
Hugh de Brehull 106
Hugh de Goldingham, Sir, steward of Rockingham Forest 196
Hugh de Loges, forester of Cannock 117
Hugh de Neville, bailiff of Pickering Forest 198
Hugh de Neville, chief forester of England 101
Hugh le Scot 191
Hugh of Eynsham, riding forester of Cannock Forest 108
Hugh of Goldingham, steward of Rockingham Forest 117
Hugh of St. Victor 229
humoral theory 70, 72, 133, 213
Hundred Years War 126–27
Hungary 34, 51, 127, 163
hunting 8, 79, 83, 207; ancient 141, 143–44, 176–78; medieval 98, 101, 110, 119, 122–23, 125, 127, 129, 131, 166, 171, 176–200, 204–6, 232; modern xvi, 191

Huntingdon, England 118–19; forest of 118; hospital of 118; prior of 120
Huntingdonshire 33, 98, 198, 208

ibex 148, 176
ibises 142
Ibn Hayyān, Jābir, *see* Geber
Ibn Jumay' 14
Ibn Khatima 53–55, 72, 210
Ibn Luqa, Qusta 13
Ibn Masawaiyh, *see* Mesue
Ibn al-Quff 17
Ibn Ridwan, Ali 14
Ibn Rushd, *see* Averroes
Ice Age 84
ice core readings 3, 10, 26, 50
Iceland 26–27
*Iliad,The,* by Homer 1
India 78, 150
Industrial Revolution 76
Ine, king of Wessex 90
Ingham, England 175
Inquisition 21, 192
inquisitions, forest 105, 120
insects 3, 26, 79, 142, 144, 146, 158, 206, 209, 216, 230; on trial 218–19, 221–24, 232
Iraq 77
Ireland 19, 34, 51–52, 84, 123–24, 184, 208
Iron Age 84–85
iron forges 34, 65, 105, 119, 122
irrigation 32–34
Isaac Israeli ben Solomon 53, 172
Isaac the Jew, *see* Isaac Israeli ben Solomon
Isaac ben Todros, Rabbi 72
Isabel de Brus, Lady 195–96
Isabella of France, queen of England 103
Ishtar 177
Isidore of Seville 18, 42–43, 45–46, 132–33, 155, 158, 166, 171–72, 229
Islam 14, 27–28, 55, 217
Israel, archangel 23
al-Isrā'īlī, Ya'qūb 14
Italy 31–32, 34, 36–37, 61, 64, 83, 85–87, 125, 151, 157, 160, 174, 180
*Iudicium Iovis* (The Judgment of Jupiter), by Paul Schneevogel 7
Ivo of Chartres 226

jackals 142
jackdaws 231
Jacme d'Agramont 53–55, 68, 71, 210, 213–14

Jacobi, John 60, 165
Jainism 48, 78, 143
James III, king of Mallorca 68
James of Panton 196
Jaret River 66
Jean de Venette 67
Jewish pogroms 68–69
Jews 66–69, 73, 171, 214–15, 218–19, 223–24, 233
Joan of Navarre, queen of England 175
John, king of England 100–102, 182
John de Beka 52
John de Chaucombe, keeper of Lynhurst Park 107
John de Shafaldon, woodward of Deepdale Springs 109
John of Burgundy 55–56
John of Saxony 59–60, 63, 209, 215
John of Speyer 211
John of Tornamira 60
John the Damascene 55
Jord 25
juniper trees 77
Jupiter 7, 20, 54, 58, 68, 78, 80, 83
justices of the forest 100, 103–6, 109, 114

Kent 33, 123, 131
Khepri 142
al-Khwarazmi 17
kid, *see* goats
Kilvington, John, keeper of forest of Pickering 117
al-Kindi 14
Kings Coughton, England 105
Kinver, forest of, England 96, 189–90
*Kitab al-Maliki* (Complete Book of the Medical Art), by Haly Abbas 13–14, 17
*Kitab al-Mansuri* (Book of Medicine for Mansur), by Rhazes 13
kites 217
Knighton, Henry 208
Konrad of Megenberg 10, 18, 44, 56–58, 63, 68, 71, 132, 166, 215
Koran, *see* Qur'an
Koralpe, Mount, Austria 20
Kramer, Heinrich 231
Kutna Hora, Czech Republic 31

lakes 12, 18–19, 22, 26, 43, 54, 210
lambs, *see* sheep
Lamme, Heinrich 215
Lancashire 116

Lancaster: earl of 117, 120, 198–99; forest of 98, 119, 121
*Lancelot*, by Chrétien de Troyes 128
Langeley, England 186
Langley Marish, England 182
larks 185
Lascaux, France 141
Latobius 20
Lausanne, Switzerland 230
Le Franc, Martin 231
leather, tanning of 32, 34, 63–66, 119, 124
Lebanon 77, 83
Ledbury, England 183
leeches 165, 218
Leicester, England 175
Leicestershire 109
Leipzig, Germany 67
Lemovices tribe 80
Lemyngton, England 107
leopards 142, 174
Lérida, Spain 53, 68, 210, 213
Leuven, Belgium 64
Leviticus, Book of 224, 226–27
*Liber de Causis Proprietatum Elementorum* (Book on the Causes of the Properites of the Elements), by Albert the Great 46, 53–54
*Liber de Natura Rerum* (Book on the Nature of Things), by Thomas of Cantimpré 42, 56, 58, 132, 147
*Liber Pantegni*, see *Kitab al-Maliki*
*Liber ruralia commodorum* (Book of Rural Benefits), by Pietro de Crescenzi 134, 161
Lichfield, England 107
Lickey, hay of, forest of Feckenham 106
Liège, Belgium 7, 55
*Life of St. Boniface*, by Willibald 83
*Life of St. Cuthbert* 24
*Life of St. Gall* 22
*Life of St. Sturm*, by Egil of Fulda 90, 128
lightning 18, 20, 25, 45–46, 59–60, 76
lime, manufacturing of 108, 119
lime trees 84–85, 106, 108, 114
Lincoln, England 34; bishop of 120
Lincolnshire 34, 50, 97
linen manufacturing 32, 63
Lindow Man 19
lions 137, 142, 147, 166, 168–69, 177–78, 213
Little Ice Age xiv, 2–4, 26, 49–51, 138, 233
Little Optimum, see Medieval Warm Period
Liutprand, king of Lombards 153
Livonia 52

*Livre de Chasse* (Book of Hunting), by Gaston Phoebus 195, 199
*Livre des Deduis, Le* (Book of Delights), by Henri de Ferrières 185, 199
*Livres du Roi Modus et de la Reine Ratio* (Book of King Method and Queen Reason), see *Le Livre des Deduis* lizards 216, 229
Llyn Cerrig Bach, Wales 19
Llyn Fawr, Wales 19
locusts, see insects
Lombard laws 83, 87, 90, 152–53, 171, 180
Lombards 87, 90, 152, 180
London, England 65–66, 164–65, 195
Loneti, Theobaldus 165
*Long Consilium*, by Gentile da Foligno 70, 209
Lorraine 76
Los Caballos, Spain 176
Lot (Old Testament figure) 58
Louis of Hungary, King 58
Louis the Pious 21
Low Countries 31, 49, 51, 96, 125
Lübeck, Germany 63, 67–68, 213, 215
Lucan 79
Lucifer 25
Lucretius 17, 86
Lucullus 217
lumber mills 34
Luttrell Psalter 175
Lucerne, Switzerland 147, 230
Lydgate, John 168
Lyndhurst, England 190; manor and park of 107
lynxes 185
Lyonnais 73
Lyons, France 20, 224

*Macbeth*, by William Shakespeare 229
Macclesfield, England: forest of 184
Macedonia 150
Machiavelli, Niccolò 49
Macrobius, Ambrosius 45
magic, see witchcraft
Magna Carta, of 1215 101–2
Mahavira, Vardhamana, founder of Jainism 78
al-Majusi, Ali ibn Abbas, see Haly Abbas
*Malleus Maleficarum* (Witches' Hammer), by Heinrich Kramer and Jacob Sprenger, 228, 230–31
Mallorca 68, 227
Mallory, Thomas 130
Malmesbury Abbey, England 52

Malthus, Thomas, economist 2–3, 29
Mandeville, John, Sir 227
Map, Walter 128
maple trees 85, 114, 119, 133
Marculus, Bishop 24
Mareschall, Walter, keeper of the bailiwick of Alta Foresta 106
Marmion, Philip, Lord 108
marmots 207
Mars 54, 58, 71, 79
Marseilles, France 66, 79
marshes 12–14, 17, 19, 33, 54, 94
martens 185
Martial 178
Martin of Braga, Archbishop 80
Martin of Tours, Saint 81–83
Massahalla 56
*Master of Game, The,* by Edward, duke of York 196, 199
masters, of the forests 125–27, 185
mastiffs 186, 196
al-Mas'udi 17
Matilda de Bruys 199
Matthew of Thurlbear, forester of Brigstock Park 198
Maurienne, France 221
Maurus, Rabanus 18, 132, 171
Maydenhith, England 131
meat, eating of 64, 138, 143, 162–65, 168, 180, 183, 185–87, 195, 213, 216, 232
Medieval Warm Period 26–27, 49, 233
Mediterranean Sea, region of 67, 86, 125, 150
Meissen, Germany 51
menhirs 20
Mercury 7
Merle, William, rector of Driby 50–51
Merlin 228
Mesolithic Age 84
Mesopotamia 11, 77–78, 150, 176, 218–19, 228
Mesue 55
metallurgy 37, 85, 121
metals 18–19, 22, 30–31, 34, 42, 45–46
*Metaphysics,* by Aristotle 142
*Meteorics,* by Aristotle 15
Metz, France 76
Meuse River 65
miasmatic theory 13, 70–72
mice, *see* rodents
Middle East 1, 149
Miguel de Cervantes 197
Milan, Italy 68, 155
milling technology 32, 34–38, 156

minerals 12, 44, 70
Minerva 7, 19
mining 10, 28, 30–31, 38, 58, 63, 101, 119, 122
missionaries, Christian 5, 23, 80–83
mistletoe 80
*Moine et la Sorcière, Le* (The Sorceress) 175
Moldava River 51
moles 218–19
Mongol Empire 149
monkeys 142, 146, 169, 175
Monmouth, England 132
Montargis, France 174–75
Montelucco, Italy 83
Montpellier, France 67, 213; barony of 68; university of 60, 66, 165
moon 43, 46, 76, 80, 132
Moravia 31
*Morte Darthur,* by Thomas Mallory 130
Moses (Old Testament figure) 11, 144
*Motion of Animals,* by Aristotle 142
Moton, Walter 183
mountains 12, 15–16, 19–20, 22, 36, 57–58, 62, 68, 76–78, 209, 212
Muhammad, founder of Islam 27
mules 149–50, 161, 166
*Mulomedicina* (Mule-medicine), by Vegetius 160
Mundus 41
Munich, Germany 76
Murdach, Henry, abbot of Vauclair 128
murrains, 138, 157–58, 161, 192, 207–8, 215
myrrh trees 133

Naiads 7
Namur, Belgium 65
Nantes, Council of 81
Naples, Italy 58; university of 71
Narbonne, France 67–68
Natura 6–7, 41, 48
*Natura, De* (On Nature), by Bernard of Frankfurt 58
*Natura Rerum, De, see Liber de Natura Rerum*
*Natural History, see Historia Naturalis*
nature xiv, xvi, 2, 4–12, 15, 17, 20, 22–23, 25, 30, 36, 41-2, 45–48, 83, 86, 93, 144–45, 168, 213, 226, 232-33
*Naturis Rerum, De* (On the Natures of Things), by Alexander Neckam 42, 132, 173
Nearing, Helen xiii
Nearing, Scott xiii

## Index

Neckam, Alexander 42, 57, 132, 158, 166, 173
necromancy, *see* witchcraft
Nemetes tribe 79
Nemetona 79
Neolithic Age 84, 86, 141, 148–49, 206
Neoplatonism 9, 41
Netherlands, *see* Holland
new ecology 2, 8
New Forest, England 96, 98, 106–7, 110–11, 116–18, 186–87, 190, 193, 195–96, 199
New Park, England 196
Newman, John 195
Newnham, England 132
Newton Bury, England 195
newts 229
Niavis, Paulus, *see* Schneevogel
Nicholas of Orton 199
*Nicomachean Ethics*, by Aristotle 142, 213
Nider, Johannes 73, 230–31
Nigel de Longchamps, *see* Wirecker, Nigel
nightingales 168
Noah (Old Testament figure) 11, 46, 145, 218
Norfolk 34, 156, 162, 175
Norman Conquest: of England 92–93, 97, 156, 181, 207; of Wales 124
Normandy 51, 97, 115, 125, 181
Normans 92, 98–99, 155
North America 148–49
North Sea 33, 49
Northamptonshire 116, 118, 190, 196
Norway 24, 27, 49, 181, 202
Norwegian laws 83, 152, 180, 226
Norwich, England, castle of 174
Nottingham, England 117, 130; castle of 109, 117
Nottinghamshire 96, 109, 190, 208
Noys 6, 41, 145
Nürnberg, Germany 165
nutmeg trees 133

oak trees 47, 79–80, 83–85, 90, 94, 98, 101, 105–11, 114–17, 119, 121–22, 124–25, 127, 133, 135, 149, 186
Oakham Castle, England 109
Oakley, England 117
oats 27, 157, 182
Obermarsberg, Germany 83
oceans 18, 27, 46, 58
Odense, Denmark 174
Odenwald, forest of, Germany 88, 90

Odin 25
Ogley, England: wood of 131
Old Testament 1, 4, 22–23, 47, 52, 77, 137, 224
Oldberrow, England 186
old-growth woodland, *see* wildwood
olive trees 90
Olympias of Epirus, mother of Alexander the Great 225
onagers 155
*On the Characteristics of Animals,* by Aelian 225
*On Contagion,* by Qusta ibn Luqa 14
*Opus agriculturae* (Work of agriculture), by Palladius 161
Ordinance of Brunoy 126
Ordinance of the Forest, of 1306 102–3
Ordinances, of 1311 103
*Ordonnances touchant les Eaux et Forêts* (Ordinances touching the Waters and Forests) 125–26
Ore Mountains 7
Origen 143–45, 147
Orleans, France: duchy of 126
Orphic school of philosophy 8, 10, 78, 143–44
Othe Forest, France 31
otters 185
outgassing 4, 58
Overton, Wales 124
Ovid 128, 132
*Owl and the Nightingale, The* 168
owls 168, 216, 229
oxen, *see* cattle

paganism 4–5, 18–23, 25, 80–83, 145, 225–26
Paleolithic Age 149, 206
Palestine 28
Palladius 161
pannage 28, 88, 90, 102, 104, 117, 124, 156, 158
Paris, France 21, 65, 127, 168, 219; university of 53–54, 61, 71–72, 167
parks, deer 104, 130, 182–83, 187–93, 204
partridges 161, 183, 185, 193, 205, 216
*Parts of Animals,* by Aristotle 142
Passavanti, Jacopo 231
peacocks 161, 179
pear trees, *see* fruit trees
peat bogs 26, 33, 84–85
Pelsall, wood of, England 108
Penates 7
*Perceval,* by Chrétien de Troyes 128

perch 204
Persia 13
Persian empire 178
Persian Wars 25
Perugia, Italy 53, 61
pestilence, *see* plague
Peter, Saint 83
Peter de Kottbus 59
Peter de Neville, warden of Rutland Forest 108–9, 119
Peterborough, England: abbot of 120, 191
Petrarch, Francesco 8
pets, *see* animals, as pets
pheasants 161, 183, 185, 193, 216
Philip II Augustus, king of France 125
Philip IV, king of France 125, 158
Philip VI, king of France 68
Philip of Stanton, sheriff of Huntingdon 118
*Philosophia Mundi* (Philosophy of the World), by William of Conches 41
Phoebus, Gaston 195, 199
*Physics*, by Aristotle 11
*Physiologus* 172
Physis 6, 41
Piacenza, Italy 1
Pickering: castle of 116–17; forest of 98, 109–10, 115–17, 120, 184, 186, 189, 196, 198–99
Pico della Mirandola 9
Pietro d'Abano 69–70
Pietro de Crescenzi 134, 161
Pietro de Tussignano 212
pigeons 142, 151, 159, 161, 163
pigs 28, 63–64, 88, 105, 124, 141–42, 148–53, 155–58, 160–63, 171, 187, 210–11, 214, 225, 230; on trial 217–20
pike 204
Pilis Forest, Hungary 127
Pillaton Hall 107
pine trees 77–78, 80, 83–84, 114, 133
Pistoia, Italy 64
Pitard, Jean, physician, 158
plague 1–5, 7, 9–10, 13, 15, 18, 43–44, 46, 52–64, 66–76, 89, 137–38, 158, 164–65, 175, 195, 207–17, 223–24, 229; *see also* Black Death
plains 15–16, 137, 209, 211
*Planctu Naturae, De* (The Complaint of Nature), by Alain of Lille 6, 42
planets 54, 56, 58, 60, 211
plants xvi, 5, 8–9, 12–13, 20, 42, 54, 62, 70–71, 78–79, 132-4, 136, 210, 213–14

Platearius, Matthaeus, of Salerno 46, 132
Plato 144
Platonism 6, 143
Pliny the Elder 17, 20, 42, 44, 46, 78, 80, 132, 158, 172
Plotinus 9
plovers 185
Pluto 7
poaching, of deer 108, 125, 182–83, 186–87, 190–99; *see also* hunting
poison 17, 44, 53, 57–58, 66–76, 80, 210–11, 214–15, 230, 233
Poitiers, France, battle of 168
Poland 51–52, 127, 163
*Politics*, by Aristotle
pollarding 111, 138
pollards 106–7, 111, 124
pollen analyses, of trees 84
pollution 3, 8, 13–14, 17, 31–32, 43, 52–53, 63–69, 76, 233
Pompeii, Italy 44
ponds, 44, 182; fish-32, 102, 117, 122, 185, 202–3; mill-32, 34
popinjays 175
poplar trees 133
Portugal 80, 185
potash, manufacturing of 107–8, 119, 124
Prague, Czech Republic 59, 214; university of 60, 68
precession 50
Prentice, Joan 231
Prestwood, wood of, England 108
primeval forest, *see* wildwood
Primus of Görlitz 61, 72, 211, 213
*Problems, The,* by Aristotle 212
*Progression of Animals*, by Aristotle 142
Prometheus 146
*Proprietatibus Rerum, De* (On the Properties of Things), by Bartholomaeu Anglicus 42, 132
Protagoras 134
Provence 68
Prudhomme, Adam 109
pruning, of trees 79, 135
Ptolemy 17, 59
purprestures 94–96, 100, 104–5, 120
Pyrenees Mountains 20
Pythagoras 8, 10, 78, 143–44

*Al-Qanun fi al-Tibb*, see *Canon of Medicine*
quail 216
Quernmore, forest of, England 116

322    Index

*Queste del Saint Graal* (Quest for the Holy Grail) 129
*Quixote, Don* by Miguel de Cervantes 197
*Quod liceat pestilentiam fugere* (That is should be permitted to flee the pestilence), by Dominico Amanti 216
Qur'an 16–17

rabbits 175, 183, 185, 193, 206
Rackham, Oliver 91–93, 110
Ragouel, archangel 23
rain 12–13, 15–16, 18, 21, 25–26, 28, 45–46, 49–51, 61, 76, 138, 207, 223
rainbows 18, 45–46, 60
Ralph de Sandwich, warden of the Forest of Dean 115
Rammelsberg, Germany 31
Ramsey, England 98; abbey of 99, 208
rangers, of the forest 121
Ranulph of Combreiton 113
Raphael, archangel 23
rats, *see* rodents
ravens 179, 207, 215–16, 228
al-Razi, *see* Rhazes
Reake, Turkey 23
red deer 180–81, 183, 186, 190–91, 193, 197–99, 205, 230
Red Sea 11
regarders 104, 106, 108, 110, 117
regards, of the forest 104, 121
Reginald of Durham 205
Rembaud, Peter 221–22
Remigia, Spain 176
Renaissance 220
*Repelling General Harm from the Human Body*, by Avicenna 17
reptiles, *see* crocodiles, lizards, snakes
Reynard the Fox, epic of 221
Reynolds, Walter, archbishop of Canterbury 52
Rhazes 13, 15, 43, 55
Rhineland 19
rhinoceros 176, 178
Ribbenitz, Heinrich 60, 63, 68, 214
Richard, earl of Cornwall 108
Richard I, king of England 101–2
Richard II, king of England 121
Richard III, king of England 193
Richard de Lyndhurst 118
Richard de Monte Viron, keeper of king's hay of Lickey 106
Richard de Muntfechet 182
Richard of Aldwinkle, verderer of Rockingham Forest 118

Ridlington, park of, England 108
Rigaud, Eudes, archbishop of Rouen 175
Rijnland, Holland 33
Ripon, England 167
rivers 12, 18–19, 27–28, 31–32, 34, 42, 44, 65, 84, 203
roach 204
Robert de Clifford, Sir 98
Robert le Baud, master of hospital of Huntingdon 118
Robert le Noble, chaplain 186
Robin Hood 8, 97, 121, 124, 130–31
*Robin Hood and the Monk* 130
Rocelin de Lyndhurst 118
Roch, Saint 175
Rockingham, England: forest of 116–18, 190, 196, 198
rocks, *see* stones
Rodbaston, England 117
rodents 3, 15–16, 48, 61, 142, 169, 173, 176, 207, 209, 211, 214, 216, 229, 231; on trial 218–19
Rodewode, Wales 124
roe deer 183, 191, 196
Roger de Chandos 131
Roger le Scot 195
Roger of Wollaston, forester of Essex 196
*Roman de Fauvel* (Story of the Fawn-Colored Beast) 233
*Roman de Renart* (Romance of Reynard the Fox) 221
Roman Empire 3–5, 18, 23, 27, 34, 80, 83, 89–90, 143, 151-52, 225
Roman gods 20
Roman law 21, 219
Romania 51
Romans, ancient 14, 19, 31, 34, 78–80, 85–86, 88, 90–93, 123, 143, 169, 178, 225, 228
Rome, Italy 7–8, 17, 150, 170, 176–78
Romsey, England 117; nunnery of 175
Roos, Margaret, Lady 175
roosters, *see* chickens
Roscommon, county of, Ireland 124
Ross, England 183
Rouen, France 175
Rous Lench, England 184
Rousillon 68
royal forest, of England 97–123, 181–83, 185–86, 189–92, 197-8
Rubicon River 228
Rufford, abbot of 109
Ruffus, Jordanus 160
Russia 150

Rusti, park of, England 111
Rutland 109; forest of, 108, 119, 196
rye 27

sacrifice: animal 20; human 19–20, 79–80
Saint-Denis, abbey of, France 151
St. Cuthbert 24
St. Denys, prior of, Southampton 118
St. John's Wood, Ireland 124
St. Julien, France 221–24
St. Mary's, abbey of, York 190
St. Mary's, prior of, Southampton 118
St. Peter, church of, Aldwinkle, England 118
St. Peter's, abbey of, Gloucester 115
St. Werbergh, abbey of, Chester 189
St. Willibald, 83
Sainte-Trinité, Falaise, France 220
Saintogne 137
Saladin Ferro de Esculo 208–9, 212–13, 223
Salerno, Italy 167
sallow trees 114
salmon 179, 200
salt manufacturing 85, 101, 107, 122
saltpeter 31
Sambre River 65
Sampson, Norman, riding forester of Huntingdon 119
Samuel, William, of Kings Coughton 105
sanctuaries: forest 19, 79–81; mountain 19–20, 80; water 18–19, 79–80
Saturn 54, 56, 68
Savernake Forest, England 116
Savonarola, Michele 215
Savoy 221
Saxony 31, 51, 68, 90, 128
Scalby Hay, forest of Pickering 115–17
Scandinavia 27–28, 51, 185
Scarborough, England, castle of 117
Schedel, Hermann 165
Schneevogel, Paul 7
science, impact of upon environment 4, 6
Scientific Revolution 11, 220
Scotland 27, 49, 84–85, 123, 138, 184, 199
Scotus, Michael 146
Scotus, Sedulius 24
Scythian kingdom 150
seals 225
seas 18, 24, 33, 42, 44
seasons, of the year 12–13, 15–16, 43, 55, 61, 80, 217
Second Barons' War, England 108, 120

*Secret of Secrets*, by Rhazes 13
Sedgeford, England 162
Seine River 19, 65
Seneca 20, 42, 44
Senlis, France 80
Sens, France 80
Sequana 19
Serapion of Alexandria 55
serpents, *see* snakes
Severus, Sulpicius 81
sewers 13–14, 31, 54, 60, 63–64, 66
Shakespeare, William 229
sharks 142
sheep 18, 20, 105, 138, 141–42, 148–53, 155–58, 160–63, 166, 168–69, 171, 177, 185, 187, 215; diseases of 157–58, 161, 206–8, 210
Sherwood Forest, England 96, 109–10, 117, 190–91, 208
Shotewi, park of, England 111
Shrewsbury, earl of 99, 196
Silvanus 79
Silvester, Bernard 6, 41–42, 48
Simon of Corvino 7
*Simplici Medicina, De* (Concerning Simple Medicine), by Galen 212
Sinquates 79
*Sir Gawain and the Green Knight* 129–30, 195, 197
*Sir Launfal* 129
*Sir Tristram* 196
Skallagrimsson, Egil 25
sky gods 20, 80, 150
snails 165, 218
snakes 15–16, 25, 61, 76, 142, 146, 173, 179, 205, 207, 209–11, 214, 216, 218–19, 225, 228–29
snow 12, 17, 45–46, 50
Sobek 142
Socrates 144
soils 12, 23, 29–30, 92, 97, 106, 110; *see also* erosion
solar radiation 50
Solinus 158, 172
Somerset 84–85, 98; forest of 119
sorcery, *see* witchcraft
South Frith, England 123
Southampton, England 32, 107, 118, 187
Spain 17, 34, 61, 68, 87, 125, 156, 160, 176, 180, 199, 210, 227
spaniels 186
sparrows 230
*Speculum Maius* (The Great Mirror), by Vincent of Beauvais 42, 132

## 324  *Index*

*Speculum Stultorum* (Mirror of Fools), by Nigel Wirecker 167
Speyer, Germany 79
spiders 216
Spoleto, Italy 83
spoonbills 185
Sprenger, Jacob 231
springs 16, 18–19, 24, 31, 42–43, 68, 78–80, 215
Spurrina 228
squirrels 175, 185, 206
Stafford, England 117
Staffordshire 96, 107, 116, 131, 189–90, 199
stags, *see* red deer
Stanion, England 117
stars 12, 14, 46, 58–60, 72, 76, 144
Stephen, king of England 99, 102, 120
Stephen of Moulton, forester 118
Stokewood, England 108
stones: precious 42, 44–46, 72; quarrying of 30–31; worship of 18–20, 81
storks 217
storm-makers 20–22, 76
storms 9, 15, 20–21, 27, 44, 49–52, 76, 116, 138, 217, 223
Strabo 79, 86, 179
Strasbourg, France 59–60, 68, 204, 209, 215
Streche, Robert, keeper of the bailiwick of Alta Foresta 106
sturgeon 200
Sturm, Saint 90–91
Sudborough 186
suckering 111
Suetonius 228
Suffolk 93, 114, 156, 180, 184
Sulis 19
Sulzfeld, Germany 209
*Summa Theologiae* (Summary of Theology), by St. Thomas Aquinas 231
sun 12–14, 45, 62, 76
sunspots 50
Surrey 111
Sussex 92, 131
swallows 172, 217
swamps, *see* marshes
swanimote courts 105
swans 179, 200, 225
Sweden 25
Switzerland 22, 76, 179, 214, 219, 230–31
Syria 20, 27, 77, 83

Tacitus 79, 86, 151, 180

Talbot, Francis, earl of Shrewsbury and chief forester of Hatfield 196
al-Tamini 14
tanning, *see* leather
Taranis 25
technology, impact of upon environment 4; *see also* watermills, windmills
Teddeslay Hay, forest of Cannock 117, 189
tench 204
Tertullian 3, 86
Teutates 19
Teutoburg Forest, Germany 86
Teutonic Knights 96
Thales of Miletus 11
Thames River 65
Theodore of Sykeon, Saint 23
Theodosius, Emperor 52, 80
Theodosius de Camilla, dean of Wolverhampton 108
Theophrastus 79
thermohaline circulation 26–27, 50
Thessaly 18
Thomas, earl of Lancaster 120
Thomas of Cantilupe, bishop of Hereford 181
Thomas of Cantimpré 18, 42, 45–46, 56–58, 132–33, 147, 158, 172–74
Thomas of Chobham 226
Thor 25
thrushes 161, 211
thunder 18, 20, 22, 25, 42, 45–46, 50, 60
tide mills 35–36
tides 18, 35, 42–43, 46, 58
tigers 174
Tiglath-Pileser I, king of Assyria 78, 150, 177
toads 165, 173, 209, 211, 216, 231
Toledo, Council of 81
torture 21, 67, 73, 119
Tradde, Walter 195
*Traité des Monitoires* (Treatise concerning Ecclesiastical Monitions), by Gaspar Bailly, 224
trapping, 184–86, 193, 201, 205
*Travels*, by Sir John Mandeville 227
*Treatyse of Fysshynge wyth an Angle* 204
tree rings, *see* dendrochronology
trees 42, 49, 78–79, 91; felling and destruction of 28, 47, 62, 80–85, 90–91, 94, 101, 103–9, 114–17, 119, 121, 124, 135; natural history of 132–36; worship of 19, 78–84
Trillek, John, bishop of Hereford 183
Trinity College, Cambridge 169

*Tristan et Yseut* 128
Trojan War 1
Trondheim, Norway 49
*True Word, The,* by Celsus 143
tunnels 32, 34
Turkey 23, 210, 225
Turnastone, England 19
turtle-doves 161
Tychefeud, England, abbot of 118

Uffington, England 150
Ulm, Germany 53
Ulster 150
Uppsala, Sweden 25
Upton, England 183
Urania 41
Uruk, Iraq 176
Usama ibn Munqidh 205

Van der Weyden, Conrad 73
vapors 44, 46, 53–54, 57–60, 62–63, 67, 70–72
Varro 150
Vauclair, abbot of 128
veal calves, *see* cattle
*Vegetabilibus, De* by Albert the Great 133
Vegetius 160
Venice, Italy 64
venison: keepers of 100, 123; offenses of 98–102, 104, 108-9, 116–17, 120, 192, 195; *see also* deer
Ventoux, Mount, France 20
verderers 104, 108–10, 115, 117–18, 120, 126
vermin, *see* rodents, worms
Vermont 61; environmental history of xiii-vi
vert: keepers of 100, 123; offenses of 98–102, 104–10, 114, 116–17, 120, 191; *see also* deer
Vienna, Austria 68
Vikings 25–27, 83, 170
Villach, Austria 57
Villiers-le-Sec, France 151, 170
Vincent of Beauvais 18, 42–44, 57, 132–33, 158, 172–73, 187
Vincent of Stanley 118
Vinck, Johann 209–10, 212
Vinland, North America 26
Virgil 128, 132
Visigothic laws 83, 90, 152–53, 180
Visigoths 90, 152
*Vision of Gottschalk* 47
Vitruvius 44
volcanoes 4, 50, 58

Vosegus 19, 79
Vosges Mountains, France 19, 79
*Vulgate* cycle 129

Wales 19, 79, 84, 122–24, 130, 138, 163, 208
Walkwood, England 106
Walsall Wood, England 108
Walter de Kent, steward of New Forest 190
Walter of Henley 160
War of the Roses 121
wardens, of the forest 103, 105, 108, 114–16, 119, 126
warrens 183, 185, 193, 199
Wash, England 33
Wasmerus, forester of New Forest 118
wasps, *see* insects
waste, *see* forest, clearance and destruction of
water xv, 8–9, 11–14, 16–19, 22, 28, 31–36, 41–49, 52–55, 58, 60, 62–67, 71–76, 132, 144, 210–11, 214–16, 233
water engineering 31–33
water wheels 34–36
watermills 28, 32, 34–38, 101–2, 117
Weald, forest of, England 88, 91–92
weasels 173
weather 13–15, 20–23, 25–28, 43, 47, 49–52, 54–55, 61–62, 73–76, 138
weather magic 73–76
weevils, *see* insects
weirs, for fishing 117, 122, 202
*Well of Life,* by Rabbi Isaac ben Todros 72
wells 31, 34, 42, 45, 54, 57–58, 60, 67–68, 214–15
Welsh laws 83, 152, 171
Weser River 137
Wessex 90, 150
West Stow, England 180, 184
wetlands, draining of 28, 32, 94–96, 185
Weybridge, England 198
whales; hunting of 181
wheat 84
White, Jr., Lynn 30
Whorlton, England 175
Widmans, Johann 195
Wight, England: countess of 199
wildcats 185
wilderness 33, 48, 77, 90–91, 93–96, 127–37, 185, 205
wildlife, *see* animals, wild
wildwood 84–86, 93, 111, 130
William de la More 196
William de Morteyn 108

William de Percehay, Sir, forester of Pickering Forest 110
William le Gale 111
William of Auvergne 230
William of Conches 41–42
William of Drakenage, forester of Hopwas Hay 107
William of Malmesbury 231
William Rufus, king of England 99
William the Conqueror, king of England 91, 97, 99, 181
William the Spenser 118
willow trees 124–25
Wiltshire 116, 132
Winchester, England 65, 164, 187; bishop of 99, 158, 208
wind turbines xv
windmills 28, 33, 37–38, 105
winds 9, 12, 15–16, 18, 20, 24, 37, 42, 44–47, 50, 52–54, 59–63, 71, 76, 106
Windsor, England: forest of 131; park of 182
wine production 27
Wirecker, Nigel 167
witch trials 228–29, 232
witchcraft 9, 21–23, 43, 73, 171, 229–32
witch-hunts 221, 232–33
Wolverhampton, England 108
wolves 130, 137, 144, 169, 180, 184–85, 187, 216, 230–32

woodland, *see* forest
woodwards 99–100, 103, 109, 182
wool industry 32, 34, 37, 152–53, 157, 163–64, 168, 207
woolly mammoths 176
Worcester, England 105; bishop of 94, 122; prior of 94
Worcestershire 85, 94, 105, 122, 138, 184, 186, 189–90
world soul 6, 41
worms 142, 146, 158, 172, 211, 216, 218, 224, 229
wrens 228
Wulvivehop, England 131
Würcker, Hans 53
Wyresdale, forest of, England 116

Xerxes, king of Persia 25

yew trees 80, 124–25
York, England 65, 174; abbey of 190; archbishop of 117; minster of 175
Yorkshire 93, 109, 175, 189, 196
*Yvain,* by Chrétien de Troyes 128

Zagros Mountains 77
Zeus 78, 146, 179, 225
Zofingen, Switzerland 214
Zuiderzee, Holland 33